THE ROUTLEDGE HANDBOOK ON ECOSOCIALISM

Building on the classical works that have propelled and shaped ecosocialist thinking and action and more recent political developments on the ground, the volume will provide a reference point for international work in the field, both directly political and academic.

The *Handbook* acquaints readers with the varied roots of and sometimes conflicting approaches to ecosocialism. It does not attempt any unification of ecosocialist currents. Rather, it aims to provide a resource that is as comprehensive as possible with respect not only to theorization and ideological framing, but also and especially to existing projects, practices, and movements and giving a sense of the geographical reach that ecosocialism so far represents. This includes scholarship that extends Marxist foundations and reflects on more recent political developments. The theoretical and practice-oriented moorings are buttressed by discussions on movements, frameworks, and prefigurative processes as well as on social struggles occurring within institutional settings. Together, the collection offers a reference point for international work in the field, in social movements, and in institutional transformations.

Providing detailed but accessible overviews of the complex, varied dimensions of ecosocialism, the *Handbook* is an essential up-to-date guide and reference not only for researchers, but also for undergraduate and graduate students in geography, environmental studies, development studies, sociology, and political science, as well as for policymakers and activists.

Leigh Brownhill is an ecofeminist scholar and writer, anti-racist abolitionist, long-time decolonial researcher, and community activist. She is Associate Editor of *Capitalism Nature Socialism*, the international, multi-disciplinary journal of ecosocialism. She teaches sociology at Athabasca University, Canada.

Salvatore Engel-Di Mauro is Professor of Geography at SUNY New Paltz and Chief Editor of *Capitalism Nature Socialism*. His research areas include socialist histories, soil contamination processes, and urban food production. His latest books are on socialist states and environment and on urban agriculture and ecosocialism.

Terran Giacomini is a PhD candidate at the University of Toronto in adult education and community development. Her community-engaged scholarship explores the gendered and ethnicized class relations that support life-affirming resistance and alternatives to capitalism. She is an associate member of Canada's National Farmers Union and a founding member of the Via Campesina movement for food sovereignty and agroecology.

Ana Isla is Professor Emerita at Brock University in Canada. She is conducting research on subsistence economies in the Peruvian rainforest and extractivism in Latin America.

Michael Löwy completed two PhDs at the Sorbonne, in 1964 and 1974. He is a recipient of the CNRS Silver Medal as the best French social scientist (1994) and the European Walter Benjamin Prize (2020). He co-authored, with Joel Kovel and Ian Angus, the Declaration of Belem (2009) at the Belem Social World Forum. He is a member of the steering committee of the Global Ecosocialist Network.

Terisa E. Turner is a revolutionary, ecofeminist Marxist trained and mentored by and in the tradition of CLR James. She worked at the UN and at nine universities. She writes on petroleum, gender, social movements, and commoning transitions.

THE ROUTLEDGE HANDBOOK ON ECOSOCIALISM

*Edited by Leigh Brownhill, Salvatore Engel-Di Mauro,
Terran Giacomini, Ana Isla, Michael Löwy,
and Terisa E. Turner*

Routledge
Taylor & Francis Group
LONDON AND NEW YORK

First published 2022
by Routledge
2 Park Square, Milton Park, Abingdon, Oxon OX14 4RN

and by Routledge
605 Third Avenue, New York, NY 10158

Routledge is an imprint of the Taylor & Francis Group, an informa business

British Library Cataloguing-in-Publication Data
A catalogue record for this book is available from the British Library

Library of Congress Cataloging-in-Publication Data
Names: Brownhill, Leigh, editor.
Title: The Routledge handbook on ecosocialism / edited by Leigh Brownhill, Salvatore Engel-Di Mauro, Terran Giacomini, Ana Isla, Michael Löwy, and Terisa E. Turner.
Description: 1 Edition. | New York, NY : Routledge, 2022. | Includes bibliographical references and index.
Identifiers: LCCN 2021030011 (print) | LCCN 2021030012 (ebook) | ISBN 9780367357290 (hardback) | ISBN 9781032134949 (paperback) | ISBN 9780429341427 (ebook)
Subjects: LCSH: Ecosocialism.
Classification: LCC HX550.E25 R68 2022 (print) | LCC HX550.E25 (ebook) | DDC 304.2—dc23
LC record available at https://lccn.loc.gov/2021030011
LC ebook record available at https://lccn.loc.gov/2021030012ed

ISBN: 978-0-367-35729-0 (hbk)
ISBN: 978-1-032-13494-9 (pbk)
ISBN: 978-0-429-34142-7 (ebk)

DOI: 10.4324/9780429341427

Typeset in Bembo
by Apex CoVantage, LLC

CONTENTS

FIGURES AND TABLES

Figures

Tables

CONTRIBUTOR BIOGRAPHIES

Samuel Alexander is a lecturer and researcher at the University of Melbourne, Australia, teaching a course called Consumerism and the Growth Economy: Critical Interdisciplinary Perspectives as part of the Master of Environment. He is also co-director of the Simplicity Institute and a research fellow at the Melbourne Sustainable Society Institute.

Elmar Altvater (1938–2018) was a political scientist who, as a university professor and a concerned political intellectual, contributed an extensive range of lectures, research, writings, and public interventions. Trained in critical political economy, his main interest was directed to developing a fruitful analysis of our future-threatening economic system.

Hans A. Baer is an anthropologist based at the University of Melbourne. His areas of research include Mormonism; African American religion; socio-political life in East Germany; critical health anthropology; medical pluralism in the US, the UK, and Australia; the critical anthropology of climate change; Australian climate politics; and ecosocialism.

Benjamin Barson is an American Society of Composers, Authors and Publishers (ASCAP) award-winning composer and protégé of the late baritone saxophonist and composer Fred Ho. Barson's academic, activist, and performing arts work has been unrelenting in its commitment to fighting racism, capitalism, patriarchy, and the destruction of the planet.

Nnimmo Bassey is Director of the ecological think tank Health of Mother Earth Foundation (www.homef.org), headquartered in Benin City, Nigeria. He received the Right Livelihood Award in 2010 "for revealing the full ecological and human horrors of oil production and for his inspired work to strengthen the environmental movement in Nigeria and globally."

Frei Betto is Dominican, an advisor of social movements, and a writer. Among his books is *Batismo de Sangue* [Baptism of Blood] (2000), which inspired a film by the same title by Helvecio Ratton (2006).

Leigh Brownhill is an ecofeminist scholar and writer, anti-racist abolitionist, long-time decolonial researcher, and community activist. She authored *Land Food Freedom: Struggles for the*

Gendered Commons in Kenya and is associate editor of *Capitalism Nature Socialism*, the international, multi-disciplinary journal of ecosocialism. She teaches sociology at Athabasca University, Canada.

Peter Burdon is Associate Professor and Deputy Dean at the Adelaide Law School.

Renán Vega Cantor has a PhD in political science from the University of Paris 8. He is a professor at the National Pedagogical University (Bogotá) and author of books on the history of Colombia, Latin America and on capitalism and imperialism. He was a student of Michael Löwy, who directed his doctoral thesis on social struggles in Colombia in the first decades of the twentieth century.

Brett Clark is a professor of sociology at the University of Utah and the associate editor of *Monthly Review*. His most recent book, with John Bellamy Foster, is *The Robbery of Nature* (Monthly Review Press, 2020).

John Clark is an eco-communitarian anarchist writer, activist, and educator in New Orleans. His most recent book is *Between Earth and Empire: From the Necrocene to the Beloved Community*. He directs La Terre Institute for Community and Ecology and is a member of the Education and Research Workers Union of the IWW.

Jacklyn Cock is Honorary Research Professor in the Society, Work and Politics Institute and Professor Emeritus in the Department of Sociology, University of the Witwatersrand, South Africa. A committed ecofeminist-socialist and environmental justice activist, she has recently written *The War Against Ourselves: Nature, Power and Justice* (2007) and *Writing the Ancestral River* (2018).

Pat Devine is a lifelong communist, now ecosocialist, who taught economics at Manchester University and is now an honorary research fellow in social science. Devine is the author of *Democracy and Economic Planning: The Political Economy of a Self-governing Society* and co-author of *What On Earth Is To Be Done?* as well as convenor of Red Green Study Group, unionist, and a member of CND and Climbers' Club.

Salvatore Engel-Di Mauro is Professor of Geography at SUNY New Paltz and Chief Editor of *Capitalism Nature Socialism*. His most recent books are *Socialist States and Environment* (Pluto Press) and, with George Martin, *Urban Food Production for Ecosocialism* (Routledge). His research areas include socialist histories, soil contamination processes, and urban food production.

John Bellamy Foster is a professor of sociology at the University of Oregon and the editor of *Monthly Review*. His most recent book is *The Return of Nature* (Monthly Review Press, 2020).

Arran Gare is Associate Professor in Philosophy and Cultural Inquiry, Swinburne University, Melbourne, Australia. He is a former Fulbright postdoctoral fellow and author of a number of books, including *Postmodernism and the Environmental Crisis*, *Nihilism Inc.*, and *The Philosophical Foundations of Ecological Civilization: A Manifesto for the Future*.

Terran Giacomini is a PhD candidate at the University of Toronto in adult education and community development. Her community-engaged scholarship explores the gendered and ethnicized class relations that support life-affirming resistance and alternatives to capitalism. She is an associate member of Canada's National Farmers Union and a founding member of the Via Campesina movement for food sovereignty and agroecology.

José Luis Haro García has a PhD in political science from the University of Barcelona and focuses on political theory through different works on environmentalism, globalization, and economic democracy. He is the author of *Environmentalism and Workplace Democracy: A Green Theory on the Democratization of Companies* (CEPC, Madrid, 2020).

Natasha Heenan is a political economist and PhD candidate at the University of Sydney. Her research interests include the political economy of climate change, just transitions, and the future of work. She is currently writing a thesis on the relationship between the Australian labour and environmental movements.

Qingzhi Huan is Professor of Comparative Politics, focusing on environmental thought and social mobilization in comparative perspective. He has authored several articles on social-ecological transformation and socialist eco-civilization in English and Chinese. He leads a research team working on this issue, in collaboration with the Rosa-Luxemburg Foundation of Germany.

Ana Isla is Professor Emerita at Brock University in Canada. She is the author of *The "Greening" of Costa Rica: Women, Peasants, Indigenous Peoples, and the Remaking of Nature*, Toronto University Press (2015), and editor of *Climate Chaos: Ecofeminism and the Land Question*, Inanna Publications & Education Inc., Toronto (2019).

Joel Kovel (1936–2018) was a prolific ecosocialist and anti-Zionist scholar and activist, as well as former psychiatrist. He authored, among other books, *The Enemy of Nature* and *Overcoming Zionism*. A main force in the global development of ecosocialism, he co-authored with Michael Löwy the 2001 International Ecosocialist Manifesto.

Michael Löwy is CNRS Emeritus Research Director (France). In 2001, he co-authored the International Ecosocialist Manifesto with Joel Kovel and *Romanticism Against the Current of Modernity* (Duke University Press) with Robert Sayre. He is the author of *Ecosocialism: A Radical Alternative to the Capitalist Catastrophe* (Haymarket, 2015).

Mary Mellor is Emeritus Professor, Northumbria University. She first advocated feminist green socialism in her 1992 book *Breaking the Boundaries*. Since then, she has published many books and articles on ecofeminism, ecosocialism, social economy, and money. She is the author of *Money: Myths, Truths and Alternatives*.

Anitra Nelson is an activist-scholar affiliated with the Centre for Urban Research at RMIT University (Melbourne, Australia), best known for advocacy and works on a post-capitalism beyond money, degrowth, housing and communal sustainability. Find out more about her here: https://anitranelson.info/.

Miguel Angel Núñez is an agroecologist, writer, teacher, and researcher at various national universities in Venezuela. Linked to international social movements, he has published on the environment, agri-food policy and ecosocialism, and agroecology, among other topics. He currently works in the office of the Ministry of Science and Technology of Venezuela.

James O'Connor (1930–2017) was a political economist and professor of sociology at the University of California, Santa Cruz, teaching sociology, economics, and environmental studies until his retirement. With Barbara Laurence, he co-founded *Capitalism Nature Socialism* and co-directed the Center for Political Ecology.

Silvia Ribeiro is the Latin American director of the Action Group on Erosion, Technology and Concentration (ETC Group). She is a well-known lecturer, writer, and educator on the impacts of new technologies on society, particularly on peoples and communities. With ETC Group, she has critically followed the debate and developments of geoengineering and its actors since 2006.

Kohei Saito, Associate Professor in the Department of Economics, Osaka City University, is the author of *Karl Marx's Ecosocialism: Capital, Nature, and the Unfinished Critique of Political Economy* (Monthly Review Press, 2017) and *Capital in the Anthropocene* (Shueisha, 2020; in Japanese) and editor of *Reexamining Engels' Legacy in the 21st Century* (Palgrave, 2021).

Ariel Salleh is a founding editor of *Capitalism Nature Socialism* (1989); Visiting Professor in Humanities, Nelson Mandela University; formerly Honorary Associate in Political Economy, University of Sydney; and Fellow in Post-Growth Societies, Friedrich Schiller University, Jena. Her work includes *Ecofeminism as Politics* (2017/1997), *Eco-Sufficiency & Global Justice* (2009), *Pluriverse* (2019), and some 200 articles at www.arielsalleh.info.

Vishwas Satgar is an associate professor of international relations and principal investigator for Emancipatory Futures Studies in the Anthropocene, University of Witwatersrand, South Africa. He also edits the Democratic Marxism series and has been an activist for the past four decades. He is the co-founder of the South African Food Sovereignty Campaign and Climate Justice Charter process.

David Schwartzman is a researcher in climate/energy and biogeochemistry and Professor Emeritus at Howard University and has a PhD in geochemistry from Brown University. His books include *Life, Temperature and the Earth* (1999, 2002); *The Earth is Not for Sale* (with Peter Schwartzman) (2019); and *The Global Solar Commons* (2020). He is on the editorial board of *Science & Society* and the advisory board of *Capitalism Nature Socialism*. Website: http://theearthisnotforsale.org.

Nadia Singh lectures in economics at Northumbria University and is a fellow of the Higher Education Academy of UK. Her research interests include ecological economics, feminist economics, and alternative pedagogies. Nadia is a socially engaged academic and regularly contributes newspaper op-eds in Indian dailies on current policy debates.

Pritam Singh got his DPhil from the University of Oxford and is professor emeritus at Oxford Brookes University. He has taught economics at Delhi University and Panjab University, Chandigarh, and has held visiting positions at the University of Oxford; Lomonosov Moscow State University; Jawaharlal Nehru University, Delhi; and the University of Uberlandia (Brazil).

Daniel Tanuro is a certified agriculturalist and Belgian ecosocialist. His book *L'Impossible Capitalisme Vert* (La Découverte, 2010) has been translated into English (as *Green Capitalism:*

Why It Can't Work, Resistance Books) and five other languages. His latest publication is *Trop Tard pour Être Pessimistes* (2020, Édition Textuel).

Seth Tobocman is a comic book artist who teaches at the School of Visual Arts in Manhattan. With Peter Kuper, he co-founded *World War Three Illustrated* in 1980 as an artist-run collective. His books include *You Don't Have to Fuck People Over to Survive* and *Disaster and Resistance*, among others.

Terisa E. Turner is a revolutionary, ecofeminist Marxist trained and mentored by and in the tradition of CLR James. She published *Oil and Class Struggle* (1980) and *Arise Ye Mighty People!* (1994) and worked at the UN and at nine universities. She writes on petroleum, gender, social movements, and commoning transitions.

Victor Wallis is the author of *Red-Green Revolution: The Politics and Technology of Ecosocialism* (2018, 2021), *Democracy Denied: Five Lectures on US Politics* (2019), and *Socialist Practice: Histories and Theories* (2020). He teaches at Berklee College of Music and is editor-at-large of *Socialism and Democracy*. www.victorwallis.com/.

ACKNOWLEDGEMENTS

First of all, we acknowledge that we, the editors of this volume, are living on stolen Indigenous lands: Leigh Brownhill lives on Haudenosaunee territory on the banks of the Grand River; Terran Giacomini on the land of the Wendat, Haudenosaunee, and Anishinaabe; and Terisa E. Turner on the lands of the Haudenosaunee and Pocumtuk. Salvatore Engel-Di Mauro, originally from Biella (Italy), lives near the Magonk (Wallkill) River in originally Esopus territory. Ana Isla, originally from the rainforest of Peru, lives in the Niagara Region of Ontario on the traditional territory of the Haudenosaunee and Anishinaabe. Michael Löwy would like to acknowledge the Tupi of the Heights of Piratininga, present-day São Paulo, where he was born. We are all deeply grateful for the privilege of being connected to these lands. We strive to build relationships of solidarity and kinship with Indigenous peoples to protect all life.

The idea for this *Handbook* arose thanks to the solicitation of Andrew Mould, Routledge publisher, who in 2018 contacted one of us (Engel-Di Mauro) to put together an edited volume that would provide an overview of and introduction to ecosocialism and its diverse origins, frameworks, and politics. Without Andrew Mould's vision, there would not have been such a handbook, at least not of the sort now before you. In this, we are especially indebted to Senior Editorial Assistant Egle Zigaite, who ably shepherded our work through the horrific time of the pandemic, ensuring that the *Handbook* would see the light of day. We are grateful to the Center for Political Ecology and their publisher, Taylor & Francis Ltd., for granting permission to reprint several foundational texts. These reprints appeared in the journal *Capitalism Nature Socialism*, a main outlet for ecosocialist thought since its founding in 1988. These reprinted works are the following:

Elmar Altvater, "Ecological and Economic Modalities of Time and Space," *Capitalism Nature Socialism* 1 no. 3 (1989): 59–70.

Joel Kovel, "The Ecofeminist Ground of Ecosocialism," *Capitalism Nature Socialism* 16 no. 2 (2005): 1–8.

Michael Löwy, "From Marx to Ecosocialism," *Capitalism Nature Socialism* 13 no. 1 (2002): 121–33.

James O'Connor, "Capitalism, Nature, Socialism: A Theoretical Introduction," *Capitalism Nature Socialism* 1 no. 1 (1988): 11–38.

We express endless gratitude to the authors who contributed to this volume. Without their efforts, diligence, and generosity of spirit, this *Handbook* could not have come into existence. The completion of this collective work benefited decisively from their support and from the comradely relationships developed over the decades among all involved.

Finally, we are deeply grateful for the people in our lives who have supported us and for our Mother Earth, to whom this work is dedicated, along with all the freedom fighters, land defenders, and water and sky protectors who are courageously standing in the way of the complete annihilation of life. We are with you.

PREFACE

An introduction to ecosocialism

Leigh Brownhill, Salvatore Engel-Di Mauro, Terran Giacomini, Ana Isla, Michael Löwy, and Terisa E. Turner

Ecosocial rationales for ecosocialism

Current struggles for egalitarian and ecologically just futures have emerged as responses to the long history of social and environmental miseries wrought in the global expansion and intensification of capitalist relations (Isla 2019; Kovel 2002; Sarkar 1999; Turner and Brownhill 2006; Wall 2010). Ecosocialism has arisen from and informs many of these struggles. It represents resistance and alternatives to the concentration and centralization in the hands of the few of the wealth produced now by most of humanity (Marx 1867 [1992], 776). Among the many sustained dire repercussions of capitalism are chronic hunger and malnutrition for more than a billion people, bloated militaries and ever-deadlier wars, mass human warehousing (incarceration, refugee camps), a third of the world population having little to no access to safe potable water (in both the global South and North), forced migrations, and the denial of shelter to millions, while vast riches and political privileges accrue to a small fraction of humanity.

Intimately tied to and undergirding this social disaster are the continuing ravages that capital is imposing on the rest of nature. Capitalism has meant the constant growth of greenhouse gas output. This has promoted more frequent and deadly extreme weather (Herring et al. 2020) and the expansion of the melting of glaciers and ice sheets, with steadily rising sea levels and the submergence of coastlines and islands. Although all aspects of the ecological crisis are linked, climate change is a political issue of primary importance because it is leading, at an inexorable pace, to a situation in which not only will most towns of human civilization disappear under water, but also the basic conditions of human life itself may be threatened. But this capitalism-caused planetary disaster is hardly the only one. The more widely acknowledged climate chaos has been historically preceded by and concurrent with the precipitous decline of many species' populations and a geologically unprecedented rate of extinctions (Ceballos, Ehrlich, and Dirzo 2017; Sánchez-Bayo and Wyckhuys 2019; WWF 2020). Just as importantly, capitalism is a mode of creating prodigious amounts of waste and of producing an increasingly more contaminated environment with mounting volumes of plastics, persistent toxic substances (PCBs, lead, radioactive waste, etc.), oil spills, and much else. Capitalism is the destruction of millions of lives to benefit the endless profit thirst and the debauched profligate lifestyle of the ruling capitalist classes.

 10.4324/9780429341427-1

We live in a capitalist world economy that some have named "fossil capitalism" because it is dependent on natural gas, oil, and coal. But endless capital accumulation, which is inherent to capitalism, means the endless extraction of resources generally. No path is available or conceivable within capitalism that would allow for the development of ecologically unsustainable modes of living. A transition away from fossil fuels (Malm 2016) will result in displacing harms onto people and ecosystems elsewhere from the communities benefiting from that energy system transition. To prevent further ecological destruction and have a chance to reduce the now decades-long climate chaos, capitalism has to go. Much of the historical and ongoing damage from capitalism, like species extinctions, world deforestation, lead contamination, and radioactive waste, is either entirely irreversible or irreversible relative to multiple human lifespans. The issue is then to prevent further social and ecological devastation. Leftists set themselves up for failure if they think that just transforming a social system solves the vast ecological destruction that has accumulated and intensified thanks to capitalism. Capitalism will only worsen and widen its destructive effects and create even more, novel forms of ecological damage, including via "renewable" energy systems.

To be ecologically sustainable means to free ourselves altogether from capitalist relations, which are exploitative by definition. This is why system change towards ecosocialism is essential. In contrast to capitalist systems, cooperation, mutualism, solidarity, and life support are among the main founding principles of ecosocialism. Lasting or irreversible ecological damage, especially at the planetary scale, can only be addressed through the cooperation of all communities dedicated to finding reciprocally beneficial solutions. This is a most practical matter. Communities assaulted and scarred by the social inequalities and multiple forms of oppression that inhere capitalism cannot tackle problems requiring close, mutually respectful collaboration and egalitarian decision-making processes. In this light, as ecosocialists, we have much to learn from existing and resisting oppressed communities, especially Indigenous peoples. Ecosocialism is already alive in ancient and Indigenous worldviews and prefigurative experiments. Indigenous struggles for a non-capitalist, anti-extractive, and solar-based commoning offer crucial lessons and inspiration for those of us seeking to building a vibrant egalitarian and ecologically sustainable world.

Capitalism's threat to planetary life is evident from the major examples cited earlier, along with the profit system's reliance on racist and misogynistic violence from officials, including law makers and law enforcement, as well as from supremacist political forces. Against and despite these threats, ecosocialist experiments are striving to overcome capitalism-induced calamities, not by reforming at the edges but by doing away with capitalism altogether. As this volume of collected works demonstrates, overcoming capitalism now may not resemble precisely the earlier visions of global revolution, led by unions, national liberation movements, and vanguard parties. Instead, capitalism's successor society is emerging right now, among us, in a multitude of prefigurative efforts, involving myriad moments and movements of transformation, building, and reinforcing the foundations of ecosocialist alternatives globally.

A brief survey of the history and multiple meanings of ecosocialism

A history of the ideas of ecosocialism is, in the historical materialist tradition, a history of the social movements whose actions, claims, and demands have shaped the theory and politics of ecosocialism. For us, ecosocialism is not primarily a theory or party line that rains from above, but rather the convergence of resistance and anti-capitalist movements from below, and their practices and critiques, that together articulate opposition to relations of exploitation and dispossession and the defense, establishment, and elaboration of praxes of an alternative political economy and way of being, rooted in social and ecological justice. Indigenous Amazonia's experience with Spanish and Portuguese colonization and evangelization, which incorporated them into global commerce, and the early twentieth-century boom of rubber, which linked them to industrial capitalism, made them among the first inhabitants rejecting and resisting colonization and capitalism to this day. They stuck steadfast to their ecologically sustainable mode of production, which is characterized by its high degree of autonomy and freedom in the organization of work. Sociability is obtained in the interaction between human beings (group work) and in the synergy with the elements of the biophysical environment, where the dialogue with the rest of nature follows the rhythm given by the cultural system based on Common Rights Usages; they do not divide life in two periods – utilitarian work and pleasant leisure; their activities bring together the useful and the pleasant, and life is built in the exercise of solidarity, according to social rights and obligations (Gashe and Vela 2012).

Ecosocialism, briefly put, is a movement, a perspective, a practice, and in some cases an institutional politics that gathers together socialist and ecological principles and objectives. It is socialist in the sense of identifying capitalist relations as the ultimate and systemic cause of structural inequalities and environmental destruction. Politically, this means struggling for social equality by establishing social control over the means of life and ending all forms of exploitation of nature, which includes people. This struggle embraces decolonization and cross-generational justice as well as overcoming the gendered and racialized class hierarchies that capitalism depends on for the exploitation of the unwaged majority of the world's people, notably women. It means developing respect for differing knowledge systems, striving to combine them to the benefit of all. Ecosocialism is environmentalist in calling attention to the biophysically destructive character of currently conventional power relations and their imposed ways of living. Ecosocialism premises the understanding of biophysical processes on diverse forms of systematic knowledge and inquiry, institutional and otherwise. Ecosocialism, in other words, stands for the development of ecologically sane egalitarian communities worldwide (Kovel 2014; Löwy 2015; Sarkar 1999; Turner and Brownhill 2006).

Environmental concern was part of early socialism, expressed in the writings of Marx, Engels, Reclus, Kropotkin, Luxemburg, and Lenin, among others. But it was not until the 1960s that socialist movements returned to and elaborated on those germinal ideas within the socialisms of the 1800s and early 1900s (Foster 2000; Gare 1993). The global 1968 revolts are a particularly important juncture in this recovery and elaboration process because they brought to the fore the ecological aspects of socialism that had been buried in much of the institutional Left. The past few decades have witnessed the increasingly explicit linkages made between development, social justice or self-determination, and environmental preservation. Ecosocialism is associated with a great diversity of Left-wing political currents involved in such linkage making, some of which contrast with each other in crucial ways. Within these currents one may find variants of Marxism, feminism, anarchism, Indigenism, decolonizationism (national liberationism), abolitionism, syndicalism cooperativism, spiritualism, and sometimes aspects of certain strands of social democracy, technocratic statism, and religious philosophies, like liberation theology, Buddhism, and Taoism (Baer 2018; Pepper 1993; Wallis 2018).

There have been socialist ideas infused to some extent in environmentalist movements as well, especially since the late 1980s. Arguably, more advanced environmental understandings in socialism emerged in areas of the world where struggles for self-determination and sheer survival (e.g., decolonization) involved the protection of ecosystems, such as forests, as means of production and subsistence and their spiritual and intrinsic value. Indigenous people's existential requirements for subsistence and needs for reproducing their ways of life have been a major source of their empowerment and resistance. The more-than-Brazilian Seringueiros movement led by Chico Mendes was among the first to combine ecological and socialist approaches in an explicit manner (Löwy 2015). Chico Mendes tied the rights of rubber-tappers together with the defense of the Amazonian forests against the rapacity of large landed and financial concerns. For such organizing against logging companies, he was assassinated in 1988, as other such activists have been in different South and Central American countries (Méndez 2018). In so-called Honduras, we remember Berta Cáceres, who was assassinated for standing up against mining and dam projects that threatened to destroy her Lenca people's ancestral rivers and homelands. She ignited a wave of solidarity and resistance as the story of her struggle and commitment to life travelled around the globe:

> Let us wake up! Wake up, humankind! We're out of time. We must shake our conscience free of the rapacious capitalism, racism and patriarchy that will only assure our own self-destruction. The Gualcarque River has called upon us, as have other gravely threatened rivers. We must answer their call.
>
> *(Cáceres 2015)*

In so-called Perù, Indigenous peoples are resisting fossil capital. The Peruvian state has declared war on Indigenous peoples living near petroleum concessions known as Blocks1AB and 8, as their territory and economy go unrecognized or invisibilized. The concept of territory as a place of dispute is part of this struggle: how oil fields have gone hand in hand with encroachment on Indigenous people's land, water, and health; how subsistence confiscation resulted in the incorporation of indigenous people into discriminatory and poorly paid jobs; how a protest against poor wages placed Indigenous people into the judicial court in Loreto in 2009; and how the court verdict recognized Indigenous people's collective rights in opposition to state criminalization (Isla 2019). These are but a few examples of resistance involved in the building of ecosocialism.

An ecosocialism for the present has germinated significantly in Indigenous resistance against capitalist violence, enclosure, and exploitation of people and nature: a resistance that was, at the same time, a defense of *already existing* as well as *aspirational* ecosocialism, or egalitarian political economies, values, and relationships built on social and ecological justice for all. Exemplary among such movements are the Zapatistas, the largely Tzotzil Mayan *Ejército Zapatista de Liberación Nacional* (Zapatista Army of National Liberation), founded in 1983 and launched into global recognition on 1 January 1994, the day NAFTA was signed, and Mayan corn farmers, who rose up in arms to stop neoliberalism and spark a global movement of social movements.

The 1990s saw the wide emergence and spread of red-green movements. In some cases, green parties have incorporated selected traditionally socialist issues as part of their platforms, as in the United Kingdom (Wall 2010), where environmental concerns have been interwoven with calls for greater social welfare provisions and more worker protection in the workplace, among other things. Already in the early 1990s, Saral Sarkar, as local Green Party secretary in Köln (Germany), was integrating the aspirations of the peace and environmental movements with socialist principles and was among the first to employ the term "ecosocialism" to that effect (Sarkar 1999). Anarchist groups and Trotskyist and other communist parties have increasingly adopted environmental platforms of their own, such as in the IV International (see chapter by Löwy and Tanuro in this volume). At a minimum, what different ecosocialist approaches share is the striving to contribute to practices and/or to establish theoretical foundations for the attainment of a society founded on the cooperative (or at least coordinated, socially planned) use of resources for the benefit of all and in ways that do not compromise the ecological conditions of our existence. This is also one way to describe the meaning and practice of the multifarious and rambunctious nature of "red-green" politics.

Movements and organizations emerging from these eclectic groundings have been among the most important in the formation and development of existing anti-systemic movements, through the World Social Forum, among other international groupings like *La Via Campesina*. The latter is constituted of small- and medium-size farmers, peasants, Indigenous peoples, landless workers, migrants, fisherfolk, and women and youth seeking to defend systems of shared control over the means of life and food production. There are also revolutionary movements and communities drawing inspiration directly from communist and socialist histories and ideas. Salient examples include the Brazilian Movimento dos Trabalhadores Rurais Sem Terra (MST, or Landless Workers Movement, formed in 1984) and the South African Landless People's Movement (founded in 2001).

In core capitalist countries, there are or there have been the likes of Occupy, the Anarchist Black Cross, and Cooperation Jackson in North America and various communities in Italy struggling to re-establish the commons (e.g., NoTav in the Piemonte region), as well as long-standing squatters' movements in many metropolitan areas (Akuno and Nangwaya 2017; Cattaneo and Engel-Di Mauro 2015; De Angelis 2017). This is in addition to the continuity of relatively small political formations and periodically erupting popular demands for more democratic-socialist provisions, such as universal basic income, workplace rights, unemployment benefits, and socialized healthcare.

Aside from movements taking up these diverse ideas, as expressed in Kovel and Löwy's *Ecosocialist Manifesto* (2001), there have been several countries where ecosocialism has been incorporated into state platforms and policies, as in Ecuador, Bolivia, Venezuela (with the latter having a ministry dedicated to ecosocialism), Iceland (Left-Green Movement), and Brazil (Partido Socialismo e Libertade). Since 2006, activists led by campesino revolutionary Hugo Blanco, who participated in the signing and propagation of the *Ecosocialist Manifesto*, have founded an organization and publication called *Lucha Indígena* (www.luchaindigena.com), dedicated to documenting and analyzing ecosocialist struggles, especially in Peru. More recently still, since 2011, Kurdish-led communities in Rojava (Northern Syria) have adopted the ecosocialist principles of democratic confederalism, according to the thought of the imprisoned Kurdish leader Abdulah Öcalan (in part by way of social ecology). These communities have devised and implemented, as much as feasible, ecologically sensitized, ecofeminist, and participatory democracy ideas in the midst of war (saed 2015). What is also novel to these popular state and institutional developments is their attentiveness to and promotion of feminist, anti-racist, and decolonial egalitarian outlooks, which remain core to ecosocialist principles and transformations (Baer 2018).

Not all of these perspectives could be encompassed in this *Handbook*. Nor would it be possible to accomplish a comprehensive sweep of a still-evolving panoply of ecosocialist movements and worldviews. Instead, what this *Handbook* does offer is a representation of key historical and more recent works that have propelled and shaped ecosocialist thinking and action over the past four decades, when self-described ecosocialist perspectives emerged. This collection can thus be a reference point for international work in the field, including in social movements and institutions of the state. Specifically, the *Handbook* is a way of acquainting people with the varied roots of and sometimes conflicting approaches to ecosocialism. There is no attempt at any unification of ecosocialist currents. Rather, the aim is to provide a resource that is as comprehensive as possible with respect not only to theorization and ideological framing, but also and especially to existing projects, practices, and movements. The Handbook also gives a sense of the geographical reach that ecosocialism so far represents.

Outline of the *Handbook*

The *Handbook* is divided into four topical parts. Each part covers ecosocialism's historical precedents and reference political philosophies, especially from ecofeminism and Marxist currents, and many of the movements and institutional efforts that underpin the varied expressions of ecosocialism. This includes scholarship that extends Marxist foundations and reflects on contemporary political developments. These theoretical and practice-oriented ties are buttressed by discussions of movements, frameworks, and prefigurative processes as well as social struggles occurring within institutional settings.

Part I, "Historical and theoretical groundings," includes works of major historical influence on the development of ecosocialism. Some of these works have been already published or harken back to already-published writings that have been crucial to the crystallizing of an ecosocialist paradigm. A foundational text is from Jim O'Connor (1930–2017), a major figure in the development of ecological Marxism, a main pillar of ecosocialism. In his 1988 introduction to the journal he co-founded, *Capitalism Nature Socialism*, O'Connor articulates a modification to classical Marxist thought that can account for the ecological crisis as well as the emergence of new social movements. He does this by delineating two kinds of contradictions characterizing capitalism: one between the relations and forces of production and a second contradiction between those processes and the conditions of production, which are undermined through the regular workings of capitalism. Succeeding O'Connor in editing the same journal, Joel Kovel (1936–2018) complements and extends the second-contradiction thesis by underlining the crucial role of specifically materialist ecofeminism in forging the necessary alliances and unity across movements to overcome capitalism and forge an ecosocialist society.

Terisa E. Turner elaborates and builds on this view by outlining a theoretical and methodological framework of "gendered, ethnicized class analysis." Key to this framing is the understanding of unwaged people as part of the working class, who, in resisting exploitation, are resisting the power of capital. This in part reiterates and extends Mary Mellor's analysis of the tight linkages between the exploitation and degradation of the natural world and the subordination and oppression of women that form a main basis of capitalism as a simultaneously patriarchal system. Ariel Salleh shows yet another way that ecosocialism must be based on ecofeminism by highlighting the importance of eco-centred practices among the unrecognized workers worldwide. The worldviews and practices inhering their lived experiences provide the means to integrate multiple forms of struggles to overcome capitalism. The importance of this standpoint is illustrated by Ana Isla, who shows how different movements, especially based on the reality of women's lives, are best placed to expose the institutional greenwashing of the brutality of extractivism in Central and South America and beyond.

Renán Vega Cantor, drawing from the work of Michael Löwy, takes us on a different yet interlinked path by exploring the complementarity of Marxist and romanticist critiques of progress, which, in its capitalist garb, implies extractivism. The section ends with an investigation into the roots of dialectical views of nature and a survey of historical and materialist dialectics among Marxists and ecosocialists, upon which he builds a dialectical ecology. He thereby also draws attention to the wealth of existing alternative philosophical frameworks useful towards ecosocialism, which pre-empt the reductionistic, ecological, and socially disconnected thinking prevalent in capitalism-friendly philosophies. Together, these works clarify how ecosocialism has been from the very beginning an intertwining of different worldviews characterized by major overlaps as well as tensions. Authors come from different starting points to understand the history and the way out of the crises of capitalism.

Part II, "Extending Marxist roots," engages with the contributions Marxists have made in the development of ecosocialist frameworks. Marxist thinkers, as evident in Part I, have from early on endeavoured to recover the ecologically oriented aspects of the writings of Marx and Engels, as well as subsequent Marxist scholarship, which have historically been buried by the socialist currents that prevailed during much of the 1900s. The diversity of ecosocialism can be more easily gleaned by reading the first as well as later writings attempting to establish the nature of ecosocialism. This is where the varied intellectual influences from the approaches represented in the previous section become more evident in terms of how they have shaped and provided diversity in ecosocialist thinking from the start.

Accordingly, Elmar Altvater (1938–2018) shows how Marx conceived of the pressure for endless accumulation that was leading to the transgression of ecological borders and threatening the very survival of the human species. In particular, the chapter brings into focus an analysis of the contradiction between capitalist economy and ecology, as found in incessant efforts to compress the space and time necessary for people and the rest of nature to exist and reproduce themselves.

Addressing further the theme of reproduction, Leigh Brownhill addresses the central importance of women's labour in producing the strategically necessary commodity of human labour power. She argues that capitalists' control over women's labour-power-producing capacities is essential to the continuity of the profit system. Under historic and contemporary capitalism, women's alienation from nature and the commons undergirds their vulnerability to violence and control by men. The chapter's ecofeminist-ecosocialist perspective affirms the transformational power and potential of popular struggles led by alliances with Indigenous women, peasants, and people of colour to establish the socially and ecologically just political economies of the future, free of the value chains and crises of capitalism.

Michael Löwy draws attention to the crises of capitalism, especially climate chaos, underlying the imperative for the development of a red-green politics of ecosocialism. He argues that socialists must heed environmentalist critiques of Marx's productivism, and, at the same time, environmentalists must come to grips with the capitalist causes of productivism and adopt Marxist critiques of capitalism. Brett Clark and John Bellamy Foster explicate further the metabolic rift scholarship that has helped, since the late 1990s, excavate the ecological foundations of classical historical materialism. Their chapter illuminates this scholarship, in particular examining how capitalism creates an alienated social metabolism, which in turn produces ecological rifts by transgressing natural limits and fatally disrupting ecosystems. Kohei Saito complements this work by challenging the conviction, held by many environmentalists and Marxists, that Marx supported hyper-industrialization and the domination of society over nature. Saito addresses recent studies, including notebooks only recently available, that illuminate Marx as an ecosocialist. With this backdrop, he shows that Marxism in the twenty-first century continues to develop ecological critiques of capitalism and to envision a sustainable society beyond capitalism.

Closing out this section of the *Handbook* is Arran Gare's tracing of the history of ecology and the concept of culture in the Soviet Union, particularly in the 1920s, to support the need for a radical socialist ecological civilization, to be developed globally and to transform every part of society, changing the way people relate to each other and to nature. He argues that with this notion being officially embraced within China, a tradition of socialist thought has been revived that has the potential to challenge and replace global capitalism.

Part III, "Movements, prefiguration and frameworks," gives examples of concrete grassroots movements and actions based on or linked to the diffusion and building of ecosocialism. This section is dedicated to discussions of ecosocialist theory and its potential application in political work, drawing from early ecosocialist theorization as well as from novel and ongoing movements that link socialism and ecology in explicit ways. The section opens with Right Livelihood Award winner Nnimmo Bassey's survey of the social and ecological destruction wrought by oil operations in Nigeria's Niger Delta. Bassey illustrates two instances of the deep trauma caused by the oil industry, pointing to the urgent need for an ecosocialist transition. He sheds new light on the anti-oil activism of Ken Saro-Wiwa and eight other Ogoni Indigenous leaders, whose courage and sacrifice have been sources of great inspiration for freedom fighters and land defenders around the world. In the same vein, Seth Tobocman, Leigh Brownhill, and Terisa E. Turner highlight the anti-oil and subsistence activism of women in the Niger Delta. The chapter illustrates, through a graphic narrative, eco-activists' use of the tactic of simultaneous, cross-border, coordinated direct action involving the blockade of specific companies' oil production sites and the boycott of their oil consumption outlets. These actions were inspired by Nigerian women, who threatened to use the curse of nakedness to impose extreme social ostracization on any men in league with Chevron and other oil companies. The simultaneous global women's nakedness demonstrations that followed starkly contrasted the life-oriented, peaceful power of ecocentric women (and the men who joined them) against the destructive and exploitative operations of the fossil fuel industry, whose markets are protected and enlarged through warfare.

The chapter by Vishwas Satgar and Jacklyn Cock provides an overview and analysis of environmental and climate justice politics emerging from struggles against South Africa's carbon capitalist economy and its "minerals-energy complex." The authors present a framework for understanding the emergence of such struggles, relating them to an expanding process of carbon capitalist accumulation. They highlight the contribution of democratic ecosocialism and address some of the challenges facing democratic ecosocialist politics in South Africa.

Any volume on ecosocialism must include a discussion of the global 200-million-strong Via Campesina movement for food sovereignty and agroecology. Terran Giacomini, a scholar-activist involved in the movement, offers an ecofeminist analysis of the food provisioning practices and activism of specific women and gender non-binary food producers in the movement. She highlights the ways in which women's life priorities and struggles are crucial to the transformative character of the movement for a "just transition" from capitalism to ecosocialism. Black liberation struggles are also crucial to a just transition towards ecosocialism. Benjamin Barson's analysis of the politics of Black jazz musicians and activists in Louisiana's Sugar Parishes in the nineteenth-century US South traces the emergence of an "eco-logic aesthetic" within the struggles of Black jazz musicians fighting for land, freedom, and the commons. Barson shows that nineteenth-century Black activists and musicians anticipated the ecological crisis which is now unfolding globally and how mass music making has served as a check against the unrestrained power of capital.

José Luis Haro García's chapter offers an analysis of workplace democracy and its role in ecosocialist transformation. He argues that workplace democracy is both a socio-economic project and strategy to achieve ecosocialist objectives. Silvia Ribeiro explores the political economy of geoengineering with a focus on resistance to it. She highlights specific instances of struggle against geoengineering, arguing that these struggles are crucial to the formation of an ecosocialist politics that prioritizes the interests and demands of grassroots peasant and Indigenous communities around the world. Pritam Singh's argument that the greening of capitalism and the rise of ecosocialism are inter-related makes an important contribution to ecosocialist praxis, particularly the role of "green reforms" and of individual action in the spheres of economy, politics, and culture. In the global North, Extinction Rebellion (XR) has emerged as one of the most active and prominent organizations fighting to defend life by challenging fossil fuels and extractivism. Samuel Alexander and Peter D. Burdon address the potential for XR, especially their practices of non-violent civil disobedience, to contribute to the realization of an ecosocialist society.

Part IV addresses "Power struggles on institutional terrains": that is to say, strategies and platforms that could be developed within and/or outside state institutional frameworks. Leftist politics have always been diverse, and ecosocialism follows this pattern, particularly owing to its multiple origins and geographical contexts. In this section, authors ponder over existing and potential political strategies useful for spreading ecosocialism, within or outside state institutions and on the basis of research and/or actual political organizing.

Based on his involvement in the climate justice movement and his life trajectories, Hans A. Baer describes guidelines that can be helpful in developing a form of ecosocialism that is integrated into a democratic-socialist perspective. Also explicitly grounding her thoughts in her life experiences, Anitra Nelson addresses the main economic instruments of capitalism and, pinpointing structural weaknesses in money-based systems, formulates an alternative, money-less economy in which what actually sustains us, socially and ecologically, is valued and prioritized. Pat Devine reinforces the importance of developing such an ecosocialist economics, one that focuses on ensuring coverage of basic needs in ecologically sustainable ways.

The ultimate goal may be expressed as ecosocialism for some. To others, it is formulated as communism, as in David Schwartzman's work, where the technical and energetic basis for a transition to ecosocialism and then to solar communism is explicated. Victor Wallis complements this by examining the general decision-making processes that would need to inform ecosocialist technological and infrastructural development and that would transform human impacts on the rest of nature in ecologically constructive directions. As Miguel Angel Núñez demonstrates, attempts at this shift are already underway in Venezuela by means of community-based planning and an iterative methodology to transition towards ecosocialist practices, or towards their strengthening and deepening.

On the mainstream institutional front, within core capitalist countries, Natasha Heenan considers the opportunities presented by the emergent politics of the Green New Deal for the wider dissemination and discussion of ecosocialist ideas and praxis. Nadia Singh cautions against bioenergy schemes that may also form part of Green New Deal policies and describes an ecosocialist framework to inform the development of sustainable bioenergy systems responsive to human and ecological well-being, rather than capital accumulation prerogatives.

In the remaining chapters, Salvatore Engel-Di Mauro discusses the major ecologically beneficial achievements of socialist states despite constant military and economic assaults from liberal democracies and calls for ecosocialists to build on and learn from those historical and current experiences. Huan Qingzhi follows with an analysis of the strides made in the People's Republic of China that can be conducive to building ecosocialism there. Analyzing the IV International's historical self-transformation, Michael Löwy and Daniel Tanuro conclude the section, and with that the *Handbook*, by demonstrating the way forward in forging concrete struggles within and outside institutional contexts, including the fight for anti-capitalist reforms that fulfil both social and environmental concerns.

The chapters described here address wide-ranging topics and cover multiple perspectives on ecosocialism, but there is no claim here of being exhaustive. As alluded to earlier, this *Handbook*, as with any work that is not encyclopaedic, can only give a partial overview of ecosocialist currents, but with the intent of whetting the reader's appetite. There are as well ecosocialist thinkers and movements who could not be represented herein, including as a result of challenges faced by invited authors but also due to the SARS-CoV-2 (COVID-19) pandemic. We would therefore like to take the opportunity here to recognize Cooperation Jackson (Akuno and Nangwaya 2017), the Red Nation (see also Red Nation 2015), ecosocialists in the US and UK Green

Parties, and members of the US-based System Change Not Climate Change, who could not complete the respective chapters for which they had been invited.

There are, moreover, not a few themes that regrettably could not figure in this collection. One is the relationship between ecosocialism and the abolitionist movements (but see saed 2012) and the Rojava revolutionary project, which has much affinity with ecosocialism (see, for example, Roelofs 2018; saed 2015). Doubtless other themes could be listed, but among them, we would like to highlight the long-standing but often under-appreciated importance of spirituality in the development of ecosocialism. That is to say, spirituality in the sense of understanding oneself as part of a larger whole of interconnected beings (Kovel 1991). As Frei Betto and Michael Löwy put it, without the spiritual dimension, which includes utopias,

> there can be no mobilizations. And without the possibility of envisaging a different, a new and better world, no hope can exist. Hope favours the upsurge of new utopias, which have to be translated into political and cultural projects that signal a new society.
> *(Betto and Löwy 2010, 98)*

May this *Handbook* be in some way helpful towards forging the necessary new utopias to overcome the current horrific capitalist impasse and build ecosocialist futures.

References

Akuno, Kali, and Ajamu Nangwaya. 2017. "Toward Economic Democracy, Labor Self-Management and Self-Determination." In *Jackson Rising: The Struggle for Economic Democracy and Black Self-Determination in Jackson. Mississippi*, edited by Kali Akuno, Ajamu Nangwaya, and Cooperation Jackson, 43–66. Cantley: Daraja Press.

Baer, Hans. 2018. *Democratic Eco-Socialism as a Real Utopia: Transitioning to an Alternative World System*. New York: Berghahn.

Betto, Frei, and Michael Löwy. 2010. "Ecosocialism and Spirituality." *Capitalism Nature Socialism* 21 (2): 87–99.

Cáceres, Berta. 2015. "Berta Cáceres Acceptance Speech, 2015 Goldman Prize Ceremony." 22 April. youtube.com/watch?v=AR1kwx8b0ms

Cattaneo, Claudio, and Salvatore Engel-Di Mauro. 2015. "Urban Squats as Eco-Social Resistance to and Resilience in the Face of Capitalist Relations: Case Studies from Barcelona and Rome." *Partecipazione e Conflitto* 8 (2): 343–366.

Ceballos, Gerardo, Paul R. Ehrlich, and Rodolfo Dirzo. 2017. "Population Losses and the Sixth Mass Extinction." *Proceedings of the National Academy of Sciences* 114 (30): E6089–E6096.

De Angelis, Massimo. 2017. *Omnia Sunt Communia: On the Commons and the Transformation to Postcapitalism*. London: Zed Books.

Foster, John Bellamy. 2000. *Marx's Ecology*. New York: Monthly Review Press.

Gare, Arran. 1993. "Soviet Environmentalism: The Path Not Taken." *Capitalism Nature Socialism* 4 (4): 69–88.

Gashe, Jorge, and Napoleon Vela. 2012. *Sociedad Bosquesina*. Tomo 1. Iquitos, Peru: Instituto de Investigaciones de la Amazonia Peruana (IIAP).

Herring, Stephanie C., Nikolaos Christidis, Andrew Hoell, Martin P. Hoerling, and Peter A. Stott, 2020. "Explaining Extreme Events of 2018 from a Climate Perspective." *Bulletin of the American Meteorological Society* 101: S1–S140.

Isla, Ana. 2019. "Indigenous Andoas Uprising: Defending Territorial Integrity and Autonomy in Perù." In *Climate Chaos: Ecofeminism and the Land Question*, edited by Ana Isla, 283–308. Toronto: Inanna Publications & Education Inc.

Kovel, Joel. 1991. *History and Spirit, an Inquiry into the Philosophy of Liberation*. Boston: Beacon Press.

Kovel, Joel. 2002. *The Enemy of Nature: The End of Capitalism or the End of the World?* London: Zed Books.

Kovel, Joel. 2014. "Ecosocialism as a Human Phenomenon." *Capitalism Nature Socialism* 25 (1): 10–23.

Löwy, Michael. 2015. *Ecosocialism: The Radical Alternative to Capitalist Ecological Catastrophe*. Chicago: Haymarket.

Malm, Andreas. 2016. *Fossil Capital: The Rise of Steam Power and the Roots of Global Warming*. New York: Verso.

Marx, Karl. 1867 [1992]. *Capital: A Critical Analysis of Capitalist Production*. Translated by S. Moore and E. Aveling. Vol. 1. New York: International Publishers.

Méndez, María José. 2018. "'The River Told Me': Rethinking Intersectionality from the World of Berta Cáceres." *Capitalism Nature Socialism* 29 (1): 7–24.

Pepper, David. 1993. *Eco-Socialism: From Deep Ecology to Social Justice*. London: Routledge.

Red Nation. 2015. "Native Liberation Struggles in North America: The Red Nation 10-Point Program." *Capitalism Nature Socialism* 26 (2): 1–7.

Roelofs, Cora. 2018. "Liberating Life: A Woman's Revolution." *Capitalism Nature Socialism* 29 (1): 128–131.

saed. 2012. "Prison Abolition as an Ecosocialist Struggle." *Capitalism Nature Socialism* 23 (1): 1–5.

saed. 2015. "Rojava." *Capitalism Nature Socialism* 26 (1): 1–15.

Sánchez-Bayo, Francisco, and Kris A.G. Wyckhuys. 2019. "Worldwide Decline of the Entomofauna: A Review of Its Drivers." *Biological Conservation* 232: 8–27.

Sarkar, Saral. 1999. *Eco-Socialism or Eco-Capitalism: A Critical Analysis of Humanity's Fundamental Choices*. London and New Delhi: Zed Books, Orient Longman.

Turner Terisa E., and Leigh Brownhill. 2006. "Ecofeminism as Gendered, Ethnicized Class Struggle: A Rejoinder to Stuart Rosewarne." *Capitalism Nature Socialism* 17 (4): 87–95.

Wall, Derek. 2010. *The Rise of the Green Left: Inside the Worldwide Ecosocialist Movement*. London: Pluto Press.

Wallis, Victor. 2018. *Red-Green Revolution: The Politics and Technology of Ecosocialism*. Toronto and Chicago: Political Animal Press.

World Wildlife Federation. 2020. *Living Planet Report 2020 – Bending the Curve of Biodiversity Loss*. Gland: WWF.

PART I

Historical and theoretical groundings

1

CAPITALISM, NATURE, SOCIALISM

A theoretical introduction[1]

James O'Connor

> Those who insist that [environmental destruction] has nothing to do with Marxism merely ensure that what they choose to call Marxism will have nothing to do with what happens in the world.
>
> –*Aiden Foster-Carter*

Introduction

In 1944, Karl Polányi published his masterpiece, *The Great Transformation*, which discussed the ways in which the growth of the capitalist market impaired or destroyed its own social and environmental conditions.[2] Despite the fact that this book is alive with insights into the problem of economic development and the social and natural environment, it was widely forgotten. The subject of the ecological limits to economic growth and the interrelationships between development and environment were reintroduced into Western bourgeois thought in the late 1960s and early 1970s. The results have been mixed and highly dubious. Polányi's work remains a shining light in a heaven filled with dying stars and black holes of bourgeois naturalism, neo Malthusianism, Club of Rome technocratism, romantic deep ecologism, and United Nations one-worldism (World Commission on Environment and Development 1987). Class exploitation, capitalist crisis, uneven and combined capitalist development, national independence struggles, and so on are missing from these kinds of accounts. The results of these and most other modern efforts to discuss the problem of capitalism, nature, and socialism wither on the vine because they fail to focus on the nature of specifically capitalist scarcity: that is, the process whereby capital is its own barrier or limit because of its self-destructive forms of proletarianization of human nature and appropriation of labour and capitalization of external nature.[3] The usual approaches to the problem – the identification of "limits to growth" in terms of "resource scarcity," "ecological fragility," "harmful industrial technology," "destructive cultural values," "tragedy of the commons," "overpopulation," "wasteful consumption," "production treadmill," etc. either ignore or mangle Marx's theories of historically produced forms of nature and capitalist accumulation and development.

This should not be surprising since Marx wrote little pertaining to the ways that capital limits itself by impairing its own social and environmental conditions, hence increasing the costs and expenses of capital, thereby threatening capital's ability to produce profits, i.e., threatening

 10.4324/9780429341427-3

economic crisis. More, he wrote little or nothing about the effects of social struggles organized around the provision of the conditions of production on the costs and expenses and variability of capital. Nor did he theorize the relationship between social and material dimensions of production conditions, excepting his extended discussion of ground rent (i.e., social relation between landed and industrial capital and material and economic relation between raw materials and industrial production). Marx was, however, convinced of at least three things. The first was that deficiencies of production conditions or "natural conditions" ("bad harvest") may take the form of economic crisis.[4] Second, he was convinced of the more general proposition that some barriers to production are truly external to the mode of production ("the productiveness of labour is fettered by physical conditions")[5] but that in capitalism these barriers assume the form of economic crisis.[6] Put another way, some barriers are "general," not "specific" to capitalism. What is specific is the way these barriers assume the form of crisis. Third, Marx believed that capitalist agriculture and silviculture are harmful to nature, as well as that capitalist exploitation is harmful to human labour power.

In sum, Marx believed that capitalist farming (for example) ruined soil quality. He was also clear that bad harvests take the form of economic costs. However, although he did state that a rational agriculture is incompatible with capitalism,[7] he never considered the possibility that ecologically destructive methods of agriculture might raise the costs of the elements of capital, which, in turn, might threaten an economic crisis of a particular type: namely, underproduction of capital.[8] Put another way, Marx never put two and two together to argue that "natural barriers" may be capitalistically produced barriers, i.e., a "second" capitalized nature.[9] In other words, there may exist a contradiction of capitalism which leads to an "ecological" theory of crisis and social transformation.

Two kinds of crisis theory

The point of departure for the traditional Marxist theory of economic crisis and the transition to socialism is the contradiction between capitalist productive forces and production relations.[10] The specific form of this contradiction is between the production and realization of value and surplus value, or between the production and circulation of capital. The agency of socialist revolution is the working class. Capitalist production relations constitute the immediate object of social transformation. The site of transformation is politics and the state and the process of production and exchange.

By contrast, the point of departure for an "ecological Marxist"[11] theory of economic crisis and transition to socialism is the contradiction between capitalist production relations (and productive forces) and the conditions of capitalist production, or "capitalist relations and forces of social reproduction."[12] Marx defined three kinds of production conditions. The first is "external physical conditions,"[13] or the natural elements entering into constant and variable capital. Second, the "labour power" of workers was defined as the "personal conditions of production." Third, Marx referred to "the communal, general conditions of social production, e.g., 'means of communication.'" (Marx 1973, 533; see also Folin 1979).

Today "external physical conditions" arc discussed in terms of the viability of ecosystems; the adequacy of atmospheric ozone levels; the stability of coastlines and watersheds; soil, air and water quality; and so on. "Labour power" is discussed in terms of the physical and mental well-being of workers, the kind and degree of socialization, toxicity of work relations and the workers' ability to cope, and human beings as social productive forces and biological organisms generally. "Communal conditions" are discussed in terms of "social capital," "infrastructure," and so on. Implied in the concepts of "external physical conditions," "labour power," and "communal conditions" are the concepts of space and "social environment." We include as a production condition, therefore, "urban space" ("urban capitalized nature") and other forms of space which structure and are structured by the relationship between people and "environment,"[14] which in turn helps produce social environments. In short, production conditions include commodified or capitalized materiality and sociality, excluding commodity production, distribution, and exchange themselves.

The specific form of the contradiction between capitalist production relations (and forces) and production conditions is also between the production and realization of value and surplus value. The agency of social transformation is "new social movements" or new social struggles, including struggles within production over workplace health and safety, toxic waste production and disposal, and so on. The social relationships of reproduction of the conditions of production (e.g., state and family as structures of social relations and also the relations of production themselves insofar as "new struggles" occur within capitalist production) constitute the immediate object of social transformation. The immediate site of transformation is the material process of reproduction of production conditions (e.g., division of labour within the family, land use patterns, education, etc.) and the production process itself, again insofar as new struggles occur within the capitalist workplace.

In traditional Marxist theory, the contradiction between production and realization of value and economic crisis takes the form of a "realization crisis," or overproduction of capital. In ecological Marxist theory, economic crisis assumes the form of a "liquidity crisis," or underproduction of capital. In traditional theory, economic crisis is the cauldron in which capital restructures productive forces and production relations in ways which make both more transparently social in form and content, e.g., indicative planning, nationalization, profit-sharing, etc. In ecological Marxism, economic crisis is the cauldron in which capital restructures the conditions of production also in ways which make them more transparently social in form and content, e.g., permanent yield forests, land reclamation, regional land use and/or resource planning, population policy, health policy, labour market regulation, toxic waste disposal planning, etc.

In traditional theory, the development of more social forms of productive forces and production relations is regarded as a necessary but not sufficient condition for the transition to socialism. In ecological Marxism, the development of more social forms of the provision of the conditions of production also may be regarded as a necessary but not sufficient condition for socialism. It should be quickly added that an "ecological socialism" would be different than that imagined by traditional Marxism, first because, from the perspective of the "conditions of production," most struggles have strong, particularistic "romantic anti-capitalist" dimensions (i.e., are "defensive" rather than "offensive") and second because it has become obvious that much capitalist technology, forms of work, etc., including the ideology of material progress, have become part of the problem, not the solution. In sum, there may be not one but two paths to socialism or, to be more accurate, two tendencies which together lead to increased (albeit historically reversible) socialization of productive forces, production relations, conditions of production, and social relations of reproduction of these conditions.

The traditional Marxist account of capitalism as a crisis-ridden system

In traditional Marxism, the contradiction between the production and circulation of capital is "internal" to capitalism because capitalist production is not only commodity production but also production of surplus value (i.e., exploitation of labour). It is a valorization process in which capitalists extract not only socially necessary labour (labour required to reproduce constant and variable capital) but also surplus labour from the working class. Everything else being the same,[15] any given amount of surplus value produced and/or any given rate of exploitation will have the effect of creating a particular shortfall of commodity demand at market prices. Or, put the opposite way, any particular shortage of commodity demand presupposes a given amount of surplus value produced and/or a given rate of exploitation. Further, the greater the amount of surplus value produced and/or the higher the rate of exploitation, the greater the difficulty of realizing value and surplus value in the market. Thus, the basic problem of capitalism is, where does the extra commodity demand which is required to buy the product of surplus labour originate? Time-honored answers include capitalist class consumption; capital investment which is made independently of changes in wage advances and consumer demand; markets created by these new investments; new investment, consumption, or government spending financed by expanded business, consumer, or government credit; the theft of markets of other capitals and/or capitals in other countries; and so on. However, these "solutions" to the problem of value realization (that of maintaining a level of aggregate demand for commodities which is sufficient to maintain a given rate of profit without threatening economic crisis and the devaluation of fixed capital) turn into other kinds of potential "problems" of capitalism. Capitalist consumption constitutes an unproductive use of surplus value, as does the utilization of capital in the sphere of circulation with the aim of selling commodities faster. New capital investment may expand faster than, or independently of, new consumer demand with the result of increasing chances of a more severe realization crisis in the future. While a well-developed credit system can provide the wherewithal to expand commodity demand independent of increases in wages and salaries, the expansion of consumer demand based on increases in consumer or mortgage credit greater than increases in wages and salaries threatens to transform a potential crisis of capital overproduction into a crisis of capital underproduction. Moreover, any expansion of credit creates debt (as well as assets) and financial speculation, instabilities in financial structures, thus threatening a crisis in the financial system. The theft of markets from other capitals implies the concentration and/or centralization of capital, hence a worsening of the problem of realization of value in the future and/or social unrest arising from the destruction of weaker capitals, political instability, bitter international rivalries, protectionism, even war. And so on. In sum, economic crisis can assume varied forms besides the traditional "realization crisis," including liquidity crisis, financial crisis or collapse, fiscal crisis of the state, and social and political crisis tendencies. However, whatever the specific forms of historical crises (the list is meant to be suggestive, not exhaustive), and whatever the specific course of their development and resolution, most if not all Marxists accept the premise based on the real conditions of capitalist exploitation that capitalism is a crisis-ridden system.

The traditional Marxist account of capitalism as a crisis-dependent system and the transition to socialism

In traditional Marxism, capitalism is not only crisis ridden but also crisis dependent. Capital accumulates through crisis, which functions as an economic disciplinary mechanism. Crisis is the occasion which capital seizes to restructure and rationalize itself in order to restore its capacity to exploit labour and accumulate. There are two general, interdependent ways in which capital changes itself to weather the crisis and resolve it in capital's own favor. One is changes in the productive forces; the second is changes in the production relations. Changes in either typically presuppose or require new forms of direct and indirect cooperation within and between individual capitals and/or within the state and/or between capital and the state. More cooperation or planning has the effect of making production more transparently social, meanwhile subverting commodity and capital fetishism, or the apparent "naturalness" of capitalist economy. The telos of crisis is thus to create the possibility of imagining a transition to socialism.

Crisis-induced changes in productive forces by capitals seeking to defend or restore profits (and exemplified by technological changes which lower unit costs, increase flexibility in production, and so on) have the systematic effect of lowering the costs of reproducing the workforce, making raw materials available more cheaply or their utilization more efficient, reducing the period of production and/or circulation, etc. Whatever the immediate sources of the crisis, restructuring productive forces with the aim of raising profits is a foregone conclusion. More, crisis-induced changes in productive forces imply or presuppose more social forms of production relationships, e.g., more direct forms of cooperation within production.[16] Examples of changes in productive forces today, and associated changes in production relationships, include computerized, flexible manufacturing systems and robotics, which are associated with the development of "creative team play" and other forms of cooperation in the workplace; profit sharing; etc. And, of course, the greatest productive force is human cooperation, and science or the social production of practical knowledge has become an almost completely cooperative enterprise (Knight 1988), partly as a result of cumulative historical economic, social, and political crises.

The second way that capital restructures itself is crisis-induced changes in production relations within and between capital, within the state, and/or between the state and capital, which are introduced with the aim of exercising more control of production, markets, and so on (i.e., more planning). Historically, planning has taken many forms (e.g., nationalization, fiscal policy, indicative planning, etc.), including, at the political level, fascism, New Dealism, and social democracy. Whatever the immediate sources of crisis, the restructuring of production relations with the aim of developing more control of labour, raw material supplies, etc. is a foregone conclusion. More, crisis-induced changes in production relations imply or presuppose more social forms of productive forces, e.g., more direct forms of cooperation. Examples of changes in production relations today are "strategic agreements" between high-tech capitals, massive state intervention in financial markets, and centralization of capital via takeovers and mergers. These changes imply sharing or socialization of high-tech secrets and technical personnel, new forms of financial controls, and restructuring of management and production systems, respectively.

To sum up, crisis forcibly causes capital to lower costs and increase flexibility and to exercise more control or planning over production and circulation. Crisis causes new forms of flexible planning and planned flexibility (even at the level of state-organized production), which increases the tensions between a more flexible capitalism (usually market created) and a more planned capitalism (usually state created). Crisis forcibly makes capital confront its own basic contradiction, which is subsequently displaced to the spheres of the state, corporate management, etc. when more social forms of productive forces and production relations are introduced, which imply or presuppose one another while developing independently of one another. In this way, capital itself creates some of the technical and social preconditions for the transition to socialism. However, whether we start from the productive-force or production-relation side, it is clear that technology and power embody one another; hence, new forms of cooperation hold out only tenuous and ambiguous promises for the possibilities of socialism. For example, state capitalism, political capitalism, and so on contain within them socialist forms, but highly distorted ones, which in the course of the class struggle may be politically appropriated to develop less distorted social forms of material and social life. But this is a highly charged political and ideological question. Only in a limited sense can it be said that socialism is imminent in crisis-induced changes in productive forces and production relations. Whether or not these new social forms are imminently socialist forms depends on the ideological and political terrain, degree of popular mobilization and organization, national traditions, etc., including and especially the particular world conjuncture. The same cautionary warning applies to the specific forms of cooperation in the workplace which emerge from the crisis, which may or may not preclude other forms which would lend themselves better to socialist practice, which cannot be regarded as some fixed trajectory but itself an object of struggle and defined only through struggle.

Nothing can be said *a priori* about "socialist imminence" except at the highest levels of abstraction. The key point is that capitalism tends to self-destruct or subvert itself when it switches to more social forms of production relations and forces. The premise of this argument is that *any given set of capitalist technologies, work relations, etc. is consistent with more than one set of production relations, and any given set of production relations is consistent with more than one set of technologies, etc.* The "fit" between relations and forces is thus assumed to be quite loose and flexible. In the crisis, there is a kind of two-sided struggle to fit new productive forces into new production relations and vice versa in more social forms, without, however, any "natural" tendency for capitalism to transform itself to socialism. Nationalization of industry, for example, may or may not be a step towards socialism. It is certainly a step towards more social forms of production and a more specifically political form of appropriation and utilization of surplus value. On the other side, quality circles, work teams, technology sharing, etc. may or may not be a step towards socialism. They are certainly steps towards more social forms of productive forces.

Towards an ecological Marxist account of capitalism as a crisis-ridden system

The point of departure for "ecological Marxism" is the contradiction between capitalist production relations and productive forces and conditions of production. Neither human labour power nor external nature nor infrastructures, including their space/time dimensions, are produced capitalistically, although capital treats these conditions of production as if they are commodities or commodity capital. Precisely because they are not produced and reproduced capitalistically, yet are bought and sold and utilized as if they were commodities, the conditions of supply (quantity and quality, place and time) must be regulated by the state or capitals acting as if they are the state. Although the capitalization of nature implies the increased penetration of capital into the conditions of production (e.g., trees produced on plantations, genetically altered species, private postal services, voucher education, etc.), the state places itself between capital and nature, or mediates between capital and nature, with the immediate result that the conditions of capitalist production are politicized. This means that whether or not raw materials, labour force, and useful spatial and infrastructural configurations are available to capital in requisite quantities and qualities and at the right time and place depends on the political power of capital, the power of social movements which challenge particular capitalist forms of production conditions (e.g., struggles over land as means of production versus means of consumption), state structures which mediate or screen struggles over the definition and use of production conditions (e.g., zoning boards), and so on.[17] Excepting the branches of the state regulating money and certain aspect of foreign relations (those which do not have any obvious relation to accessing foreign sources of raw materials, labour power, etc.), every state agency and political party agenda may be regarded as a kind of interface between capital and nature (including human beings and space). In sum, whether or not capital faces "external barriers" to accumulation, including external barriers in the form of new social struggles over the definition and use of production conditions (i.e., "social barriers" which mediate between internal or specific and external or general barriers),[18] whether or not these "external barriers" take the form of economic crisis, and whether or not economic crisis is resolved in favor of or against capital are political and ideological questions first and foremost, economic questions only secondarily. This is so because production conditions are by definition politicized (unlike production itself) and also because the whole corpus of Marx's work privileges labour power as a production condition; access to nature is mediated by struggles while external nature has no subjectivity of its own.[19] Labour power alone struggles around the conditions of its own well-being and social environment broadly defined.

An ecological Marxist account of capitalism as a crisis-ridden system focuses on the way that the combined power of capitalist production relations and productive forces self-destruct by impairing or destroying rather than reproducing their own conditions ("conditions" defined in terms of both their social and material dimensions). Such an account stresses the process of exploitation of labour and self-expanding capital, state regulation of the provision of production conditions, and social struggles organized around capital's use and abuse of these conditions. The main question – does capital create its own barriers or limits by destroying its own production conditions? – needs to be asked in terms of specific use values, as well as exchange value. This is so because conditions of production are not produced as commodities; hence, problems pertaining to them are "site specific," including the individual body as a unique "site." The question – why does capital impair its own conditions? – needs to be asked in terms of the theory of self-expanding capital; its universalizing tendencies, which tend to negate principles of site specificity; its lack of ownership of labour power; external nature; and space: hence, without state or monopolistic capitalist planning, capital's inability to prevent itself from impairing its own conditions. The question – why do social struggles against the destruction of production conditions (which resist the capitalization of nature: for example, environmental, public health, occupational health and safety, urban, and other movements) potentially impair capital flexibility and variability? – needs to be asked in terms of conflicts over conditions defined both as use values and exchange values.

Examples of capitalist accumulation impairing or destroying capital's own conditions, thus threatening its own profits and capacity to produce and accumulate more capital, are well known. The warming of the atmosphere will inevitably destroy people, places, and profits, not to speak of other species life. Acid rain destroys forests and lakes and buildings and profits alike. Salinization of water tables, toxic waste, soil erosion, etc. impair nature and profitability. The pesticide treadmill destroys profits as well as nature. Urban capital running on an "urban renewal treadmill" impairs its own conditions and thus profits: e.g., congestion costs, high rents, etc.[20] The decrepit state of the physical infrastructure in this country may be mentioned in this connection. There is also an "education treadmill," a "welfare treadmill," a "technological fix treadmill" a "health-care treadmill," etc.[21] This line of thinking also applies to the "personal conditions of production . . . labour power" in connection with capital's destruction of traditionalist family life as well as the introduction of work relations which impair coping skills and the presently toxic social environment generally. In these ways, we can safely introduce "scarcity" into the theory of economic crisis in a Marxist, not neo-Malthusian, way. We can also introduce the possibility of capital underproduction once we add up the rising costs of reproducing the conditions of production. Examples include the health bill necessitated by capitalist work and family relations; the drug and drug rehabilitation bill; the vast sums expended as a result of the deterioration of the social environment (e.g., police and divorce bill); the enormous revenues expended to prevent further environmental destruction and to clean up or repair the legacy of ecological destruction from the past; monies required to invent, develop, and produce synthetics and "natural" substitutes as means and objects of production and consumption; the huge sums required to pay off oil sheiks and energy companies (e.g., ground rent, monopoly profit, etc.); the garbage disposal bill; the extra cost of congested urban space; and the costs falling on governments, peasants, and workers in the Third World as a result of the twin crises of ecology and development. And so on. No one has estimated the total revenues required to compensate for impaired or lost production conditions and/or to restore these conditions and develop substitutes. It is conceivable that total revenues allocated to protecting or restoring production conditions may amount to one-half or more of the total social product – all unproductive expenses from the standpoint of self-expanding capital. Is it possible to link these unproductive expenditures (and those anticipated in the future) to the vast credit and debt system in the world today? To the growth of fictitious capital? To the fiscal crisis of the state? To the internationalization of production? The traditional Marxist theory of crisis interprets credit/debt structures as the result of capital overproduction. Ecological Marxism would interpret the same phenomena as the result of capital underproduction and unproductive use of capital produced. Do these tendencies reinforce or offset one another? Without prejudging the answer, the question clearly needs to be on the agenda of Marxist theory.

Towards an ecological Marxist account of capitalism as a crisis-ridden system and the transition to socialism

Neither Marx nor Marxists have developed a theory of the relationship between crisis-induced changes in the conditions of production and the establishment of the conditions of socialism. In traditional Marxism, crisis-induced changes in productive forces and relations are determined by the need to cut cost, restructure capital, etc. Forces and relations are transformed into more transparently social forms. In ecological Marxism, like traditional Marxism, capitalism is also not only crisis ridden but also crisis dependent. Crisis-induced changes in production conditions (whether crisis itself originates in capital overproduction or underproduction) are also determined by the need to cut costs, reduce ground rent, increase flexibility, etc. and to restructure conditions themselves (e.g., expanding preventive health, reforestation, reorganization of urban space, etc.).

There are two general, interdependent ways in which capital (helped by the state) changes its own conditions to weather the crisis and to resolve it in capital's favor. One is changes in conditions defined as productive forces. The other is changes in the social relations of reproduction of conditions. Changes in either typically presuppose or require new forms of cooperation between and within capitals and/or between capital and the state and/or within the state or more social forms of the "regulation of the metabolism between humankind and nature" as well as the "metabolism" between the individual and the physical and social environment. More cooperation has the effect of making production conditions (already politicized) more transparently political, thereby subverting further the apparent "naturalness" of capital existence. The telos of crisis is thus to create the possibility of imagining more clearly a transition to socialism.

Crisis-induced changes in conditions such as productive forces with the purpose of defending or restoring profit (exemplified by technological changes which lower congestion costs, increase flexibility in the utilization of raw materials, etc.) have the systemic effect of lowering the costs of reproducing the workforce, making raw materials available more cheaply, etc. Whatever the immediate sources of the crisis, restructuring production conditions with the aim of raising profits is a foregone conclusion. More, crisis-induced changes in production conditions imply or presuppose more social forms of the social relations of reproduction of production conditions (e.g., more direct forms of cooperation within the sphere of production conditions). An example of a change in production conditions today, and the associated change in the social relations of reproduction of production conditions, is integrated pest management, which presupposes not only more coordination of farmers' efforts but also more coordination of training and education programmes.[22] Another example is preventative health technology in relation to AIDS and associated changes in community relations in a more cooperative direction.

The second form of restructuring is crisis-induced changes in the social relations of reproduction of production conditions introduced with the aim of exercising more control over production conditions: i.e., more planning. Historically, planning has taken many forms, e.g., urban and regional transportation and health planning, natural resource planning, etc.[23] Whatever the immediate sources of crisis, the restructuring of these social relations with the aim of developing more control over production conditions is also a foregone conclusion. More, crisis-induced changes in the social relations of reproduction of production conditions imply or presuppose more social forms of production conditions defined as productive forces. An example of such a change today is "planning" to deal with urban smog, which presupposes coalitions of associations and groups (i.e., political cooperation) to legitimate tough yet cooperative smog-reduction measures (Daggett 1988). Another example is the proposed restructuring of the US Bureau of Reclamation, which new technical changes in water policy presupposes.[24]

To sum up, crisis forcibly causes capital and state to exercise more control or planning over production conditions (as well as over production and circulation of capital itself). Crisis brings into being new forms of flexible planning and planned flexibility, which increase tensions between a more flexible capitalism and a more planned capitalism – more so than in the traditional Marxist account of the restructuring of production and circulation because of the key role of the state bureaucracy in the provision of production conditions. Crisis forcibly makes capital and state confront their own basic contradictions, which are subsequently displaced to the political and ideological spheres (twice removed from direct production and circulation), where more social forms of production conditions are introduced, defined both materially and socially (e.g., the dominance of political bipartisanship in relation to urban redevelopment, educational reform, environmental planning, and other forms of provision of production conditions which exemplify new and significant forms of class compromise). However, it is clear that technology and power embody one another at the level of conditions as well as production itself; hence, new forms of political cooperation hold out only tenuous promises of socialism. Again, nothing can be said *a priori* about "socialist imminence" excepting at a high level of abstraction. The key point is that capitalism tends to self-destruct or subvert itself when it switches to more social forms of the provision of production conditions via politics and ideology. The premise of this argument (like the argument of the present interpretation of traditional Marxism) is that any given set of production condition technologies, work relations, etc. is consistent with more than one set of social relations of reproduction of these conditions and that any given set of these social relations is consistent with more than one set of production condition technologies, work relations, etc. The "fit" between social relations and forces of reproduction of production conditions is thus assumed to be quite loose and flexible. In the crisis in which the future is unknowable, there is a kind of two-sided struggle to fit new production conditions defined as forces into new production conditions defined as relations, and vice versa, into more social forms without, however, any "natural" tendency for capitalism to transform itself into socialism. Urban and regional planning mechanisms, for example, may or may not be a step towards socialism. They are certainly a step towards more social forms of the provision of production conditions, making socialism at least more imaginable. On the other side, regional transportation networks, health-care services, and bioregional water distribution, for example, may or may not be a step towards socialism. They are certainly a step towards more social forms of the provision of production conditions.

In the modern world, the list of new social and political forms of reproduction of production conditions is endless. It seems highly significant, and also theoretically understated within Marxism, that the world crisis today appears to result in more, and require many additional, social forms of not only productive forces and relations but also production conditions, although the institutional and ideological aspects of these forms are confusing and often contradictory, and although these forms should not be regarded as irreversible (e.g., reprivatization, deregulation, etc.). Yet it is conceivable that we are engaging in a long process in which different yet parallel paths to socialism occur; hence, Marx was not so much wrong as he was half right. It may be that the traditional process of "socialist construction" is giving way to a new process of "socialist reconstruction," or the reconstruction of the relationship between human beings and production conditions, including the social environment. It is at least plausible that in the "first world," socialist reconstruction will be seen as first, desirable and second, necessary; in the "second world" as equally desirable and necessary; and in the "third world" as first, necessary and second, desirable. It is more plausible that atmospheric warming, acid rain, and pollution of the seas will make highly social forms of reconstruction of material and social life absolutely indispensable.

To elaborate somewhat, we know that the labour movement "pushed" capitalism into more social forms of productive forces and relations (e.g., collective bargaining). Perhaps we can surmise that feminism, environmental movements, etc. are "pushing" capital and state into more social forms of the reproduction of production conditions. As labour exploitation (the basis of Marxist crisis theory, traditionally defined) engendered a labour movement which, during particular times and places, turned itself into a "social barrier" to capital, nature exploitation (including exploitation of human biology) engenders an environmental movement (e.g., environmentalism, public health movement, occupational health and safety movement, women's movement organized around the politics or the body, etc.), which may also constitute a "social barrier" to capital. In a country such as Nicaragua, the combination of economic and ecological crises and political dictatorship in the old regime has engendered a national liberation movement and eco-development planning.

Concrete analysis of concrete situations is required before anything sensible can be said about environmentalism, defined in the broadest sense, and capital's short- and long-term prospects. For example, acid rain causes ecological and economic damage. The environmental movement demands cleanup and restoration of the environment and protection of nature. This may restore profits in the long run or reduce government cleanup expenses, which may or may not be congruent with the short- and medium-term needs of capital. Implied in a systematic programme of politically regulated social environment are kinds of planning which protect capital against its worst excesses, yet which may or may not be congruent with capital's needs in particular conjunctures. One scenario is that "the destruction of the environment can lead to vast new industries designed to restore it. Imagine, lake dredging equipment, forest cleaning machines, land revitalizers, air restorers, acid rain combatants" (Correspondence, Saul Landau). These kinds of super-tech solutions would be a huge drain on surplus value unless they lowered the reproduction cost of labour power, yet at the same time help to "solve" any realization problems arising from traditional capital overproduction. Vast sums of credit money would be required to restore or rebuild the social environment, however, which would displace the contradiction into the financial and fiscal spheres in more or less the same ways that the traditional contradiction between production and circulation of capital is displaced into the financial and fiscal spheres today.

This kind of technology-led restructuring of production conditions (including technique-led restructuring of the conditions of supply of labour power) may or may not be functional for capital as a whole or individual capitals in the short or long run. The results will depend on other crisis prevention and resolution measures, their exact conjuncture, and the way in which they articulate the crisis of nature, broadly defined. In the last analysis, the results will depend on the degree of unity and diversity in labour movements, environmental movements, solidarity movements, etc. And this is a political, ideological, and organizational question.

In any event, crisis-induced changes in production conditions necessarily lead to more state controls, more planning within the bloc of large-scale capital, and a more socially and politically administered or regulated capitalism: hence, a less nature-like capitalism, one in which changes in production conditions would need to be legitimated because they would be more politicized and one in which capitalist reification would be less opaque. The combination of crisis-stricken capitals externalizing more costs, the reckless use of technology and nature for value realization in the sphere of circulation, and the like must sooner or later lead to a "rebellion of nature": i.e., powerful social movements demanding an end to ecological exploitation. Especially in today's crisis, whatever its source, capital attempts to reduce production and circulation time, which typically has the effect of making environmental practices, health and safety practices, etc. worse. Hence, capital restructuring may deepen, rather than resolve ecological problems. Just as capital ruins its own markets, i.e., realized profits, the greater the production of surplus value, so does capital ruin its own produced profits, i.e., raise costs and reduce capital flexibility, the greater is the production of surplus value based on the destructive appropriation of nature, broadly defined. And just as overproduction crises imply a restructuring of both productive forces and relations, so do underproduction crises imply a restructuring of production conditions. And just as restructuring of productive forces implies more social forms of production relations and vice versa, so does restructuring of production conditions imply a twofold effect – more social forms of production conditions, defined as productive forces, and more social forms of the social relationships in which production conditions are reproduced. In sum, more social forms of production relations, productive forces, and conditions of production together contain within them the possibilities of socialist forms. These are, in effect, crisis induced not only by the traditional contradiction between forces and relations but also by the contradiction between forces/relations and their conditions. Two crises – not one – are thus inherent in capitalism; two sets – not one – of crisis-induced reorganizations and restructurings in the direction of more social forms are also inherent in capitalism.

Conclusion

Some reference needs to be made to post-Marxist thought and its objects of study, "post-industrial society," "alternative movements" or "new social movements," and "radical democracy."[25] This is so because post-Marxism has practically monopolized discussions of what Marx called "conditions of production." No longer is the working class seen as the privileged agent of historical transformation nor is the struggle for socialism first on the historical agenda. Instead, there is the fight for "radical democracy" by "new social movements" in a "post-industrial society."

These basic post-Marxist postulates deserve close scrutiny, especially given post-Marxist readings of Marx and Marxism and the political implications therein.[26] So does the declaration by radical bourgeois feminists, ecofeminists, deep ecologists, libertarian ecologists, communitarians, etc. that Marxism is dead. In the present discussion, however, it is possible only to point out that in ecological Marxist theory, the struggle over production conditions has redefined and broadened the class struggle beyond any self-recognition as such, at least until now. This means that capitalist threats to the reproduction of production conditions are threats not only to profits and accumulation but also to the viability of the social and "natural" environment as a means of life. The struggle between capital and "new social movements," in which the most basic concepts of "cost" and "efficiency" are contended, has two basic "moments." The first is the popular and nearly universal struggle to protect the conditions of production, or means of life, from further destruction resulting from capital's own inherent recklessness and excesses. This includes needs and demands for the reduction of risks in all forms. This struggle pertains to the form by which "nature" is appropriated: as a means of reproduction of capital or a means of reproduction of civil and human society. The second is the struggle over the programmes and policies of capital and state to restructure the production conditions, i.e., struggles over the forms and contents of changes in conditions. Put another way, new social struggles are confronted with both the impairment and the crisis-induced restructuring of production conditions at the same time. Both "moments" of struggle occur outside the state and also within and against the state: i.e., they pertain to "public administration" (in Carlo Carboni's words). Seen this way, the demand for radical democracy is the demand to democratize the provision and reconstruction of production conditions, which in the last analysis is the demand to democratize the state, i.e., the administration of the division of social labour.[27] In truth, in the absence of struggles to democratize the state, it is difficult to take the demand for "radical democracy" seriously.

In post-Marxist thought, great stress is placed on "site specificity" and the "integrity" of the individual's body, a particular meadow or species life, a specific urban place, etc.[28] The word "difference" has become post-Marxism's mantra, which, it is thought, expels the word "unity," which in the post-Marxist mind is often another way to spell "totalitarian." In the well-thought-out versions of post-Marxist thought, the "site specificity" on which new social movements base themselves are considered to make any universal demands impossible,[29] at least any universal demand beyond the demand for the universal recognition of site specificity. This is contrasted with the bourgeois revolution, which universalized the demand for rights against privilege and the old working-class struggle, which universalized the demand for public property in the means of production against capitalist property. However, our discussion of production conditions and the contradictions therein reveals clearly that there is a universal demand implicit or latent in new social struggles: namely, the demand to democratize the state (which regulates the provision of production conditions), as well as the family, local community, etc. In fact, no way exists for diverse social struggles defending the integrity of particular sites to universalize themselves – hence, win – and at the same time retain their diversity, excepting through struggles for the democratic state and also by uniting with the labour movement, recognizing what we have in common – cooperative labour – thereby theorizing the unity of social labour.[30]

Moreover, post-Marxism, influenced by the "free rider problem" and problems of "rational choice" and "social choice" (all problems which presuppose bourgeois individualism), states or implies that struggles over production conditions are different than traditional wage, hours, and working conditions struggles because conditions of production are to a large degree "commons," clean air being an obvious example, urban space and educational facilities being somewhat less obvious ones. The argument is that struggles against air pollution (or capitalist urban renewal or racist tracking in the schools) do not have an immediate "payoff" for the individual involved; hence, in Offe's account, the phenomenon of cycles of social passivity and outrage owing to the impossibility of combining individual and collective action around goals which "pay off" for both the individual and group. Again, this is not the place for a developed critique of this view, one which would begin with an account of how the process of social struggle itself changes self-definitions of "individuality." It needs to be said, however, that labour unions, if they are anything, are disciplinary mechanisms against "free riders" (e.g., individual workers who try to offer their labour power at less than the union wage are the object of discipline and punishment by the union). Further, it should be said that the "free rider" problem exists in struggles to protect the "commons" only insofar as these struggles are ends in and of themselves, not also means to the specifically political, and hence universal, end of establishing a democratic state.

Also in relation to the problem of the "commons," and beyond the problem of the relation between the individual and the group, there is the problem of the relationship between groups and classes. Specifically, the struggles of "new social movements" over conditions of production are generally regarded in the self-defined post-Marxist universe as non-class issues or multi-class issues. "Transformative processes that no doubt go on in our societies are very likely not class conflicts . . . but non-class issues" (Offe 1987, 234). Especially in struggles over production conditions (compared with production itself), it is understandable that these appear as non-class issues and that agents define themselves as non-class actors. This is so not only because the issues cut across class lines (e.g., urban renewal, clean air, etc.), but also because of the site specificity and "people" specificity of the struggles (i.e., because the fight is to determine what kind of use values production conditions will in fact be). But, of course, there is a class dimension to all struggles over conditions (e.g., tracking in the schools, urban renewal as "people removal," toxic waste dumps in low-income or poor districts and communities, the worker as the "canary" in the workplace, the inability of most unemployed and many workers to access "wilderness areas," etc.). Most problems of the natural and social environment are bigger problems from the standpoint of the poor, including the working poor, than for the salariat and the well-to-do. In other words, issues pertaining to production conditions are class issues, even though they are also more than class issues, which becomes immediately obvious when we ask who opposes popular struggles around conditions. The answer is, typically, capital, which fights against massive public health programmes, emancipatory education, controls on investments to protect nature, even adequate expenditures on childcare and certainly demands for autonomy or substantive participation in the planning and organization of social life. What "new social movements" and their demands does capital support? Few, if any. What "new social movements" does labour oppose? Certainly, those which threaten ideologies of male supremacy and/or white race supremacy, in many instances, as well as those which threaten wages and jobs, even some which benefit labour, e.g., clean air. Hence, the struggle over conditions is not only a class struggle, but also a struggle against such ideologies and their practices. This is why it can be said that struggles over conditions are not less but more than class issues. And that to the degree that this is true, the struggle for "radical democracy" is that much more a struggle to democratize the state, a struggle for democracy within state agencies charged with regulating the provision of production conditions. In the absence of this perspective and vision, "new social movements" will remain at the level of anarcho-communalist and related struggles which are bound to self-destruct in the course of their attempts to "deconstruct" Marxism.

Acknowledgements

I am grateful to Carlo Carboni, John Ely, Danny Faber, Bob Marotto, and David Peerla for their encouragement and helpful criticisms and comments.

Notes

1 Editors' Note: The original was published as James O'Connor. 1988. "Capitalism, Nature, Socialism. A Theoretical Introduction." *Capitalism Nature Socialism* 1 (1): 11–38. The original publication has been modified to conform to the style adopted for *The Routledge Handbook on Ecosocialism*, and the cited references have been amended by completing missing information and removing any redundancies.
2 Polányi's focus was altogether on capitalist markets, not exploitation of labour.
3 The closest anyone has come to a "Marxist" account of the problem is Schnaiberg (1980). This is a path-breaking and useful work. The relation between the capitalization of nature and political conflict between states is another, albeit closely related, question (Timberlake and Tinker 1985).

4 "In the case of bad harvests, *the value of the raw material . . . rises;* its *volume* decreases. . . . More must be expended on *raw material,* less remains for *labour,* and it is not possible to absorb the same quantity of labour as before. Firstly, this is *physically impossible*. . . . Secondly, it is impossible because a greater *portion of the value of the product* has to be converted into raw material. . . . Reproduction cannot be *repeated* on the same scale. A part *of fixed capital* stands idle and a part of the workers is thrown out into the streets. The *rate of profit* falls because the value of constant capital has risen as against that of variable capital. . . . The fixed charges – interest, rent – which were based on the anticipation of a *constant* rate of profit and exploitation of labour, remain the same and in part *cannot be paid*. Hence *crisis*. . . . More, although the *rate of profit* is decreasing, there is a *rise in the price of the product.* If this product enters into the other spheres of reproduction as a means of production, the rise in its price will result in the same disturbance in *reproduction* in these spheres" (Marx 1968, 515–516).

5 "Apart from the degree of development, greater or less, in the form of social production, the productiveness of labour is fettered by physical conditions" (*Capital I*). In *Theories of Surplus Value*, Marx states that the precondition for the existence of absolute surplus value is the "natural fertility of the land" (Marx 1968, 449).

6 Lebowitz (1982) includes as "general" barriers the supply of labour and the availability of land and natural resources. However, he does not distinguish between the supply of labour per se and the supply of disciplined wage labour. As for natural resources, he does not distinguish between "natural" shortages and shortages capital creates for itself in the process of capitalizing nature nor those created politically by ecology movements.

7 *Capital III*, Chapter 6, 215 [Editor's note: no further information provided in original].

8 We can therefore distinguish two kinds of scarcity: first, scarcity arising from economic crisis based on traditional capital overproduction (i.e., a purely social scarcity) and second, scarcity arising from economic crisis based on capitalistically produced scarcity of nature or production conditions generally. Both types of scarcity are ultimately attributable to capitalist production relations. The second type, however, is not due to "bad harvests," for example, but to capitalistically produced "bad harvests" as a result of mining, not fanning, land; polluting water tables; etc.

9 There are two reasons Marx ran from any theory of capitalism and socialism which privileged any aspect of social reproduction besides the contradiction between production and circulation of capital. One was his opposition to any theory which might "naturalize" – hence, reify – the economic contradictions of capital. His polemics against Malthus and especially his rejection of any and all naturalistic explanations of social phenomena led him away from "pulling two and two together." Second, it would have been difficult in the third quarter of the nineteenth century to argue plausibly that the impairment of the conditions of production and social struggles therein are self-imposed barriers of capital because historical nature was not capitalized to the degree that it is today (i.e., the historical conditions of the reproduction of the conditions of production today make an "ecological Marxism" possible).

10 State-of-the-art accounts of the problematic categories of productive forces and production relations are Sayer (1987) and Marotto (1984).

11 Murray Bookchin deserves most of the credit for developing the theory of "social ecology" in the USA. The basic impulse of his method and theory is libertarian not Marxist, "social ecology" not "socialist ecology." To my knowledge, "ecological Marxism" was coined by Ben Agger (1979, 316–339). Agger's focus is "consumption" not "production." His thesis is that the ever-expanding consumption required to maintain economic and social stability impairs the environment and that ecological crisis has replaced the economic crisis as the main problem of capitalism. This chapter may be regarded as, among other things, a critique of Agger's often-insightful views.

12 According to Carlo Carboni, who also uses the expression "social reproductive conditions." I use "conditions of production" because I want to reconstruct the problem using Marx's own terminology and also because I want to limit my discussion mainly to crisis tendencies in the process of the production and circulation of capital, rather than to the process of social reproduction, i.e., reproduction of the social formation as a whole. This means that I will follow Marx's lead and interpret "production conditions" in "objective" terms, excepting in the last section, which suggests that these conditions are increasingly grasped as "subjective" today.

13 External physical conditions include "natural wealth in means of subsistence" and "natural wealth in the instruments of labour" (Marx 1906, 562).

14 In a conversation with David Harvey, who pioneered the theory of the spatial configurations and barriers to capital (Harvey 1982), tentative "permission" was granted the author to interpret urban and other forms of space as a "production condition."

15 The following is a deliberate "Smithian" simplification of the traditionally defined economic contradiction of capitalism, which altogether neglects Marx's critique of Smith: namely, that it is the rising organic composition of capital, not a falling rate of exploitation, which causes the profit rate to fall even though capitalism "presents itself" otherwise. To be absolutely clear, the following account is not meant to review Marx's critique of capital fetishism or Adam Smith, et al. I put the contradiction of capitalism in its simplest terms with the two-fold aim of (a) preparing a discussion of crisis-induced restructuring of the productive forces and production relations and (b) setting up a standard by which we can compare the "traditional" with the "non-traditional" or "second" contradiction of capitalism based on the process of capitalist-created scarcities of external and human nature.

16 "Cooperation" (e.g., "work relations") is both a productive force and a production relationship, i.e., ambiguously determined by both "technological necessity" and "power."

17 This kind of formulation of the problem avoids the functionalism of the "state derivation school" of Marxism as well as political sociological or Weberian theories of the state, which are not grounded in material existence.

18 So-called external barriers may be interpreted as internal barriers. In fact, if we assume that (a) external nature being considered is commodified or capitalized nature, and
(b) new social struggles organized under the sign of "ecology" or "environmentalism" have their roots in the class structure and relations of modem capitalism, e.g., the rise of the new middle class or salariat, which is the backbone of environmentalism in the USA.

19 "External and universal nature can be considered to be differences within a unity from the standpoint of capital accumulation and state actions necessary to assure that capital can accumulate. Yet the difference is no less significant than the unity from the standpoint of social and ecological action and political conflict. The reason is that laborpower is a subject which struggles over health and the (natural) conditions of social health broadly defined, whereas the 'natural elements entering into constant and variable capital' are objects of struggle" (Robert Marotto, Correspondence).

20 "Economists and business leaders say that urban areas in California are facing such serious traffic congestion that the state's economic vitality is in jeopardy" (*The New York Times*, 5 April 1988).

21 "If schools cannot figure out how to do a better job of educating these growing populations and turn them into productive workers and citizens, then the stability of the economy could be threatened" (Fiske 1988). Fisk is referring to minorities which today make up 17 percent of the population, a figure expected to jump to one-third by 2020. In the USA, health-care costs as a percentage of GNP were about 6 percent in 1965; in 2000 they are expected to be 15 percent. "Health care has become an economic cancer in this country," screams a *San Francisco Chronicle* headline writer (14 March 1988).

22 The well-known IPM [Editor's Note: Integrated Pest Management] programme in Indonesia reportedly increases profits by reducing costs and also increasing yields. It depends on new training and education programmes, coordination of farm planning, etc. (Postel 1988, 4).

23 For example, West German organized industry and industry-scale coordination successfully internalizes many externalities or social costs. This occurs without serious harm to profits because the Federal Republic of Germany produces such high-quality and desirable goods for the world market that the costs of protecting or restoring production conditions can be absorbed while industry remains competitive (Conversation, Claus Offe).

24 The idea that crisis induced by inadequate conditions of production results in more social forms of production and production relations is not new in non-Marxist circles. Schnaiberg linked rapid economic expansion to increased exploitation of resources and growing environmental problems, which in turn posed restrictions on economic growth, hence making some kind of planning of resource use, pollution levels, etc. essential. He interpreted the environmental legislation and control policies of the 1970s as the start of environmental planning (Schnaiberg 1980). More, the idea that crisis induced by unfavourable production conditions results in more social productive forces, as well as production relationships (which is also Schnaiberg's thesis, since planning is a form of cooperation – hence, both a force and relation of production), can be found in embryonic form in works such as that of Wilkinson (1973), which argues that epoch-making technological changes have often resulted from ecological scarcities, and Sunkel and Leal (1986, 413), which argues that depletion of resources and scarcity increases the costs of economic growth because of declines in natural productivity of resources – hence, new energy resources and technological subsidies (implying more planning) are needed.

25 The most sophisticated post-Marxist text is Laclau and Mouffe (1985). A home-grown version is Albert et al. (1986).

26 For example, Laclau and Mouffe's discussion of what they call Marxist "essentialism" violates both the spirit and substance of Marx's theory of capital.

27 See O'Connor (1978). It should be noted that in the entire post-Marxist literature, it is impossible for me to find any reference to the division of social labour, so obsessed are the "theorists" with the division of industrial labour, division of labour within the family, etc. This absence or silence permits us to grasp post-Marxism as recycled anarchism, populist-anarchism, communitarianism, libertarianism, etc.

28 Accordingly, to Carboni, "the challenge of specificity is propelled by all new social actors in advanced capitalist societies. It is an outcome of the complex network of policies, planning, and so on which are implemented by both capital and the state in order to integrate people while changing production conditions. On the one hand, this specificity (difference) represents the breakage of collective and class solidarity. On the other hand, it reveals both new micro-webs of social solidarity and the universalistic network of solidarity based on social citizenship" (Communication to the author).

29 This and the following point were made by Claus Offe in conversation with the author, who is grateful for the chance to discuss these issues with someone who gracefully, and in a spirit of scientific collaboration, presents a post-Marxist point of view.

30 David Peerla explains the issue thus: "The issue in dispute is the post-Marxist claim that we have multiple social identities against the present claim that there exists a theoretical unity in these identities in the unity of the conditions of production and capital production and realization. On the level of appearances, it is true that we have multiple identities, but in essence the unity of our identity sterns from capitalism as a mode of production. The trick is to make the theoretical unity a reality. An environmental struggle may be an unintentional barrier to capital in the realm of accumulation while not being ideologically anti-capitalist. The question is how to make environmentalists conscious of the fact that they are making the reproduction of the conditions of production more social. The post-Marxists do not want to find a unity in the fragmented social identities we have. But even to build alliances between social movements some unity must be constructed. In the absence of an agreed upon telos of struggle, or any common definitions, dialogue cannot take place. If we are unable to agree on any terms and objects of struggle in what sense can we say new social movements are reconstructing the public realm as a realm of dialogue. It is agreed that we have to struggle over what socialism means but in some sense we are required to struggle for a common language which will necessarily obscure particular differences. As capitalism abstracts out the social nature of labor in the exchange of commodities, it obscures what we have in common, cooperative labor, thereby fragmenting our identity. What is disturbing is the lack of any move on the part of the post-Marxists to theorize the unity of social labor" (Communication, David Peerla).

References

Agger, Ben. 1979. *Western Marxism, an Introduction: Classical and Contemporary Sources*. Santa Monica, CA: Goodyear Publishing.

Albert, Michael, Mel King, Robin Hahnel, Leslie Cagan, Holly Sklar, Lydia Sargent, and Noam Chomsky. 1986. *Liberating Theory*. Boston: Southend Press.

Daggett, Christopher J. 1988. "Smog, More Smog, and Still More Smog." *The New York Times*, 23 January.

Fiske, Edward B. 1988. "US Business Tums Attention to Workers of the Future." *International Herald Tribune*, 20–21 February.

Folin, Marin. 1979. "Public Enterprise, Public Works, Social Fixed Capital: Capitalist Production of the 'Communal, General Conditions of Social Production'." *International Journal of Urban and Regional Research* 3 (1–3): 333–360.

Harvey, David. 1982. *The Limits to Capital*. Chicago: University of Chicago Press.

Knight, David. 1988. *The Age of Science: The Scientific World View in the Nineteenth Century*. Oxford: Basil Blackwell.

Laclau, Ernesto, and Chantal Mouffe. 1985. *Hegemony and Socialist Strategy: Towards a Radical Democratic Politics*. London: Verso.

Lebowitz, Michael. 1982. "The General and the Specific in Marx's Theory of Crisis." *Studies in Political Economy* 7: 5–24.

Marotto, Robert. 1984. "Forces and Relations of Production." PhD diss., University of California, Santa Cruz.

Marx, Karl. 1906 [1867]. *Capital: A Critical Analysis of Capitalist Production*. Vol. 1. New York: Random House.
Marx, Karl. 1968. *Theories of Surplus Value*. Moscow: International Publishers.
Marx, Karl. 1973. *Grundrisse*, 533. New York: Harmondsworth, Penguin.
O'Connor, James. 1978. "The Democratic Movement in the United States." *Kapitalistate* 7: 15–26.
Offe, Claus. 1987. "Panel Discussion with Robert A. Dahl, Claus Offe, and Alain Touraine." *Scandinavian Political Studies* 10 (3): 223–237.
Polányi, Karl. 1967. *The Great Transformation*. Boston: Beacon.
Postel, Sandra. 1988. "Indonesia Steps Off the Pesticide Treadmill." *World Watch*, January–February.
Sayer, Derek. 1987. *The Violence of Abstraction: The Analytical Foundations of Historical Materialism*. Oxford: Basil Blackwell.
Schnaiberg, Allan. 1980. *The Environment: From Surplus to Scarcity*. New York: Oxford University Press.
Sunkel, Osvaldo, and José Leal. 1986. "Economics and Environment in a Developmental Perspective." *International Social Science Journal* 38 (3): 411–427.
Timberlake, Lloyd, and Jon Tinker. 1985. "The Environmental Origin of Political Conflict." *Socialist Review* 16 (6): 57–75.
Wilkinson, Richard G. 1973. *Poverty and Progress: An Ecological Perspective on Economic Development*. New York: Praeger.
World Commission on Environment and Development. 1987. *Our Common Future*. New York: Oxford University Press.

2
THE ECOFEMINIST GROUND OF ECOSOCIALISM

Joel Kovel[1]

A twofold mission

The ecological crisis is too serious to be left to greenwashers, technocrats, and reformers; it demands, rather, the "ruthless criticism" Marx set forward in 1843, criticism that fears neither the powers that be nor its own findings. We believe that the scale of what is now transpiring should fire the radical imagination to rethink the whole of existence and the entirety of our relations with nature. Such is the credo of the journal *Capitalism Nature Socialism* (*CNS*), which has since 1988 been home to radical thought about the crisis.

Much of the work the journal publishes consists of analytic studies that track the intricate workings out of the crisis across the innumerable boundary points that define our relation to nature. Nothing human is alien to us, to reprise an ancient and still valid ideal. But we are also open to work that might be called expressive and is variously aesthetic, exhortatory, even prophetic – work that announces where we want to be going. These two types are also dimensions that appear in various proportions within each individual *CNS* article and immanently imply one another, inasmuch as any work's basic intentions are embedded in its results. The *New York Times*'s "all the news that is fit to print" is news that fits into a worldview defined by investment in the status quo and desire to preserve its class structure. Consequently, mainstream "environmental" reporting takes on a non-ecological cast: it is about what is outside us and never questions the essential assumptions behind the facts. Never making the vital connections that could point to a basic change, the mainstream presents the world as needing to be controlled and manipulated.

The *CNS* worldview is distinctly different because it is grounded in the desire to preserve the integrity of life. Accordingly, if the dominant system turns out to be a menace to life, then to Hell with it. We strive, therefore, to be ecocentric in our work, by preserving interconnections, seeing the whole, not isolating the inside from the outside nor the observer from what is observed. Fully realized, this vision becomes dialectical and emancipatory. It seeks to supersede what is, not to patch the given state of affairs and preserve it for further plunder.

We call the given system capitalism, as in *Capitalism Nature Socialism*, and claim that our reality makes the most sense when regarded as the workings out of an apparatus that subordinates all reality to the goal of accumulation. We reject postmodern ways of thought as evasions of this hard truth. And we also try to go beyond the half measures that would locate the core problem as inhering in derivatives of capital, such as the corporations, or the industrial system. We take

10.4324/9780429341427-4

these phenomena very seriously but regard them as essentially the instruments of capital. For capital is the monster[2] that drives the ecological crisis and is our real quarry.

This commits us to a position that includes but also surpasses the discourses of political economy. We reject the notions of ecological economics, that oxymoron created out of the pathological need to commodify nature, but we are also open to work that goes beyond the framework of the economy as such. For capital is a way of being as well as an economy. It is the outcome of an ancient disorder, congealing into an overarching economism that represses all other aspects of existence. The critique of capital should therefore undo this repression and open onto a wider range of inquiry than is ordinarily comprehended by political ecology. This point is essential for the imagining of where we want to be going, as it opens our imagined "better world" to a wider and deeper realization than that comprising the simple extension of legal mechanisms, regulatory codes, alternative currencies, etc. We see these as means to an end, posted at various distances along the way to a telos, that, like all objects of human labour, must be imagined in the mind before it is realized in the world.

Ecosocialism as telos

If we would consign capitalism to Hell, it is for the sake of some heaven: a heaven, moreover, on the earth and not pie in the sky. Thus, our thinking is not merely utopian, but prefigurative. This destination beyond capital, this ecologically rational society, remains an open project. Nevertheless, to reject the capitalist system as life threatening demands a working name in order to focus the mind on its alternative. It is in this spirit that we speak of socialism, as in *Capitalism Nature Socialism*. The absence of commas between the three words of our title was the inspired choice of James O'Connor, founder of *CNS*. It opens the trio of terms, standing in ambiguous sequential splendor like the monoliths at Stonehenge, to the whole ambit of possibilities inherent in the bracketing of nature with capitalism and socialism.

I call socialism a "working name" because of all the words in our language, it is best suited to signify capital's antagonist and successor because it summates a history variously glorious and doomed, a history full of lessons to be learned and turns both false and true on the road to ecological integrity. However, the remarkable example of Cuba notwithstanding, no one should regard those actual socialisms that arose over the last century through the struggle to emancipate labour as nearly adequate to define the ecologically rational society of the future. And if the socialism of the "first epoch" was not able to encompass the ecological crisis, then there needs be a "next epoch" socialism that does. For this notion, we reserve the word *ecosocialism*, to signify the "where we want to be going." Ecosocialism is socialism made ecologically rational. It is both destination and the road to be built as we travel.

Ecosocialism, it seems to me, will take shape out of three interwoven lines, the first two of which will simply be mentioned, and the third explored a bit:

- Ecosocialism requires preserving and extending the core of socialism, which is to allow human beings to freely express themselves in self-determined, social labour. This means doing what is necessary to overcome the separation of the producers from the means of production, which in turn means, as ever, undoing the class system and the capitalist state that enforces that class system – and it means facing the dire implications of this. If we want peace, which is another word for ecological sanity, we need justice, which is another word for bringing down the system in which certain people accumulate wealth by exploiting others. A really just world still needs to be a classless world, beyond imperialism, militarism, and the other accoutrements of accumulation.

- Second, ecosocialism requires the development of a post-capitalist, ecological mode of production, one that replaces capitalist production with that freely expressed by emancipated labour. This entails a recentering of production away from the commodity and towards the creating of flourishing, integral ecosystems. In practice, this begins with the restoration of use values within production (associated with the devaluation of exchange value), and it extends to the recognition of an intrinsic value beyond use value, for which end we need to comprehend ourselves in nature and nature in ourselves. The notion of an integral ecosystem is allied with that of a "moral economy"; it incorporates the human being as an active, transformative agent and not a demiurge standing over mute nature. Human ecosystems entering into the building of ecosocialism might include an organic garden, a "wilderness" park, a schoolroom, a labour union, a factory setting, or indeed, an ecosocialist movement. As these become integral, they develop relations of wholeness, which, encompassing ever-wider regions of existence, become emancipated and develop revolutionary potential.
- The third foundational line of ecosocialism is its gendered ground in human existence and is comprised under the name of ecofeminism, which signifies that branch of the feminist movement that regards nature and gender as deeply inter-related so that the one cannot be emancipated without the other.

The gendered bifurcation of nature

Gender is the point at which human being interacts with nature according to the notion of difference. So long as humans have had self-consciousness and language, they have recognized two versions of themselves according to body plan. No other structure of difference is so profound, enduring, and grounded in the natural reality of sex.

We can – and should – imagine a world without class. And while many of the various constructions of race are related to nature through some physical sign or other, this is no real foundation for what divides people, but a false difference (a "splitting") spelled out with the bad biology of pseudo-speciation. Hence, we also can – and should – imagine a post-racist world, in which complex networks of social difference develop outside the ground of racialization.

We cannot, however, imagine a world without gender, as some postmodernists would have it – though we can – and should – imagine one with better gender relations. Gender remains the form of difference authentically grounded in nature, insofar as sex is a basic natural category. It is always constructed, yet always out of real materials – and the "always" reaches back to the origins of each person and the building blocks of the species itself. Hence, it is the ground of difference and, like a proper ground, endures and holds up what rests upon it.

Our capacity to transform nature, i.e., to produce, is itself gendered, as is the language with which we account for reality. When we produce, we interact with nature to bring forth new form, which being new, is also a difference. Since thought is always interconnected through the endless plasticity of language, gender difference, however constructed under various historical conditions, is transferred onto all other differences. In one direction the process includes aspects of the human, produced world, and in the other, those of nature itself. The primary division of labour is therefore organized by gender; it "incorporates" body plan and connects male and female functions with the sky, the earth, the sun, the moon, the wind, the waters, and the numberless creatures inhabiting nature.

None of this is inherently tied to domination and violence. Yet domination and violence also entered the world as gendered. I should think this happened through the mutation of male hunting bands[3] into forcible expropriators of other humans and particularly of the threefold prizes embodied in the female – her productive labour, her reproductive power (also with "labour"), and her sexuality. From this point of origin, patriarchy, class, and the state derived through processes far too complex to be summarized, much less developed, here.

In the realm of gender domination, women stand between men and nature and mediate the two. In this scheme of things, the true human being became considered male (which is why the received and reactionary inventory of human nature is a simulacrum of male psychology, rapacity and all, and why Freud scarcely ever took mothers into account in drafting his psychology); meanwhile, women, though formally recognized as human beings, experience their status as a tenuous and constantly undermined right. In civilization, then, women are sunk into nature, and nature and women are alike devalued, while men soar freely. The work of the genders is valued correspondingly. The woman is a mixture of earth and humanity, while the sky is reserved for the male. And the earth, or nature, is the devalued part of the female principle, variously considered wild and threatening or passive, inert, and exploitable (for paired, split-apart opposites characterize this delusional way of being). Hence, the gendered bifurcation: male is to female as humanity/civilization is to nature. Accordingly, nature forever remains outside, split-off from Man. In terms of our notions of ecological production, the gendered bifurcation imbues nature with the Otherness of repressed gender and blocks the recognition of intrinsic value. So long as this "conception" (for at a very deep level, it is also a male parthenogenetic fantasy) is maintained, the ecological crisis is not only inevitable but irremediable.

From its point of origin, all aspects of nature become stained with Eve's curse, what Engels termed the "world-historical defeat of the female sex." The curse entered into the history of capital and was further transformed by it. Capitalism, far from being a rational system of market relations, is a phase in the evolution of un-free, gendered labour. The new system carried forward the intense fear and loathing of the body developed in earlier stages of Christianity and transmuted this into the torments of Puritanism. It further widened the split between the sexes through the intense degree of rationalization imposed by the new regime of generalized exchange and monetization and became stained by hatred of the female principle to a degree scarcely ever matched in history. All this burst forth in modernity's grim underside, the great witch-hunt.[4]

Yet capital also ushered in another turn in the history of gender through the generalized commodification of labour power. In doing so, it paradoxically created the conditions for both socialism and feminism – the former by the growth of wage labour, which could be organized along socialist lines, the latter through the fact that the wage relation, being an extension of abstract exchange, is indifferent to body plan. Because exchange value is immaterial and a mere mental figment, as capital expands, it breaks down gender boundaries and opens up a path to a potential equalization of the sexes. It also does much else, by weakening communal ties, fracturing the family, and, through the consumerism that pervades its "advanced" stages, flooding society with desire.

The effects on gender are chaotic, with every conceivable tendency swirling across the map, including certain feminist advances. But there is a joker in this deck, which derives from the fact that first-epoch feminism, like first-epoch socialism, remains dependent on the integrity of capitalism rather than its supersession. Therefore, the equalization of the sexes was, and remains, equality under bourgeois relations, just as the advancement of workers has been an advanced incorporation into the alienated ways of consumerism, debt, and the money economy.[5]

The soccer mom balancing domesticity and high-tech employment and the superexploited[6] Chinese woman torn from the countryside who solders together the computer the first-world woman uses to organize her busy schedule and make her PowerPoint presentations, occupy different spots along the continuum of female subsumption into capital, with varying degrees of repression within the current division of labour. The same applies to the growing flood of women sucked into the sex industries that feed from the abovementioned insertion of desire into social existence. In the midst of it all, bourgeois feminism languishes, ultimately, I should think, because it has never been able to imagine an alternative that encompasses a better world than that of capital and therefore succumbs to the system's extremely potent ways of absorbing its partial negations.

The necessity of ecofeminism

Long ago, one could answer the question posed by the limitations of bourgeois feminism by pointing to the hope of socialism. But well before first-epoch socialism degenerated and collapsed, it demonstrated that its gender relations had done little to surpass the example of the *ancien regime*. The bold hopes of an Alexandra Kollontai and the genius of Rosa Luxemburg became associated with many remarkable stories of women who rose out of feudal bondage. But soon enough, these became eclipsed by a bureaucratic reality in which women would swell the ranks of the technical professions but rarely ascend further. The same has proven the case for Cuba, socialism's finest outcome.

First-epoch socialism, despite various initiatives towards an ecocentric position (Gare 2002),[7] failed to overcome the gendered bifurcation of nature and fell back with a thud on the capitalist metaphysic that regards both nature and the female as a passive reserve of resources with no intrinsic value. The dreadful results in terms of ecosystem integrity are well documented. But we also need to learn to recognize the profound connection between a society's treatment of nature and its treatment of women. There needs to arise, then, a "next-epoch" feminism that avoids both the pitfalls of bourgeois feminism and (first-epoch) socialist feminism: an ecofeminism. And as ecofeminism requires a socialist development if it is to break free of the fetters binding bourgeois feminism, then this means that it needs to take place as an ecosocialism, which avoids the errors of first-epoch socialism. And since one of the greatest of these occurred at the nexus of gender and nature, an ecosocialist outcome needs to also be ecofeminist, for this is predicated on overcoming the estrangement between gender and nature, freeing both in the process. Thus the two mutually recognize each other – and mutual recognition is the sign of an integrally human ecosystem.

Ecological production posits the object produced as an ecosystem. At its centre, it is neither a thing nor a fetish, but a living form, inter-related with others, made from the bringing together and holding together of mutually recognizing yet distinct elements.

This does not mean that nothing will be produced for exchange or that commodities will somehow disappear from the world, a quite unthinkable outcome. It does mean that as exchange becomes subordinated to use and intrinsic value, a different set of needs and productive relations will appear and take ecosystemic form: process will supersede product; direct sensuous appropriation will be valorized; subject-subject relations will proliferate; aesthetic considerations will be foregrounded. In the changed need structure that emerges, mutual obligation, respect for limits, and feelings of solidarity and spiritual wholeness arise, and these can overcome the hunger of possession and domination that propels the capitalist economy.

The labour entailed in this kind of production embodies the rising of what has been subjugated and segregated into mere woman's work since the rise of male dominion. Ecological production is not the activity of a producer standing over dumb nature as a chaotic set of inert objects. Still less is it the labour of the alienated worker deprived of human power and following orders as a tiny wheel in the gigantic machine of accumulation. It is, rather, a revalorization of what had been degraded. The female principle, being differentiated rather than split from nature and immemorially regarded in terms of caring for, or holding, or gathering, or providing, or of staying with, or of weaving together, is now restored, no longer marginalized, no longer the background to civilization, but dialectically brought to full being as the groundwork of integral ecosystems. We know empirically that the most vital ecopolitical movements arise spontaneously as collectives of women, often in the South and often working and living under conditions of subsistence. Now we can declare these ventures as prefiguring and announcing a new society.

On the ecofeminist path, women cease being object and Other to male desire and regain themselves as full human beings. Correspondingly, men regain connection to the earth as the Other becomes others. Thus, the bifurcation is undone. Neither of these emancipatory motions within gender will happen except as isolated and disconnected instances, however, so long as capital and the capitalist state rule. Thus, ecosocialism and ecofeminism need to be in the thick of the struggle and will have to prove their bona fides by alliances and associations across the whole range of social and political movements.

The task for *CNS*

None of this can be taken up within the perimeter of this brief and necessarily abstract presentation. But as we began by considering the work of the journal *CNS*, so may we end there as well. It goes without saying that neither ecofeminism nor ecosocialism looms large on the current landscape. We do believe, however, that we have something to offer a world tormented by impending ecological breakdown and barren of revolutionary hope. And so we press on, yet not in isolation from one another. I wrote this editorial, more substantially theoretical than previous ones, to express some ideas that have been brewing inwardly but also to underscore certain editorial priorities. These are first, that *CNS* will continue to strive for a vital mix of analytic and expressive work, the former addressing the workings of this world, the latter indicating in some manner, a better one, and second, that the ecofeminist side of things will be welcome in these pages, not as an affirmative action programme but as an integral aspect of the making of the better, ecologically rational world.

Notes

1 Editors' Note: The original was published as Joel Kovel. 2005. "The Ecofeminist Ground of Ecosocialism." *Capitalism Nature Socialism* 16 (2): 1–8. The original publication has been modified to conform to the style adopted for *The Routledge Handbook on Ecosocialism*. The content has been slightly amended by completing missing information and removing allusions to other portions of the journal where the work was originally published.

2 Heeding Marx, who repeatedly and wonderfully refers to capital as a werewolf or vampire sucking the blood of living labour.

3 Hunting is a primary mode of the gendered division of labour, deriving from the momentous evolutionary development that underlies the special role played by sex and gender in human existence: in particular, the replacement of the annual oestrus that regulates sexuality in other mammals with year-round oestrus. Along with deregulated sexuality, this brought in its wake menses, their associated blood loss, and a chronic threat of iron-deficiency anemia for women, the prevention of which accentuated the survival value of hunting and meat eating and its subsequent potential for organized violence as carried out by bands of hunters. (See Shlain 2004.)

4 For the definitive account of the intersection between gender and the rise of capitalism, the reader is advised to consult Silvia Federici's *Caliban and the Witch* (2004), excerpts from which appeared in *CNS* 58 and 59. See also Carolyn Merchant (1980) for an exploration of how gender bifurcation corrupted the history of modern science.

5 Within the sphere of gender, this would incorporate postmodern developments, inasmuch as this wrinkle in capitalist relations is predicated on the unfettered ascendance of exchange over use value. As for

workers, we can grimly see the working out of the system logic by the Bush II administration, as when the president urged replacement of Social Security with private investment accounts for workers, such being the glorious road to an "ownership society."

6 Because the patriarchal economy sees to it that female labour is remunerated at a lower level, while patriarchal culture sees to it that women workers are also more docile and easily controlled.

7 Cuba again provides a partial exception with its impressive ecological development after the collapse of the USSR. But this came too late in its development to constitute a complete social transformation.

References

Gare, Arran. 2002. "The Environmental Record of the Soviet Union." *Capitalism Nature Socialism* 13 (3): 52–72.

Merchant, Carolyn. 1980. *The Death of Nature*. Berkeley: University of California Press.

Shlain, Leonard. 2004. *Sex, Time and Power*. New York: Penguin.

Silvia, Federici. 2004. *Caliban and the Witch*. New York: Autonomedia.

3

ETHNICIZED, GENDERED CLASS ANALYSIS

A theoretical-methodological framework for analyzing ecofeminist, ecosocialist praxis

Terisa E. Turner

Introduction: Women are prominent in struggles to keep fossil fuels and uranium in the ground

Given that we are facing ecocide mainly because of hydrocarbon combustion, it is essential that we keep fossil fuels and uranium in the ground and protect the forests (Ekwurzel et al. 2017; Griffin 2017). This imperative is captured in the poetic ditty, chanted at protests globally:

> Keep the gas under the grass,
> Keep the coal in the hole,
> Keep the tar sand in the land,
> Keep the oil under the soil.

Women, especially Indigenous women and women of colour, have long been at the forefront of civil society mobilizations to keep the gas, coal, tar, and oil out of capitalists' value chains and under the soil and sea (Temper et al. 2013; The Red Nation 2021). Why women fight Big Oil is no mystery. A 2021 study by WECAN found an "indisputable connection between fossil fuel capitalists' practices and negative impacts on African American/Black/African Diaspora, Indigenous, Latina/Chicana, and low-income women's health, safety, and human rights." The study reported that fossil fuel–derived air, water, and soil pollution impact women's fertility, mental health, and daily work and responsibilities. The negative effects, specifically from coal, oil, and gas, stem from

> direct pollution of communities and through the role fossil fuel companies play as the biggest contributors to industrial carbon dioxide and methane. Additionally, temporary male housing sites used for fossil fuel pipeline construction and oil field work, referred to as "man camps," have been extensively linked to increased levels of abuse and safety threats to Indigenous women, girls, and two-spirit people, contributing to the Missing and Murdered Indigenous Women, Girls, and Two-Spirit People (MMIWG2S) epidemic.
>
> *(WECAN 2021, 6–7)*

10.4324/9780429341427-5

An important gendered class dimension of these anti-hydrocarbon actions is that very frequently, it is women at the bottom of what we could call the global "hierarchy of labour power" who take initiatives and are joined by their allies. Among the examples of women mobilizing to "keep the oil under the soil" are the Ecuadorian initiative from 1997 to 2014 to keep one billion barrels of oil under the Yasuni Indigenous people's land if global donors contribute one-half of this oil's market value to pay for alternative energy and other essentials. In Costa Rica, a massive campaign resulted in the 2002 government ruling that banned oil and gas exploitation on shore and offshore. This prohibition made Costa Rica the world's first country to completely exclude hydrocarbon extraction in favor of biodiversity, eco-tourism, and the artisanal fishing industry. Indigenous women and their allies have for decades been fighting the highly toxic Athabasca tar sands in Northern Alberta and the pipeline infrastructures that this landlocked extraction requires as a precondition of its profitability (Germanos 2020). Indigenous Akii Kwe women water protectors in Walpole Island south of Sarnia, Ontario, blocked a 350,000-barrel-a-day Shell tar sands oil refinery expansion project (Kicknosway 2008; Stephens 2009). These instances of women's action to stop oil could be multiplied many-fold (Temper et al. 2013; Simmons 2020). To anti-oil resistance can be added hundreds of other actions worldwide, the numbers of which increase over time, to keep gas, fracked gas, and coal from being extracted and to block construction of fossil fuel infrastructure. How can we theorize these concerted actions against Big Oil? The next section addresses this question. But first we make a short digression.

In the midst of this nascent new world order that is being incubated by a growing world movement of movements, we have growing racism and sexism in the reformist, mainstream, environmental "green capitalism" organizations at governmental, corporate, and civil society levels. Before we turn to theorizing the revolutionary, ecofeminist content of authentic ecosocialist class struggle, it is useful to quote Lohmann's historical materialist critique of toxic reformism:

> International popular movement-building around issues of energy and climate tends to be hamstrung by the lingering belief that energy is a universal, nonracial, genderless substance craved by generic humans from time immemorial. Now might be a good time to explore what it might mean to follow some of the lessons of recent scholarship and try to understand energy instead as the post-18th century reorganization of peasant, Indigenous and urban territories around the mass interconvertibility of motion, heat and electricity, together with the resulting wastes, under the rule of patriarchal empire, labour control and white supremacy. Energy scholars and energy activists arguably can benefit not only from confronting this reality but also from trying to work out what relationship they want to have to the whiteness and white history of the thermodynamics that they rightly take so seriously. That entails a lot of loving work of many kinds. Not the least of which will be learning how to deal with the unreasoning and identifiably white rage that often spurts forth among white environmentalists and energy specialists when, to adapt the words of James Baldwin, they "continue to hold on to their innocence long after that innocence has gone."
>
> *(Lohmann 2020)*

Theoretical concepts in ethnicized, gendered class analysis

The theoretical-methodological framework described here may be of particular interest to those who wish to analyze the global drive to address climate chaos by keeping fossil fuels in the ground from an historical, materialist perspective. The ethnicized, gendered class analysis outlined here is useful for examining hydrocarbon-related mobilizations and, within these, especially the transition to a post–fossil capitalist system. The framework can be described, at least in this iteration, as being composed of five elements. These are (a) the hierarchy of labour power, the male deal, and ethnicized, gendered class alliances; (b) the fight for fertility; (c) the four sites of confrontation between male dealers and those in a gendered class alliance: the sites of production, consumption, social reproduction, and nature; (d) the circulation of struggles and; (e) the implications of the struggle to capture or 'free' the 'oil slaves' in of a barrel of oil.

The hierarchy of labour power, the male deal, and ethnicized, gendered class alliance

All the world's people are included in this hierarchy of capitalist labour power, illustrated in Figure 3.1. We are divided, most importantly, by the fundamental class division whereby a few thousand capitalists in the state and private sectors own and control almost all the means of production, the land, the intellectual property rights, and the wealth in all its forms. Capitalists' ownership and control expand constantly, driven by the growth imperative: corporations either grow or die or are absorbed by competitors. The growth imperative is taking a very heavy toll on nature, a reality that the climate crisis has driven home.

The class relation between these capitalists and the rest of us is one of exploitation. Because we are excluded from the means of production, we have little option other than to work for the capitalists. About a fifth of us may have a wage contract whereby we sell our labour directly on a daily basis. But the other four-fifths of us are also exploited, despite not receiving a wage, because we produce labour power and cheap wage goods. Hundreds of millions constitute the immense reserve army of labour: that is, the world's poor, refugees, migrants, internally displaced people, prisoners, slum dwellers, the elderly, and the unhoused. Labour power

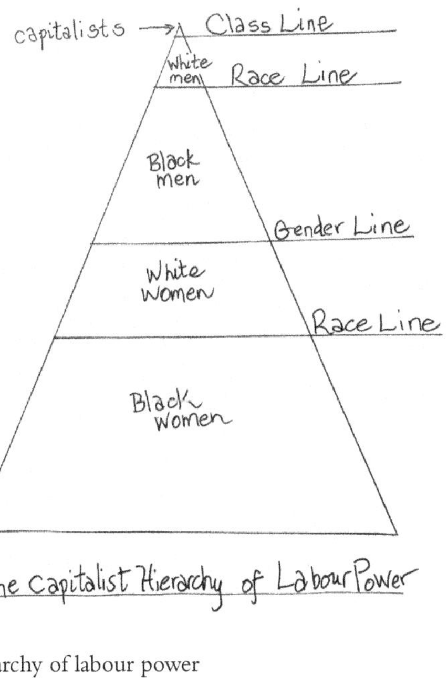

Figure 3.1 The capitalist hierarchy of labour power

production within this reserve army is accomplished at rock-bottom cost to the capitalists. It has the added attraction of being highly flexible.

The evolution of the capitalist system, especially since the late 1700s, and its transition to the use of (production-expanding) coal energy by the 1820s have been dominated by accelerating dispossession, colonization, and the shaping of a global working class that is stratified on the basis of ethnicity and gender as indicated in Figure 3.1. Of the three horizontal lines on the figure, the class line is most fundamental because, under capitalism, the class relation is the most fundamental relationship. This is because the only way to eliminate the class line is to eliminate the relationship that capitalists have with the rest of us and with extra-human nature. In order to block such elimination, capitalists have instituted racist divisions while reconfiguring highly diverse gender relations worldwide into strict patterns of male dominance and female subordination. This female subordination was affected by systematic terror and dispossession through the witch hunts between 1415 and 1650 in Europe, and on an ongoing basis elsewhere in the world. Additional forms of dispossession reduced women to labour power producers, whereas previously they had had control over myriad means of production and social power.

As illustrated in Figure 3.1, above the class line are a few hundred billionaires, or the 1 percent of the 1 percent. A mere handful of these billionaires wield overwhelming dominance. Perhaps as few as a dozen in Big Oil (the "Carbon Majors") are directly responsible for climate chaos (Griffin 2017). The vast majority of big capitalists are white men, but there is a sprinkling of people of colour and white women (Oprah, Queen Elizabeth). The capitalist system is highly unstable and draws some compensating stability from coopting Black and female showcase tokens. Such cooptation enables capitalists to assume the posture that excellence, not racism and sexism, dictates who rules the world.

In fact, most of the wealth and power of these super billionaires comes from inheritance, not innovation or entrepreneurial brilliance (Picketty 2014, 18). This is a reminder of the historically deep entanglements that capitalists have with state, legal, military, cultural, and other kinds of institutional and social power. These entanglements bolster the abilities of the capitalist class to reproduce itself and its relationships of exploitation with the billions of dispossessed people worldwide and with extra-human nature.

Below the class line are the other 7.8 billion of us, including small- and medium-scale farmers, peasants, Indigenous peoples, workers, migrants, women, youth, and virtually every other conceivable category within the human community. The 99 percent are divided most fundamentally by the gender line. The fundamental and core character of this gendered division arises from capital's need for women to produce and maintain workers – that is, labour power – for virtually no recompense. Nature, on which most women depend, in part, for their work and lives, is under dramatic corporate attack. Consequently, land and water struggles multiply. The gender line also draws attention to the extreme dispossession that capitalist enclosures and neo-colonialism have visited on women and against which they are mobilized.

Women below the gender line produce labour power with no or almost no wage. They produce cheaply available goods (food, textiles, clothing, microchips) for corporate value chains for a miniscule wage that cannot support life. In the categories of men and women, we include those who are born into these genders as well as all those who identify as such (Lewis 2016).

Above the gender line are men. So-called white (European-descendent) men, typically with salaried professional and waged jobs, are at the top of the hierarchy, immediately under the class line. It is from this stratum that are drawn most of the collaborators with capitalists in a male deal, partly because of a race-based and gender-based solidarity between largely white capitalists and largely white male actors. Official culture encourages relatively property-less and dispossessed white men directly under the class line to expect and aspire to copycat wealth and power that is a faint echo of the wealth and power of the capitalist class (Moore 2001).

In the stratum immediately below that of European-descendent men are Black, Brown, Indigenous, and other men of colour. Then we have, under the gender line, all women, with white women above Black, Brown, Indigenous, and other women of colour. European-descendent white women are, under capitalism, accorded certain privileges while nevertheless being exploited and dispossessed. While there is some overlap and blurring of these categories, with, for example, some women of colour in the higher strata of the hierarchy and transgender people present in all categories (Lewis 2016), this is an historically grounded conceptualization intended to be an approximation of the world's people as exploited labour power.

It is important to note that the hierarchy describes people, but it is located in extra-human nature. It is primarily this nature that is the focus of the fight between women and their allies on the one hand and capitalists and their working-class, collaborator men on the other hand.

The class character of the hierarchy is familiar to many. But it is not so readily grasped that the racialization and gendering of the totality of the world's people as represented in the hierarchy is set up by capitalists as essential to their continued control and profitability. Hence, we see that capitalists spend a tremendous amount of effort in maintaining and enforcing sexism and racism. They do so through official culture. They are countered by popular culture. While the first divides, the second unites. The capitalists' objective is to keep the working class divided against itself and weak in the face of capitalist power. The need for divisiveness, from capitalists' perspective, increases as people become more tightly enmeshed in the web of the world market and have at their disposal the internet and the enhanced expertise in using it acquired by people throughout the globe as a result of the social distancing required by the discipline of the COVID-19 pandemic.

The hierarchy of labour power facilitates the construction by capitalists of many global, for-profit value chains. At each node in the value chain, it is necessary for capitalists to impose and activate what we here call the male deal. The subsidiaries of corporations and the contractors linked to them must be managed and made profitable by a combination of local (relatively dispossessed) people (almost always men) and corporate officers themselves. These local men who participate in the male deals are almost always a target of fightback by those lower down in the hierarchy of labour power. The resistance that comes up from below is focused on building alternative, commoning, not-for-profit but for-life value chains. Giacomini (see her chapter in this volume) calls these commoner relationships "value matrices" to distinguish them from the exploitative "value chains" of transnational corporations. The value chains are based on exchange value for profit. In contrast, the value matrices are based on use value and intrinsic value. While the first is destructive of life, the second is supportive of life.

We now turn to a specific kind of class analysis that focuses on the gendered, ethnicized class alliance versus the male deal. Within the hierarchy of labour of power, we have two opposed categories of actors. The first consists of those who are part of a "male deal," which is a cross-class and often a cross-race or multiethnic arrangement or deal, almost always between men (although women can be involved in what might be described as a "female male deal"). The male deal operates to build the for-profit value chains. It does so by using enclosure and dispossession as well as sexism and racism to pump profits up from the lower echelons of the hierarchy and from "more than human" nature to the corporate bottom lines of those capitalist billionaires above the class line.

The second key category of actors in the hierarchy are women at the bottom who take action to defend life. If we look at who pursues efforts to stop fossil fuels and uranium extraction, it is usually, for example, Indigenous women, La Via Campesina women, unwaged women, housewives, and community activists such as the women of Quebec who, throughout some 700 villages, oppose fracking. It is women at the bottom, sometimes autonomously and sometimes with their allies from strata above them, who break with that male deal, thereby expanding the possibilities of building ecofeminist, ecosocialist commons (Gago 2018). What do these two categories of actors in the alliance and deal fight over? This takes us to the second aspect of the framework: the "fight for fertility."

The fight for fertility

What I call the "fight for fertility" is shorthand for a fight for control over the capacity to produce and maintain life, whether plant, human, or non-human animal life. It is a fight for control over the capacity to bring forth and sustain life and to have access to the prerequisites of life: most importantly, land or housing, clean air, water, and food. Female actors are clearly centralized here. For example, the California-based but global organization Women's Earth and Climate Action Network (WECAN) mobilizes mainly with Indigenous women from all over the world against Big Oil and its financiers. Here we have predominantly women of colour directly confronting predominantly white men in oil companies and banks. We have those at the very bottom of the hierarchy of labour power trying, with some success, to deny to those above the class line – a few billionaires – access to subsoil hydrocarbons and fossil fuel infrastructure associated with their exploitation.

From the point of view of those under the class line, this fight entails resisting further enclosures and dispossessions. It involves those dispossessed, exploited people struggling to secure or regain access to the means of production and the means of life that have been appropriated or destroyed by capital. One example is the fight for reparations in the US by those descended from captive Africans. From capital's point of view, fighting to control the essentials of life entails commodifying them along with grabbing them for free via conquest or under the protection of capitalist laws such as eminent domain (Barbier 2010).

Four sites of struggle for control over the means of life

The fight for fertility is evident at four different sites: production, consumption, social reproduction, and nature (Dyer-Witherford 1999). First is the site of production, where we see strikes in which waged workers confront capitalists. Second is the site of consumption, where, for example, consumers campaign to have GMO labels on food products. Or consumers partner with small-scale food growers to insist on having street markets and community-supported agriculture. After the 2008 global crash, we saw consumers in more than 40 countries taking over food supplies in what were mistakenly and prejudicially called food riots (Schneider 2008). Third, we see the fight for fertility at what some call the site of "social reproduction" (and what I described as labour power production), where there are multiple, diverse, anti-commodification struggles, for example, to defend the national healthcare system, or free, high-quality public education from corporate capture. Finally, we have this fight for fertility at the site of nature, where, for example, women and their allies resist Big Oil, as in Standing Rock in the US (Estes 2019), in Nigeria (see chapter by Tobocman et al. in this volume), and throughout the world.

In practice, every struggle has dimensions that draw from each of these four sites. But a careful examination of the class struggle in question will reveal elements of the fight to control the capacities to bring forth and sustain life that are evident in all three other sites of struggle. This is especially the case in those global, mass actions that circulate and build on each other, thereby bringing in resistors and aspiring commoners from a range of countries. What CLR

James called "strange eruptions from below" are, today, increasingly multi-issue struggles that underline the emergence of a new consciousness about the interconnectedness of all and, therefore, the need for system change.

The circulation of struggles

The fourth aspect of ethnicized, gendered class struggle has to do with the ways in which these struggles feed on each other and circulate. Resistance to injustice has always circulated. This dynamic has become more pronounced with the elaboration of the world capitalist system (Linebaugh 2021; Turner 1994). At the four sites, (production, consumption, social reproduction, and nature), a circulation of struggles is evident. This circulation is very similar to solidarity strikes. When people confront capital at one point in the corporate, for-profit value chain, it weakens that global value chain. That confrontation opens the door for people to act in another place. We do see this kind of domino effect, both horizontally and vertically, along corporate value chains and integrated capitalist operations (Bhatnagar 2021; Harvey 2021). The circulation of struggles is facilitated by what Marx called the imperative of large corporations to expand in size and globalize. "One capitalist always kills many" (Marx 1867 [1999]). As these corporations get fewer but bigger, they entangle all peoples in the net of the world market. In the process, they destroy and cause hardship and environmental degradation. "[B]ut with this too, grows the revolt of the working-class, a class always increasing in numbers, and disciplined, united and organized by the very mechanism of the process of capitalist production itself" (ibid). Marx was writing about waged workers or proletarians. Here, we extend that analysis to all the world's people, both waged and unwaged, who, through the process of corporate expansion, are transformed from a class in itself to a class for itself.

"Oil slaves," or the capacity of a barrel of oil to do work

The fifth and final element in the framework for ethnicized, gendered class analysis is the oil slave. The idea of oil slaves arises from a labour theory of oil or energy: that is, a labour theory of fossil fuels under capitalism. A barrel of oil can do a lot of work. It can do work equal to that of some 14 person years (based on 40 hours of manual work a week). This means that the capitalist who has appropriated that barrel of oil has the potential to profit from production equivalent to approximately 14 years of manual work.

We are referring here to oil that's been extracted from deep in the earth. We are not referring to a different kind of carbon emissions that are part of a natural cycle and that are generated and absorbed in a balanced manner over time. This naturally cycled carbon is living carbon. In contrast, oil corporation oil slaves embody dead carbon. Extracted oil, gas, and coal are hydrocarbons from tiny diatoms that were alive millions of years ago before the period of the dinosaurs. The exploitation of oil and the capturing of oil slaves in barrels of fossil fuel are causing ecocide. Ending extractivism is a move to free the oil slaves. This is the focus of women-led initiatives.

What do these oil slaves do for capitalists? They do two things. One, oil is a commodity like any other commodity. Capitalists get it; they sell it; they make a profit. Two, and much more importantly, the capitalists use oil slaves to engage in class struggle and in the process reveal how absolutely indispensable fossil fuels are to the capitalist epoch (Malm 2016). It follows that to adequately address the climate crisis, it is necessary to eliminate capitalist relations in favor of commoning. This imperative is denied by anti-ecofeminist, anti-ecosocialist, would-be saviors of a broken capitalist regime, represented for instance by the analyses and reforms generated by United Nations policy wonks (Barbier 2010).

Capitalists use the energy in fossil fuels to pursue two kinds of class warfare: compete with other capitalists and smash up the working class. First, to compete with other capitalists, they invest in technologies. Then they must fuel those technologies. This allows them to produce widgets or whatever more cheaply than their competitors, so they can get more market share and wipe out the less technologically advanced producers. It is evident that capitalists compete with each other using the oil slaves in fossil fuels. This competition undergirded two world wars in the twentieth century as the oil-slave have-not countries of Germany, Italy, and Japan confronted countries whose corporations commanded massive ranks of oil slaves. In the twenty-first century, militarism and in particular US military expansionism is driven by a war for oil and by oil for war.

Second, capitalists respond to militant waged workers by automating and increasing their use of technology. Malm (2016) shows that the Industrial Revolution's shift from water and wind power to coal-fired steam power at the beginning of the nineteenth century was more expensive but allowed capitalists to control workers more effectively. These early capitalists were driven to avoid costly shutdowns by workers at site-specific, hydro-powered factories. In South Africa's coal and diamond mines, strikes by Black workers were resisted by Anglo American and other transnational mining corporations by cutting the workforce and installing more machinery (Christie 1980).

Capitalists get rid of many workers by automating and employing artificial intelligence. But then they have to fuel or plug in these machines. In so doing, capitalists become more dependent on oil slaves. This dependency confronts abolitionism. Consequently, capitalists and their male deals become increasingly vulnerable to the mobilizations of ethnicized, gendered class alliances that are fighting to keep fossil fuels in the ground. Capitalists become increasingly reliant on misogynist and racist ideologies (see chapter by Brownhill in this volume) to maintain the male deal that is their bulwark against the power of the alliances that are fighting ecocide by blocking access to hydrocarbons. This is fertile ground for fascism.

Conclusion: Prefigurations of ecofeminist ecosocialism on the horizon

How can we actually make the transition from this epoch to a post-capitalist epoch? One of the significant powers of ethnicized, gendered class analysis is its capacity to illuminate and facilitate an actual transition from this to a successor epoch. This transition was prefigured, for instance, by the women of Nigeria's Niger Delta who, in 2002, denied Chevron a large portion of its petroleum production by occupying the Escravos oil export terminal for 11 days, threatening to use the curse of nakedness (see Tobocman et al. in this volume) and sparking a nationwide general strike. They did so at the same time that thousands of people internationally refused to consume Exxon and Chevron petroleum products. They did so as the US military launched its first invasion of Iraq and met global resistance in the form of demonstrations by naked women, who were also performing versions of Aristophanes's 411 BC play, *Lysistrata*, in which women go on a sexual and other domestic services strike until men stop violence and war.

Because Nigerian women shut down petroleum production and export just as hundreds of thousands of people stopped buying gasoline, Chevron faced a global production-consumption strike. The transnational oil major was simultaneously denied part of its production and part of its consumption. When a corporation doesn't have inputs and outputs, that corporation disappears. That is the end of it. This is a non-violent path to system change. Progress along this path is greatly facilitated by the COVID pandemic–induced great leap forward in online savvy, especially among the young worldwide. The dispossessed below the class line are now able to coordinate more effectively on a global scale. More fundamentally, the pandemic has made many more people question the system and recognize that we live or die together (see chapter by Tobocman et al. in this volume). These dynamics indicate that many more production-consumption strikes are on the horizon.

A second instance that prefigures a system transition draws attention to the replacement relationships made possible by production-consumption strikes. This example comes from Yemen in the midst of the 2011–2012 global uprising that included the Arab Spring. Yemeni women and their allies were trying to take control of the country's petroleum production and infrastructure. In early 2011, these alliances called for any parties in the international community to engage in barter trade with them. They offered oil, over which they had some control, in exchange for food, which was in very short supply. Potential barter partners would be those who had control over some exportable food so that they could engage in its exchange for oil. Direct exchanges such as the one proposed by Yemenis cut out capitalists. The most powerful corporations like Monsanto and Cargill in food or Chevron and Exxon in oil slaves are denied their exploitation role and hence their profits. This denial locks in the dissolution of corporate power.

Direct exchanges are a kind of commoning that have the special characteristic of exponential expansion: as more offers to engage in direct exchange arise, more efforts are made by aspiring commoners to take control over the means of life in a fight for fertility so that some of these means of life can be channeled into direct trade. Such trades prioritize use and intrinsic value over exchange value. They are at the centre of a global, post-capitalist political economy based on solar energy, cooperation, reparations, and the free association of all in the pursuit of life support, not profits.

References

Barbier, Edward B. 2010. *The Global Green New Deal, Report Prepared for the Green Economy Initiative of UNEP*. Nairobi: United Nations Environment and New York: Cambridge University Press. https://doi.org/10.1017/CBO9780511844607.017

Bhatnagar, Dipti. 2021. "African Activists: The Earth Is in Peril If Wealthy Nations Don't Slash Emissions and Pay Climate Debt." *Democracy Now*, 23 April. www.democracynow.org/2021/4/23/climate_summit_emission_cuts_global_south

Christie, Renfrew. 1980. "Why Does Capital Need Energy?" In *Oil and Class Struggle*, edited by Petter Nore and Terisa E. Turner, 10–25. London: Zed.

Dyer-Witherford, Nick. 1999. *Cyber Marx: Cycles and Circuits of Struggle in High-Technology Capitalism*. Urbana: University of Illinois Press.

Ekwurzel, B., J. Boneham, M.W. Dalton, R. Heede, R.J. Mera, M.R. Allen, and P.C. Frumhoff. 2017. "The Rise in Global Atmospheric CO2, Surface Temperature, and Sea Level from Emissions Traced to Major Carbon Producers." *Climatic Change* 144: 579–590. https://doi.org/10.1007/s10584-017-2042-9

Estes, Nick. 2019. *Our History Is the Future: Standing Rock Versus the Dakota Access Pipeline, and the Long Tradition of Indigenous Resistance*. London: Verso.

Gago, Veronica, and Raquel Gutierrez Aguilar. 2018. "Women Rising in Defense of Life: Tracing the Revolutionary Flows of Latin American Women's Many Uprisings." *NACLA Report on the Americas* 50 (4): 364–368. https://doi.org/10.1080/10714839.2018.1550978

Germanos, Andrea. 2020. "Indigenous Women to Financial CEOs: Stop Abetting 'ClimateWrecking' Tar Sands Industry." *Common Dreams*, 20 October. www.ecowatch.com/indigenous-women-financial-ceos-tar-sands-2648409004.html

Griffin, Paul. 2017. "The Carbon Majors Database: CDP Carbon Majors Report 2017." *CDP Worldwide*. https://b8f65cb373b1b7b15feb-c70d8ead6ced550b4d987d7c03fcdd1d.ssl.cf3.rackcdn.com/cms/reports/documents/000/002/327/original/Carbon-Majors-Report-2017.pdf?1499691240 (accessed 20 April 2021).

Harvey, Fiona. 2021. "Biden's Pledge to Slash US Emissions Turns Spotlight on China." *The Guardian*, 23 April. www.theguardian.com/environment/2021/apr/23/bidens-pledge-toslash-us-emissions-turns-spotlight-on-china

Kicknosway, Myrna. 2008. "'Akii Kwe: Earth Women of Bkejwanong Territory Demand a Panel Study with First Nations Representatives! Tar Sands Crude Oil Refinery Proposed for the St. Clair River." http://donmarketing.com/PDFs/akii_kwe_brochure_final.pdf (accessed 3 April 2021).

Lewis, Holly. 2016. *The Politics of Everybody: Feminism, Queer Theory, and Marxism at the Intersection*. London: Zed Books.

Linebaugh, Peter. 2021. *Red Round Globe Hot Burning: A Tale at the Crossroads of Commons and Closure, of Love and Terror, of Race and Class, and of Kate and Ned Despard*. Berkeley: University of California Press.

Lohmann, Larry. 2020. "Is Energy White? And if so, What Does that Mean for Climate Movement Strategy?" Paper presented to Durham University Centre for Culture and Ecology, Facebook Live, October 1.

Malm, Andreas. 2016. *Fossil Capital: The Rise of Steam Power and the Roots of Global Warming*. London: Verso.

Marx, Karl. 1867 [1999]. "Chapter Thirty-Two: The Historical Tendency of Capitalist Accumulation." *Capital*. www.marxists.org/archive/marx/works/1867c1/ch32.htm (accessed 20 April 2021).

Moore, Michael. 2001. *Stupid White Men*. New York: HarperCollins.

Picketty, Thomas. 2014. *Capital in the Twenty-First Century*. London and Cambridge, MA: The Belknap Press of Harvard University Press.

The Red Nation. 2021. "The Red Deal: Indigenous Action to Save Our Earth." *Common Notions*. www.commonnotions.org/the-red-deal

Schneider, Mindi. 2008. "We Are Hungry! A Summary Report of Food Riots, Government Responses, and States of Democracy in 2008." www.researchgate.net/publication/237113525_We_Are_Hungry_A_Summary_Report_of_Food_Riots_Government_Responses_and_States_of_Democracy_in_2008 (accessed 20 April 2021).

Simmons, Jordan. 2020. "Overt Bastardization of the Truth: Valve Turner Listed as 'Extremist' by U.S. Government Faces Upcoming Trial." *EcoWatch*. www.ecowatch.com/valveturners-climate-direct-action-2645054934.html

Stephens, Christianne V. 2009. "Toxic Talk at Walpole Island First Nations." Unpublished PhD thesis, McMaster University, Ontario, Canada.

Temper, Leah, Ivonne Yanez, Khadija Sharife, Godwin Ojo, and Joan Martinez-Alier, eds. 2013. "Toward a Post-Oil Civilization: Yasunization and Other Initiatives to Leave Fossil Fuels in the Ground." *EJOLT*, 21 May. www.ejolt.org/2013/05/towards-a-post-oil-civilizationyasunization-and-other-initiatives-to-leave-fossil-fuels-in-the-soil/

Turner, Terisa E. 1994. "Rastafari and the New Society: Caribbean and East African Feminist Roots of a Popular Movement to Reclaim the Earthly Commons." In *Arise Ye Mighty People! Gender, Class & Race in Popular Struggles*, edited by Terisa E. Turner, 9–56. Trenton, NJ: Africa World Press.

WECAN. 2021. "Gendered and Racial Impacts of the Fossil Fuel Industry in North America and Complicit Financial Institutions: A Call to Action for the Health of Our Communities and Nature in the Climate Crisis." *WECAN International*. https://www.wecaninternational.org/2021-divestment-report (accessed 28 August 2021).

4
ECOFEMINIST ECOSOCIALISM

Mary Mellor

Ecofeminism and ecosocialism emerged alongside each other in the latter part of the twentieth century in response to the growing global ecological challenge. While they share a common concern about environmental damage, ecosocialism and ecofeminism have tended to follow different paths. Ecosocialists have concentrated on capitalism and its destructive activities in the sphere of production, while ecofeminists have focused on what capitalism tends to ignore, the sphere of reproduction. In this, ecofeminists identify a link between the subordination and subjugation of women and environmental exploitation and damage. For socialist ecofeminists, this created a political and theoretical link between feminist, green, and socialist analysis (Mellor 1992a).

Very early on Rosemary Radford Ruether stressed the need for feminists to address wider social and ecological relations:

> Women must see that there can be no liberation for them and no solution to the ecological crisis within a society whose fundamental model of relationships continues to be one of domination. They must unite the demands of the women's movement with those of the ecological movement to envision a radical reshaping of the basic socioeconomic relations and . . . underlying values of . . . society.
>
> *(Ruether 1975, 204)*

Ruether advocated a "communitarian socialism" under which women's subordination is to be overcome by "transforming the relationship among power, work and home" (1975, 207). Women's work would be communalized and collectivized under local communal control. All forms of production would return to the local level. Work would be craft based and non-alienating, organic and non-waste-generating: "human society . . . would be consciously integrated into its environment" (1975, 209).

In a later reflection on the development of the relationship between ecofeminism and ecosocialism, Carolyn Merchant also argued for the need to embrace all three movements:

> Socialist feminism views change as dynamic, interactive and dialectical, rather than as mechanistic, linear and incremental. . . . A socialist feminist environmental ethic

 10.4324/9780429341427-6

> involves developing sustainable, non-dominating relations with nature and supplying all peoples with a high quality of life.
>
> *(Merchant 1990, 105)*

Merchant argues that socialist feminist environmental theory would give as much emphasis to reproduction as to production. She also stressed the importance of seeing reproduction in a wider context:

> Weaving together the many strands of the ecofeminist movement is the concept of reproduction construed in its broadest sense to include the continued biological and social reproduction of human life and the continuance of life on earth.
>
> *(Merchant 1992, 209)*

She advocates a "radical ecology" as the basis of a socialist position in which social movements such as bioregional movements, grassroots struggles, and mainstream environmental campaigning would largely replace social class as political agents. While environmental campaigning has grown, it is arguable how far it can be seen to have achieved Merchant's vision of replacing class politics.

Connecting the two spheres

An early theorization of the connection between the subordination of women, the degradation of nature, and capitalist economics was put forward by Maria Mies. Her book, *Patriarchy and Accumulation on a World Scale*, first published in 1986, was subtitled *Women in the International Division of Labour.* Mies, a German sociologist, drew on her years of living in India to knit together the experience of women as paid and unpaid labourers in both the North and the South. She argued that a capitalist patriarchy was condemning women to "housewifization" and thereby responsibility for the sphere of reproduction. By adopting the concept "capitalist patriarchy," she aimed to bring together the concerns of feminism and socialism. She saw housewifization as just one aspect of a system of exploitation that embraces the low-paid or unpaid labour of women, destruction of the natural environment, and the colonization of the resources and knowledge of Indigenous peoples around the world.

Mies's solution was for women in the North and South to realize and challenge their common subordination by the capitalist patriarchy. However, she put more responsibility on the relatively more "prosperous" Northern women to forgo voluntarily the benefits of the exploitation of the women of the South. Similarly, Mies argued that men needed to withdraw their labour from the capitalist patriarchy:

> [I]n an alternative economy men have to share the responsibility for the immediate production of life . . . all work so far subsumed under the term "housework." . . . Only by *doing* this life-producing and life-preserving work *themselves* will they be able to develop a concept of work which transcends the exploitative capitalist patriarchal concept.
>
> *(Mies 1998 [1986], 222, emphasis in original)*

By situating her analysis in a global context, Mies expanded a critique of capitalism and women's inequality to other victims of economic exploitation such as subsistence farmers, Indigenous people, and colonized people in general. In her later work, she argued that Western capitalist consumerism should be replaced by a return to subsistence production (Bennholdt-Thomsen and Mies 1999).

The importance of integrating analysis of the two spheres was taken up by some male ecosocialists, most notably James O'Connor. Launching the journal *Capitalism Nature Socialism* in 1988, he called for "discourses between feminism, social ecology, and Marxism" (O'Connor 1996 [1988], 5). He sought to theorize the relation between them by building on Marx to explore the "second contradiction of capital." The first contradiction of capital is the classic Marxist analysis of the crisis of production. The second contradiction refers not to the crisis within capitalist production, but the structural conditions within which the production process exists, i.e., reproduction and the natural environment.

Ecofeminists have also adapted the Marxian framework to explore the dynamic between production and reproduction. In a debate about the relationship between ecosocialism and ecofeminism, Ariel Salleh adopted the language of Marx to explore "embodied materialism," seeing the link between nature, women, labour, and capitalism as a deeper contradiction than that between capital and labour (Salleh 1994, 106). In an earlier work she argued that:

> "caring" however despised . . . is nevertheless the kind of unpaid service/labor that women under capitalist patriarchy are required to put in. While society denigrates the worth of such work, social reproduction would not occur without it.
>
> *(Salleh et al. 1991, 134)*

Salleh argued that ecosocialism without ecofeminism was incomplete and vice versa (Salleh et al. 1991, 129). It is necessary to understand the dynamics of the mode of production and the social construction of gender, which Salleh described as "reified naturalism." She saw ecofeminism as seeking to understand how such "patriarchal gender images become enmeshed in social institutions in a hegemonic way" (Salleh et al. 1991, 130). At the same time, ecofeminists have to deal with the fact that human beings inhabit sexed bodies that are embedded in the natural world. The ideological constructs of patriarchy and capital such as "mother" nature, woman as "feminine," and "human nature," as well as ideas of economic value and ownership, also relate to biological and material processes. Salleh called upon socialists to:

> examine the social, political and economic consequences of biological sex . . . to come to terms with the material conditions of women's lived experience. . . . Politicians cannot thrust "the biological" aside. That is precisely what has brought Western capitalist patriarchy to its present ecological impasse.
>
> *(Salleh et al. 1991, 131)*

She argued that the ecological impasse could be overcome if ecofeminists and ecosocialists joined together to dismantle "the ideological artifice which divides 'humanity' from 'nature'." This could not be achieved if ecosocialists continued to hold on to Enlightenment concepts of transcendence over "bodily embeddedness in place and in relationships" and failed to understand the "fundamental premise of ecofeminism that in patriarchal cultures, men's assumed right to exploit nature parallels the use/s they make of women" (Salleh et al. 1991, 131).

What emerges from these early writings is a multilevel analysis that raises key themes in the relationship between production and reproduction. Salleh sees capitalism as owing a "debt" at three levels: social debt to exploited labour, embodied debt to reproductive labour, and ecological debt for damage to the natural metabolism (2009, 24).

Materialism not essentialism

Ecofeminism is always in danger of being accused of essentialism: seeing women as being "naturally" closer to nature. How ecofeminism theorizes the relationship between women and the natural world is critical if the accusation is to be rejected. If ecofeminists see the relationship as one of affinity, arguing that women as mothers and nurturers have a unique identification with the natural world, they are open to the criticism of being essentialist. As ecofeminism emerged, socialist feminists expressed alarm at what seemed to be a retrograde movement that would trap women in mothering and nurturing roles (Coote and Campbell 1982; Jackson 1995). Socialist ecofeminists countered the critique of essentialism by arguing that they were not promoting some essential ideal of women as mothers and nurturers, but the material fact of the need for humanity to confront the reality of human existence in nature (Mellor 1992b).

The ecofeminism that is compatible with a materialist ecosocialist perspective does not focus on women as such, but on the sphere of reproduction and women's responsibility for it. That is, an analysis of the social relationship of reproduction and its categorization as "women's work." A materialism that is only concerned with the injustices and inequalities in the so-called "productive" process ignores its impact on the environment and the injustices and inequalities in the reproductive sphere.

Ecofeminism's materialism stresses the immanence (embodiedness and embeddedness) of human existence. The material starting point is that humans exist as beings in a natural environment. As embodied beings, humanity cannot avoid the biological cycle of birth, maturation, and death. Human development is also embedded within a particular ecological framework. In the process, humanity acts on nature with material consequences for the ecological framework.

Ecofeminists argue that modern economies are so destructive because they have lost touch with the overall reality of human existence in nature. Economies are fundamentally gendered in a way that marginalizes the life of the body, together with the rest of nature. As a result of this gendered division of labour, activities that represent only a very partial aspect of human existence have become the driving force and focus of modern economies. Not only are economies gendered, but the relationship between humanity and nature is also gendered. Capitalism and industrialism, compounded by patriarchy, create economies that are disembedded from local communities, local environments, and the whole of human daily life and the earth's life cycles.

Profit-based economies only want exploitable labour and take no responsibility for the rest of the human and non-human life cycle. From this perspective, the so-called wealth-creating "economy" is parasitical on the regeneration and renewal capacity of the environment, unpaid work in homes and communities mainly done by women, socially provided infrastructure and laws, and the accumulated work and knowledge of previous generations. The formal economy, particularly the profit-seeking sector, is described by materialist ecofeminists as the tip of an iceberg with the other supportive sectors lying below the surface (Bennholdt-Thomsen and Mies 1999) or as the icing on a many-layered cake (Henderson 1988).

Immanence not transcendence

Materialist ecofeminism builds on the feminist analysis of gendered work and its critique of the marginalisation and exploitation of women's labour. Far from an essentialist view of women as "naturally" caring, reproductive work can be seen as an "imposed altruism" (Mellor 1992a, 252). As a result of the gender inequality that feminism has identified, the inconveniences of human existence as part of nature is marginalized as low-paid or unpaid "women's work." The link between "women's work" and damage to the non-human environment exists in the way these processes enable dominant groups to distance themselves from the reality of their embodiment. Those who rely on others to maintain their bodies can appear to "transcend" their material conditions. Women and other groups that carry out body work, refuse work, repair work, regeneration, restitutive health work, etc. act as a buffer between dominant elites and the conditions and limitations of their embodiment and embeddedness in the cycles of life that sustain them. However, this does not mean women and other people who do "women's work" are *essentially* more rooted in nature than men; it is just that dominant men (and women) are less rooted in practice.

To put it another way, (some) men (and some women) have used their power to escape the consequences of their embodiment. As Biesecker and Hofmeister argue, if ecological sustainability is to be achieved, it is vital to look at the processes of mediation between society and nature, in particular the work of (re)productivity that would recognize that "the processes involved in the regeneration and restoration of human and non-human life are intrinsic to each and every process involving the production of goods and services" (Biesecker and Hofmeister 2010, 1707).

Materialist ecofeminism argues that the gendered nature of modern economies leads to ecological destructiveness because they have been constructed on the false basis of excluding much of the materiality of human and non-human lives (Mellor 1997). Dominant classes can behave as if human activities are without social or ecological consequences because the destruction of life cycles has no immediate consequences for *them*. As a result, they create socio-economic structures that do not take account of social and ecological realities. Failure to address the basic conditions of human existence leads to what Plumwood called the dangerous illusion of "mastery" of nature:

> After much destruction, mastery will fail, because the master denies dependency on the sustaining other: he misunderstands the conditions of his own existence and lacks sensitivity to limits and to the ultimate points of Earthian Existence.
>
> *(Plumwood 1993, 195)*

Materialist ecofeminism directly challenges the sustainability of contemporary societies and their economic structuring. This has led to the emergence of ecofeminist political economy frameworks that challenge modern economic systems.

Ecofeminist political economy

Ecofeminist political economy sees the assertion of a material link between "women's work" and the natural environment as posing a challenge to the very foundation of modern economies and conventional economics. The starting point is a critique of the "externalization" of both unpaid "women's work" and un/undervalued natural resources. That is, they are not accounted for or incorporated into economic value. In modern economies, the externalized aspects of human embodiment ("women's work") and human embeddedness (relying on the resilience of the natural environment) are both treated as free "goods." As a result, the benefits of women's unpaid work, like the cost of damage to the non-human world, are not acknowledged.

Defining an economy as activities represented by money value fails to acknowledge its true resource base and the way money value is parasitical upon sustaining systems of unpaid social labour and the natural world (Mies 1986; Salleh 2009). As a result, these are exploited and damaged. The money-framed economy can operate as it does because it can exploit the unpaid, or underpaid, caring work that is mainly done by women, together with the resources of the natural world. This leads to an unsustainable construct, "economic man," who appears not to be embodied or embedded (Mellor 1997). "He" operates according to financial calculation, whether as producer or consumer, worker or boss. In an economic context, "he" has no hinterland, domestic responsibilities, needs, or insecurities. "He" is not very young or very old; "he" is not sick or troubled. "He" consumes without any concern for the use or abuse of resources. "He" is gendered but not necessarily male; women participating in the formal economy also need to adopt the mantel of "economic man."

Ecological sustainability and gender justice would mean recognizing the material structures and relationships that enable the false construct "economic man" to seemingly transcend "his" material conditions. The question then becomes how the economic marginalization of both women's work and the sustainability of the environment are to be challenged. The challenge must be to monetary value (or rather market value) itself, with the need to identify other ways of measuring value. In a ground-breaking analysis, Marilyn Waring brought concern about unpaid labour together with a concern for the environment and wanted to allocate both a special value in national accounting (1989). She argued strongly for women's unpaid work and the unrecognized value of the natural environment to be included in the UN system of national accounts. However, writing in 2009, Waring saw the United Nations System of National Accounts (UNSNA) as "failing women miserably as a policy instrument" (Waring 2009, 178).

Silvia Federici has recognized that capitalist wealth relies on the primary accumulation of the unpaid work of housewives, indentured servants, and colonized peoples (Federici 2009, 57). Reproducing humanity is the most basic level of "production" on which everything else rests. Federici (2012) is optimistic about the political potential of women's domestic role. She argues that in the most adverse of circumstances, reproductive and solidaristic work must be done if people and communities are to survive at all. Far from being neutralized by their role within families, women are organizing within and through collectives and commons. Drawing inspiration from women's activities in social economies in the global South and North, including in urban gardening, local money systems, and credit unions, Federici sees women as key to many grassroots social movements, particularly those challenging globalization's attack on the local commons. However, she argues that there is still need for a collective struggle over social reproduction even as cooperative and collective approaches to reproductive work are creating space for experimenting with new ways to restructure human relationships.

Challenging capitalism

The contribution of ecofeminism to ecosocialism is to see the sphere of reproduction as mounting a direct challenge to capitalism. Echoing the work of Maria Mies, many ecofeminists see a resistance to globalized capitalism in the campaigns of women, peasants, and Indigenous people to preserve local economies and common resources. As Leigh Brownhill and Terisa E. Turner argue, the prominence of women in defending the commons against commodification has been evident in Africa for many decades:

> African women have faced and resisted enclosure of their commons and collectively maintained indigenous knowledge, seed, practices, food production, and energy technologies that offer clear alternatives to oil and petro-chemical reliant food and energy systems.
>
> *(Brownhill and Turner 2019, 1)*

They define ecofeminist ecosocialism as "a global, horizontal, subsistence-oriented, decolonized commoning political economy" (Brownhill and Turner 2019, 5). It is a politics of resistance against a neoliberal corporate globalization that exploits women, Indigenous and colonized people, and nature. It seeks to defend the commons against commodification through building economies based on self-provisioning, revitalization of subsistence systems, and local production and distribution.

The concepts of the commons and provisioning are important for ecofeminists. Seeing the environment as a commons is in opposition to its privatization and commodification. Provisioning is a challenge to the separation of the two spheres of production and reproduction and the conventional notion of an economy (Mellor 2016). Feminist economists use the concept of provisioning to dramatically expand the means of human sustenance. Provisioning embraces both paid and unpaid work covering the full range of activities, from love and care to food, shelter, and social and leisure activities. The concept also opens up the distinction between wants and needs and helps focus economic decision-making on the needs of human beings in all aspects of their lives. In meeting human needs, the environment is also seen as a provisioning system with the need to achieve a balance between human needs, the needs of other species, and the need of the environment to reprovision itself.

Why ecosocialism must be ecofeminist

There is a danger that ecosocialism without ecofeminism will assume that getting off the treadmill of productivism and consumerism will enable realization of Marx's "kingdom of freedom" (Löwy 2018). That abolishing exploitative labour and profit seeking will open up an era of choice and fulfilment in harmony with nature. The lesson from ecofeminism is that while it is possible to abandon the socially constructed sphere of production, it is not possible to abolish the labour required in the sphere of reproduction. Marx famously foresaw an era when it would be possible to hunt in the morning, fish in the afternoon, and be a critic after dinner. There is no mention of who cooks the dinner.

If ecosocialists are to absorb the insights of ecofeminism, they must pay attention to the relationship between the sphere of production and the sphere of reproduction. It is at least theoretically possible to deconstruct socially constructed systems such as capitalism or industrialization, but it is not possible to deconstruct the realities of human existence as beings in nature. Humans exist in biological time. They are born, grow/learn, and die. They must eat and drink, keep warm. They need help if sick or infirm. The environment also has its time scale and needs. It needs time to grow, regenerate, react. It may be possible to abolish or minimize production, but it is impossible to ignore reproduction if humanity is to survive.

An ecofeminist ecosocialist "provisioning" economy would start from a commitment to social justice and human well-being as well as the sustainability of the natural world. Patterns of work and consumption would be sensitive to the human life cycle and the replenishing needs of the planet. Provisioning of necessary goods and services would be the main focus of the economy, and the activities of production and exchange would be fully integrated with the dynamics of the body and the environment. Socialists have followed Marx in looking at a politics around a struggle over the means of production. Socialist ecofeminism makes the case for a politics around the means of sustenance: socially just and ecologically sustainable provisioning. This would embrace the right to life and flourishing not only for humans but also for the rest of nature.

There is no "natural" way for humanity to live within its environment. Ecofeminist ecosocialism is about analyzing the sources of inequality and ecological destruction and looking for new ways of living that would enable people to control democratically their means of sustenance in a way that minimizes human impact on the natural world and enables individuals to flourish in peace.

References

Bennholdt-Thomsen, Veronika, and Maria Mies. 1999. *The Subsistence Perspective: Beyond the Globalised Economy*. London: Zed Books.

Biesecker, Adelheid, and Sabine Hofmeister. 2010. "Focus: (Re)productivity. Sustainable Relations Both Between Society and Nature and Between the Genders." *Ecological Economics* 69: 1703–1711.

Brownhill, Leigh, and Terisa E. Turner. 2019. "Ecofeminism at the Heart of Ecosocialism." *Capital Nature Socialism* 30 (1): 1–10.

Coote, Anna, and Beatrix Campbell. 1982. *Sweet Freedom*. London: Picador.

Federici, Silvia. 2009. "The Devaluation of Women's Labour." In *Eco-Sufficiency and Global Justice: Women Write Political Ecology*, edited by Ariel Salleh, 43–57. London: Pluto Press.

Federici, Silvia. 2012. *Reproduction at Point Zero*. Oakland, CA: PM Press.

Henderson, Hazel. 1988. *Politics of the Solar Age*. New York: Knowledge Systems Inc.

Jackson, Cecile. 1995. "Radical Environmental Myths: A Gender Perspective." *New Left Review* 210: 124–140, March–April.

Löwy, Michael. 2018. "Why Ecosocialism: For a Red-Green Future." *Great Transition Initiative*. https://greattransition.org/publication/why-ecosocialism-red-green-future (accessed 14 April 2021).

Mellor, Mary. 1992a. *Breaking the Boundaries: Towards a Feminist, Green Socialism*. London: Virago.

Mellor, Mary. 1992b. "Eco-Feminism and Eco-Socialism: Dilemmas of Essentialism and Materialism." *Capitalism Nature Socialism* 3 (2): 43–62.

Mellor, Mary. 1997. "Women, Nature and the Social Construction of 'Economic Man'." *Ecological Economics* 20 (2): 129–140.

Mellor, Mary. 2016. *Debt or Democracy: Public Money for Sustainability and Social Justice*. London: Pluto.

Merchant, Carolyn. 1990. "Ecofeminism and Feminist Theory." In *Reweaving the World: The Emergence of Ecofeminism*, edited by Irene Diamond and Gloria Feman Orenstein, 100–105. San Francisco: Sierra Club Books.

Merchant, Carolyn. 1992. *Radical Ecology*. London: Routledge.

Mies, Maria. 1998 [1986]. *Patriarchy and Accumulation on a World Scale: Women in the International Division of Labour*. London: Zed Press.

O'Connor, James. 1996 [1988]. "The Second Contradiction of Capitalism." In *The Greening of Marx*, edited by Ted Benton, 197–222. New York: Guilford.

Plumwood, Valerie. 1993. *Feminism and the Mastery of Nature*. London: Routledge.

Ruether, Rosemary Radford. 1975. *New Woman, New Earth*. New York: The Seabury Press.

Salleh, Ariel. 1994. "Nature, Woman, Labor, Capital: Living the Deepest Contradiction. In *Is Capitalism Sustainable?*, edited by Martin O'Connor, 106–125. New York: Guilford.

Salleh, Ariel. 2009. "From Eco-Sufficiency to Global Justice." In *Eco-Sufficiency and Global Justice: Women Write Political Ecology*, edited by Ariel Salleh, 219–304. London: Pluto Press.

Salleh, Ariel, Martin O'Connor, James O'Connor, and Daniel Faber. 1991. "Discussion: Ecosocialism/Ecofeminism." *Capitalism Nature Socialism* 2 (1): 129–140.

Waring, Marilyn. 1989. *If Women Counted*. London: Palgrave Macmillan.

Waring, Marilyn. 2009. "Policy and the Measure of Woman." In *Eco-Sufficiency and Global Justice: Women Write Political Ecology*, edited by Ariel Salleh, 165–176. London: Pluto Press.

5

AN ECOCENTRIC EPISTEMOLOGY FOR ECOSOCIALISM

Ariel Salleh

This essay urges ecosocialists to reframe their analysis in a way that articulates human identity as nature-in-embodied-form.[1] It adopts Marx's focus on internal relations, which theoretical approach moved between vantage points in order to amplify one problem or another. As the dialectician Bertell Ollman says:

> Society has many levels of internally related qualities that we get to understand by using different kinds of lenses, some intimate, others more general, *with each lens or vantage point having a time scale of its own.* This means that seemingly contradictory views can often be true.
>
> *(1992, 93, emphasis added)*

Given the multiple crises generated by corporate globalization in the twenty-first century, an understanding of internal relations is indispensable to political economy. Capital's devastating environmental impact now calls radicals to reposition *History* inside of *Nature,* so to speak. From now on, socialists need to access several discursive lenses to grasp the interplay of ecology with feminism, decoloniality, and their traditional socialist constituency in labour.

Boundary conditions

While Marx's sociology of knowledge grounded the perceptions of each class in its habitual field of praxis (Ollman 1992, 69), in some respects, his own *episteme* remained complicit with the Promethean ethos of the capitalism that he rejected. In his time, the emergent Industrial Revolution was reinforced by a culture of scientific exuberance for manipulating Nature. The Newtonian physics of seventeenth-century Europe had created an illusion of patriarchal control over material processes, spatializing lived temporal flows as measurable units. In this world of "appearance," change was linear, calculated as an extensional relation between position x and position y. Absolute time, somewhere in the repressed realm of *Mother Nature*, was a necessary ground of science, but it remained untheorized. The specular order of Enlightenment men depended on a clean break from lived embodiment, inconvenient limits, and boundary conditions that might threaten the powerful imaginary of reason.

10.4324/9780429341427-7

Capitalist patriarchal value was dissociated from bio-productive processes, and standard readings of Marx have tended to repeat that defect. The preoccupation with production – as distinct from bodily and social reproduction – celebrates the moment of man-to-man exchange, equivalence, one to one. It overlooks the fact that the mode of production relies on structures of sexual and social reproduction – and, in turn, that these consist of environmental energies drawn from nature's metabolism. What gets to be accepted as "value" under capitalism is matter raped of its kinetic pulse. This commonsense is not helpful. It simplifies a complex libidinally assured territory of exchange through measurement of the unitary object (Salleh 1997 [2017]).[2] In economics, science, and political decision-making, artificial monocultures displace a humanly grounded "sense of place." The gentle material reciprocity of, say, plant photosynthesis is inconceivable to the man-made logic of industry. Conversely, in a true "economy," which should be simultaneously an "ecology," "value" would represent the capacity for life regeneration.

Marx's perspective on those who labour with nature largely outside of mechanized production – peasants, women, animals – reflected the instrumental rational discourse of masculinist eurocentrism. Thus, his work was marked by a number of patently ideological features:

- Dualisms, such as *Man* versus *Nature*, speciesism, sexism, ethnocentrism
- A separation of necessity and freedom
- A growth-oriented view that production beyond need is "properly human"
- Too much faith in rational human control and a linear idea of progress
- A very undialectical treatment of technology
- A too-narrow theory of value as men's productivist labour objectified.

Marx's standpoint is what today's nature ethicists call anthropocentric – and this vision has continued to inspire much Left thinking, even when it tries to address "the environment question." For example, Werner Grundmann maintained that

> Anthropocentrism and mastery over nature, far from causing ecological problems, are the starting points from which to address them. . . . Freedom, for Marx, can be gained only in human objectifications, in second nature. The more first nature is transformed into second nature, the more its laws are understood and the more mankind [*sic*] is able to free itself.
>
> *(1991, 2–11)*

Looking for some cultural reflexivity here, an ecological feminist might ask: "Free itself from what, exactly?"

Marxists typically emphasize the labour theory of value to explain the capitalist "laws of motion." The argument is that value derives from labour time expended and spatialized in making a commodity. According to received theory, the use value of a worker's labour is invested in the product, which in turn is appropriated by his employer for sale. What the worker receives in return as a wage is only his exchange value or price in the labour market. The worker, notionally a man – even in a time when women, children, and horses worked the mines – uses his earnings to subsist and maintain his labour power. But he is never fully reimbursed for the labour time expended. In this shortchanging, the difference remaining in capitalist hands is called surplus value.

Twentieth-century socialist feminists from Mariarosa Dalla Costa and Selma James (1972) to Lise Vogel (1983 [2013]) observed a parallel order of extraction between man and wife in the domestic sphere. This is why capitalism is fundamentally a patriarchal system. What is yielded up for free by the woman in her obligatory role as worker's personal caregiver is the labour time of bodily nurture. Further, this may involve many kinds of work – growing food, washing bodies, sewing clothes. Most significantly, a woman, as mother, manufactures the next generation of "labour power" inside her very body. By convention, she is entitled to monetary support from her partner's wage, but neither capitalist entrepreneur nor husband necessarily identifies her as "a subject" with the right to a wage of her own. It is almost half a century since the socialist feminists drew attention to this theoretic anomaly in what has become known as "the domestic labour debate," but the argument fell on deaf ears. Curiously, a woman's adult sons, as products of her embodied labour, are not sold by her for gain but go off to seek their own wage. In some cultures, daughters are taken by their father and exchanged in marriage as a form of wealth. The produce of a woman's labouring body may have various kinds of value, but she herself has none.

Bio-energetics

Appropriation of "the gendered surplus" remains a boundary condition of both Right and Left economics, see David Pepper (1993), for example. In fact, the daily subsistence work that women provide under capitalism is sometimes described as "imported labour" brought in from some "other" mode of production. Socialist James Devine explains the dualist logic this way:

> The exclusion of surplus arising from household production . . . is a common simplifying assumption of many Marxian analyses. . . . [Footnote] After all, that labour does produce use-values that are quite important, indeed totally necessary to human existence as sane and sentient beings.
>
> *(1993, 138)*

A similar theoretic "simplification" applies to "nature's economy"; quoting Devine again:

> Just because nature, in the Marxian view, produces no surplus value does not imply that nature is (or should be) either ethically or empirically unimportant to socialists or capitalists . . . all it says, is that the relationship between capitalism and nature is not a relation among people [*sic*].
>
> *(1993, 152)*

In fact, the relationship between capital and nature is very much a relation among people: unspoken people. As "economic man" gained rhetorical assurance, indeed power, from the model of classical physics, field conditions such as an ideologically constructed nature, including its corollaries in "women" and "natives," were hived off as externalities.[3]

The economic conversation about labour and value is culturally embedded in such repressive metaphorical hierarchies – *Humanity* over *Nature, Man* over *Woman, White* over *Black*. Yet if socialists took seriously the international statistics as to who actually constitutes the global proletariat, the first premise of their analysis might shift. In a pioneering survey using United Nations data, Marilyn Waring (1988) demonstrated how it is women – from global North and South – who do two-thirds of the world's work. Moreover, most of their labours do not rely heavily on technological mediation, so they tend to be environmentally benign. Turning from economic to ecological models, the hubris of capitalist patriarchal physics might have been exposed by the Second Law of Thermodynamics. This faulted the linear imaginary of infinite progress through commodity production by pointing to the possibility of non-reversible matter-energy transformations leading to entropy. That caution resonates with the experience of mothers, peasants, and others whose everyday labour is designed to protect living material processes from disintegration. A healthy ecosystem is a continuously circulating metabolism of energetic matter exchanges, and humans are an intrinsic part of these elemental trades. But when the senses are severed from nature by mechanized production, material appropriation runs wild; nature's symbiotic counterpoint of giving and taking, extracting and restoring is broken. Industrialization cuts across life-giving flows, but Eurocentric dissociation from its embodied ground simply responds to global environmental breakdown with more "modernist" policies and tech fixes.

Would it be possible for ecosocialists to move beyond this dilemma by rounding out Marx's insight into surplus extraction using a bio-energetic theory of value? *The labour nexus where direct humanity-nature energy transfers occur is the real material site of life-affirming value.* Arran Gare has noted that various arguments for bio-energetics have surfaced over the years. Podolinsky, a Narodnik who corresponded with Marx and Engels, wanted to reformulate the theory of surplus value as appropriation of usable energy so as to account for the exploitation of peasants and their lands. Later, in the Soviet era, Stanchinski researched the energy budgets of biotic communities, convinced, as Gare argues

> by studying the energy flows in a whole range of biocenoses, humans would be able to calculate the productive capacities of these natural communities and would be able to structure their own economic activity *in conformity with* them.
>
> *(Gare 1993, 87, emphasis added)*

Sadly, the opportunity for an ecocentric Marxism passed. Nicholas Georgescu-Roegen (1966) introduced thermodynamics into ecological economics, but for the most part, energetics was taken over by scientism and systems theory in the West. At home, Stanchinski's project perished under Stalin's Five-Year Plan. In the 1940s USA, Wilhelm Reich (1999 [1942]) pursued a bio-energetic psychoanalysis of the West's repressive measurement-obsessed 1/0 culture – but he, too, was soon marginalized.

Against the Promethean trend in Marxism, William Leiss (1972) would opt for liberation through a minimalist "mastery over nature." By contrast, naturalistic aspects of Marx's thought are elaborated in Ted Benton's critique of human domination over other animals. However, Benton's insightful discussion of humanity developing powers that go against its own needs begs gendered scrutiny. His ecosocialism proposed a "trans-species ethic" to oversee the temporal phases of organic growth, sexuality, and care; cooperation in the meeting of needs: and management of social integration (1993, 53–54). Benton does not point out that in almost every culture this "meta-industrial skill-set" derives from women's traditionally accorded labour role. Another early ecosocialist, Tim Hayward (1992), was critical of Benton, preferring Richard Lichtman's (1990) focus on the acculturation by which "natural-born" infants become properly human. Yet Lichtman and Hayward also sideline the labours of women, who turn children into social beings. It is not "culture" but the unvalued agency of care work that enables an individual to talk, think, and empathize in a way that feral children cannot. The "species powers" that women, and the exceptional man, elicit through this energetic outlay are trans-cultural, if discursively mediated. While the theoretic invisibility or devaluation of "feminine" contributions is ubiquitous, ecosocialist theory and strategy could benefit by recognizing how women are well placed to take an ecocentric *episteme* forward.

Non-identity

In the making of Europe's scientific revolution, women's customary expertise in herbal medicine and midwifery was eliminated systematically by establishment-condoned witch burnings. Research by Carolyn Merchant (1980) demonstrated how the old idea of the body as part of nature was replaced by the modern idea of the body as machine – signifier of modern men's power and carrier of commercial promise. Joseph Needham's (1954) study of ancient Chinese civilization reveals a quite different template: a scientific logic responsive to internal links and transfers throughout nature. An integrative ecology was not formulated in Europe until Ernst Haeckel (1876) came along in the nineteenth century. Women marvel that he named it from the Greek word *oikos*, meaning communal household. But today, the academic disciplines still reduce materiality to the single "variable" – to be studied in controlled isolation or to be administered as single-issue social policy. On every side, the one-dimensional calculus can be seen to undermine life-giving flows. As ecofeminist activist Vandana Shiva (1989) recounts it, in the name of "development," engineered expertise breaks apart the interactive cycles of air, water, and soils, thereby killing off the habitat of trees, grasses, fungi, bacteria, insects, birds, fish, and mammals.

The source of twenty-first-century dilemmas lies far deeper than private ownership and pursuit of profit. Socialism is certainly necessary to resolving environmental crises but not sufficient to check the instrumental rational mindset that drives overgrazing, deforestation, and mining into disorder and entropy. If physics hypostatized matter as a hierarchy of elements suspended in space, process thinkers from Alfred North Whitehead (1929) to Herbert Marcuse (1972) have envisaged nature as "a living subject" in its own right. Ilya Prigogene and Isabelle Stengers (1984), inspired by quantum science, would trace the generative potential of dissipative structures. Whereas machines are imagined by men to be self-referential entities, they draw heavily on natural systems. And natural systems are themselves far from equilibrium. But paradoxically, material stability can increase with openness and relational complexity. Each biophysical habitat has a characteristic field of tensions that oscillate within and against others: identity and difference (Prigogene and Stengers 1984). Other interrogations of the classic *Humanity/Nature* schism, coming from biologists Richard Lewontin and Richard Levins (1990), document active historicity – "regenerative labour" – in the *episteme* of animals:

> [A]nt "workers" switch among a variety of tasks over the course of the day in response to changing environmental circumstances. . . . There is, moreover, a "daily round" of slowly changing colony activities from morning to night and a long-term change in the activities of a colony as it occupies the same site over years.

This description evokes the multi-tasking and "spatio-temporal scatter" observed by Meike Spitzner (2009) in the daily labour of mothers and housewives.

Just as plants and animals have specific time rhythms, so according to Barbara Adam (1990), humans engaging with them learn to adapt. For this reason, the modernist attitude of positivizing reality is the ultimate political totalization. When people, including Marxist economists, take words and labels as representing fixed identities – essences – they adopt a naive realism whereby mutually transforming processes in nature or society are held artificially still and constant. Reductive measurement and closed-systems theory invariably go hand in hand. But thinking systemically by manipulating numerical indicators as substitutes for constructs like environment, population, and organization cancels out internal relations and, with them, contradiction as the possibility of change. For Marx's dialectic, also, it is the lines between the points, the interactions, that are more significant than the points themselves. Dialectical reasoning is by definition counter-essentialist. Or rather, it might be said that

> the particular ways in which things cohere become essential attributes of what they are.
> *(Ollman 1992, 37)*

Growing out of labour with everyday material processes, dialectical thought moves historically; it defies received semantic boundaries that attempt to organize the living world as if their logic were "given." This is why Theodor Adorno's (1973) "principle of non-identity" is an ecological tool as much as it is a philosophic one.

Since Newton's optic, the image of the lens guided men in focusing on discrete objects. The art of aiming the cannon in war, perspectival drawing, and causal argument in philosophy are each guided by the linear will to mastery. Analytic philosophy has also served the thought police of modernity by keeping debate locked into the synchronic grid with its either/or, theory/praxis, fact/value. This argument by reason always proceeds not by joining, but by "drawing a clear line" between entities. This versus that. When it comes to envisaging complex weblike patterns, in which each part resonates information from the whole, non-sequentially, David Bohm (1983) says that the holograph is a more useful metaphor than the lens, even if it still bends to the privileged patriarchal organ of sight. Marx's dialectic is almost a holographic *episteme* in that it traced mutually referring internal relations back and forth across a multi-dimensional field. Several of these relations have been encountered already:

- Identity and difference – for example, when women and men, or men and animals, share common features and yet are also unlike
- Interpenetration of opposites – for example, when women's gendered labours are a source of both their oppression and privileged political insight
- Quantitative increase to the point of qualitative change – for example, when waste in soil at the base of a tree stirs the growth of new fruit
- Contradiction; the negation of the negation; a spiral – for example, when antagonistic developments inside one relation cause a rearrangement of internal relations

A postmodern Marx?

Marx's sinuous analysis of natural and historical linkages held on to subversive transformations. Ecofeminist and "third-world" activists like bell hooks (1984) and Patricia Hill Collins (1990) also make good use of decentred oscillation and the generative resonances of time lived simultaneously inside and out of *Humanity/Nature*. In fact, humans create order out of chaos precisely by calling different internal relations into focus. Women do this as they labour to mediate conflicts in family life. Peasants pacify biological systems by catalyzing exchanges between hens, cows, and orchard plots. People who are privileged to work with all their senses together, come to a kinaesthetic awareness of the multiple timings embedded in what is handled. They learn about "holding labour": that is to say, synchronizing their actions with systems of Earth growth. Would-be feminist cyborgs notwithstanding, the time scales of pregnancy, washing small bodies, planting and gathering, laying out the dead, remain largely the epistemological province of women. Most of these "meta-industrial labour" activities involve intergenerational needs, sensitive to consequence.

Given that these invisible pulsations are integrated with social practices, "situated" humans have typical ethnic, class, or gendered time frames. Here, Marx's sociology of knowledge resonates with phenomenologist Alfred Schutz (1962), who called for an understanding of consciousness as crossed by "multiple realities." As Schutz noted, the paramount reality of everyday life is "bracketed out" to construct the time horizon of science. Thus, as mechanism became the bearer of white masculinist subjectivity, a separation of action from consequence became the social norm. This disastrous fragmentation was institutionalized in capitalist patriarchal economies as a division of mental versus manual labour. But technocratic efficiency did not "save time"; it simply displaced matter and disturbed the motility of life forms. In short: the abstract systems of instrumental rationality speak nothing but ecological "noise."

If the content of Marx's theory is limited by the anthropocentric values of an industrializing era, his method certainly drove hard against prevailing Western thought habits. Marx's opus exemplifies many different discursive lenses in use, just as with complementarity in physics, where the object may be treated as either light wave or particle. In sociology, a worker may be conceptualized as an individual husband, a sample of an economic class, or a member of the human species. Each distinction describes objective activities in the material world, and each has a theoretical vantage point with conceptual tools honed for different purposes. Ollman points out that the canvas of social analysis can be intimate or distancing in its level of generality: wide or narrow, diachronic or synchronic. In conventional positivist jargon. one might say that the discursive scoping which gives boundaries to an investigation is called its extension, and this has both spatial and temporal aspects. With a diachronic or temporal extension, Marx might typically

> abstract a particular group to include where they seem to be heading, together with the new set of relations that await them but which they have not yet fully acquired. In the case of peasants who are rapidly losing their land, and of small businessmen who are being driven into bankruptcy, this translates into becoming wage-labourers.
>
> *(Ollman 1992, 47)*

Marx's thought processes constantly moved back and forth between levels of generality, just as ecological feminists must do in deconstructing masculinist attitudes. Thus, using one lens, Marx could paint a compassionate psychological portrait of the *individual person*, as in his heart-rending descriptions of the miserable proletarian man or child. At other times, Marx would angle his view of the individual through a *sociological role*, like worker or peasant. This lens emphasizes the functional status of an individual in the system as a whole. Marx might use another lens to provide an economic account of the worker as part of the capitalist *economic system* and its colonizing history. This vantage point would focus on questions of exploitation and equity. It would become the preferred level of generality for both socialist feminists like Lydia Sargent (1981) from the 1970s and for ecosocialists who emerged in the 1990s. But using another lens, Marx could pitch his text against a historical view of capitalism compared with other possible *political traditions*. The lens of Marx's early work offers a naturalistic reading of history as *species development*. But equally, he recognized that a naturalized ethic was used for purposes of social control by the Church with its Great Chain of Being schema, later adapted by bourgeois thinkers like Hobbes and Locke. Ecofeminist work on patriarchal accumulation and related epistemology critiques offer critical readings of ideological naturalization. Finally, the bio-energetic ground of social life, the lens of Frederick Engels' (1883 [1974]) dialectical materialism, allows a *biological account* of the social suggesting that the same laws of transformation may run through all matter.

A meta-industrial *episteme*

An ecosocialist application of internal relations theory must be sensitive to the discursive lenses of women North and South. From the outset, feminism has needed dialogue with the political mainstream: hence, its ongoing *liberal moment.* This involves the struggle of women as individual citizen subjects for equal social and economic entitlements alongside men. Other feminists make sense of their experiences through socialist theory, only to find that here, too, patriarchal stereotyping has objectified them as domestic role embodiments. Attempts to remedy this lapse constitute the *socialist moment* in feminist history. Building on both these analyses, an *ecofeminist moment* explores the multiple lived effects of Eurocentric dissociation with its *Humanity/Nature* dualism. Reading feminism as a continuum in this way, with each discourse or moment having a specific site of political intervention, gives the movement its full power. Ecofeminism as the deepest critique demands a particularly complex dialectical praxis. Here, meta-industrial activists tread a zigzag course between their liberal and socialist task of establishing the right to a political voice; their radical deep structural and decolonial task of interrogating that same validation; and their ecofeminist task of demonstrating how women, peasants, indigenes – men too – may live differently as nature-in-embodied-form. Can ecosocialist theory ground its consciousness in this way?

Ironically, perhaps, it can be argued that Marx's and Engels's dialectic of internal relations readily converges with an ecological feminist standpoint:

- A view of social institutions as alienated second nature manifest in the ideology of linear reductionism
- An ontology of nature as multi-dimensional flow and epistemology that is dialectical and transformative
- A claim that human labours with nature are self- and life-affirming and that sensuous praxis is the basis of valid knowledge
- An urgent global political focus on gender, decolonial, and species struggles

So too, the young Marx envisaged a communist society expressing

> the consummate oneness in substance of man and nature – the true resurrection of nature – the naturalization of man and the humanism of nature both brought to fulfillment.
>
> *(1844 [1964], 104)*

Against the vanities of a dominator philosophy, that sees "everything from nowhere," *an ecofeminist epistemology embraces the ecocentric vantage point of the working majority living "the deepest contradiction" – right inside in the nexus of human exchange with nature.* Indigenous knowledges at the geographic periphery of economic globalization and housekeeping skills at its domestic periphery both typify meta-industrial connections between people and ecologies. In speaking about that doing and knowing, ecofeminists honour the metabolic bridging of *History* and *Nature*. As opposed to entropic activities like mining, smelting, and genetic engineering, subaltern labours give life back to the biosphere. Such labours are mixed transactions; non-identical, human and natural, light wave and particle. But everyday language fails dialectics, for talk of reciprocity is, itself, limited by the ancient dualist fracture of material life. In what might be called an "embodied materialism," these catalyzing moments are not exactly identical to the alienated labour that turns "first" nature into "second," nor are they strictly biology. The sphere of meta-industrial holding is relatively autonomous.

An ecocentric *episteme* is a material condition and yet unrealized intention, for human perception of nature will always be limited by species-specific capacities. More so as new generations of humans interacting with machines have their bodies and minds unconsciously colonized and dulled by fantastical solipsistic "laws of motion." That said, human exposure to the so-called "technologically developed" world is sociologically uneven as Ashish Kothari et al. (2019) emphasize. The nameless class of women, peasants, gatherers, and fishers continues to reproduce the organic enfoldment of nature's internal relations. And this hands-on conversation with the Earth is invaluable because kinaesthetic knowing helps mend ruptured ecological energies. Here, intrinsic value and the precautionary principle are embodied. Meanwhile, across continents, this class is speaking a political vision far removed from corporate greenwash, ecological ethics, or even socialism as we have known it.

Notes

1 This is an abbreviated version of chapter 10 in Salleh (1997 [2017]), "As Energy/Labour Flows." The terms *Humanity* and *Nature, History* and *Nature* are capitalized and italicized when ideological meaning is implied.

2 The reference here is to the discussion of object relations theory in the Salleh (1997 [2017]), "Body Logic: 1/0 Society."

3 For an extension of the "domestic labour debate" across to ecological and postcolonial politics see Luxemburg (1913 [2003]), Mies and Shiva (1993), and Salleh (2009).

References

Adam, Barbara. 1990. *Time and Social Theory*. Cambridge: Polity.

Adorno, Theodor. 1973. *Negative Dialectics*. Translated by E. Ashton. London: Routledge.

Benton, Ted. 1993. *Natural Relations: Ecology, Animal Rights and Social Justice*. London: Verso.

Bohm, David. 1983. *Wholeness and the Implicate Order*. London: Ark.

Collins, Patricia Hill. 1990. *Black Feminist Thought*. Boston: Unwin.

Dalla Costa, Mariarosa, and Selma James, eds. 1972. *The Power of Women and the Subversion of the Community*. Bristol: Falling Wall Press.

Devine, James. 1993. "The Law of Value and Marxian Political Ecology." In *Green on Red: Evolving Ecological Socialism*, edited by Jesse Vorst et al. Halifax: Fernwood.

Engels, Frederick. 1883 [1974]. *Dialectics of Nature*. Moscow: Progress Publishers.

Gare, Arran. 1993. "Soviet Environmentalism: The Path Not Taken." *Capitalism Nature Socialism* 4 (4): 69–88.

Georgescu-Roegen, Nicholas. 1966. *Analytical Economics: Issues and Problems*. Cambridge, MA: Harvard University Press.

Grundmann, Reiner. 1991. *Marxism and Ecology*. Oxford: Clarendon Press.

Haeckel, Ernst. 1876. *The History of Creation, or, the Development of the Earth and Its Inhabitants by the Action of Natural Causes: Doctrine of Evolution in General, and of That of Darwin, Goethe, and Lamarck in Particular*. Translated by E. Ray Lankester. London: Henry King.

Hayward, Tim. 1992. "Ecology and Human Emancipation." *Radical Philosophy* 62: 3–13.

Hooks, Bell. 1984. *Feminist Theory: From Margin to Centre*. Boston: South End.

Kothari, Ashish, Ariel Salleh, Arturo Escobar, Frederico Demaria, Alberto Acosta, eds. 2019. *Pluriverse: A Post-Development Dictionary*. New York: Columbia University Press; New Delhi: Tulika & AuthorsUpFront.

Leiss, William. 1972. *The Domination of Nature*. New York: Brazillier.

Lichtman, Richard. 1990. "The Production of Human Nature by Means of Human Nature." *Capitalism Nature Socialism* 1 (4): 13–51.

Luxemburg, Rosa. 1913 [2003]. *The Accumulation of Capital.* Translated by A. Schwarzschild. London: Routledge.
Marcuse, Herbert. 1972. *Counter Revolution and Revolt.* London: Allen Lane.
Marx, Karl. 1844 [1964]. *The Economic and Philosophic Manuscripts of 1844.* Edited by Dirk Struik. New York: International Publishers.
Merchant, Carolyn. 1980. *The Death of Nature: Women, Ecology, and the Scientific Revolution.* San Francisco: Harper.
Mies, Maria, and Vandana Shiva. 1993. *Ecofeminism.* London: Zed Books.
Needham, Joseph. 1954. *Science and Civilization in China.* Cambridge: Cambridge University Press.
Ollman, Bertell. 1992. *Dialectical Investigations.* New York: Routledge.
Pepper, David. 1993. *Eco-Socialism.* London: Routledge.
Prigogene, Ilya, and Isabelle Stengers. 1984. *Order Out of Chaos: Man's New Dialogue with Nature.* London: Heinemann.
Reich, Wilhelm. 1999 [1942]. *The Function of the Orgasm.* Translated by V. Carfagno. London: Souvenir Press.
Salleh, Ariel. 1997 [2017]. *Ecofeminism as Politics: Nature, Marx, and the Postmodern.* London: Zed Books.
Salleh, Ariel. 2009. "Ecological Debt: Embodied Debt." In *Eco-Sufficiency and Global Justice: Women Write Political Ecology*, edited by Ariel Salleh, 1–40. London: Pluto Press.
Sargent, Lydia, ed. 1981. *Women and Revolution: A Discussion of the Unhappy Marriage of Marxism and Feminism.* Boston: South End Press.
Schutz, Alfred. 1962. "On Multiple Realities." In *Collected Papers*, Vol. 1. The Hague: Martinus Nijhoff.
Shiva, Vandana. 1989. *Staying Alive: Women, Ecology and Development.* London: Zed Books.
Spitzner, Meike. 2009. "How Global Warming Is Gendered." In *Eco-Sufficiency and Global Justice: Women Write Political Ecology*, edited by Ariel Salleh, 218–2129. London: Pluto Press.
Vogel, Lise. 1983 [2013]. *Marxism and the Oppression of Women.* New York: Brill.
Waring, Marilyn. 1988. *Counting for Nothing: What Men Value and What Women Are Worth.* Sydney: Allen and Unwin.
Whitehead, Alfred North. 1929. *Process and Reality.* New York: Palgrave Macmillan.

6

"GREENING," THE HIGHEST STAGE OF EXTRACTIVISM IN LATIN AMERICA

Ana Isla

Introduction

In the Americas, extractivism – economic activities that remove huge amounts of natural resource – has existed since colonial times. These extractions grew exponentially in tandem with the neoliberal framework and engendered the debt crisis of the 1980s that then imposed the Stabilization and Structural Adjustment Programme by the IFM and the World Bank, respectively. After the Earth Summits (1992, 2002, and 2012), extractivism opened up new areas of global intervention for capital when responsibility for sustainable development (SD) was ceded to the World Bank.

The World Bank has contributed to the increase and intensification of extractive industries since taking charge of the Global Environmental Facility to pursue a so-called "green economy" through environmental non-government organizations (NGOs). The role of most international NGOs, such as the World Wildlife Fund, has been to act as brokers between corporations and indebted states and to advance the language of the seemingly more ecologically friendly economic policies and programmes. In the SD paradigm, despite ongoing debates and the search for alternatives, economic growth remains a dominant objective. Consequently, the ongoing destruction of subsistence economies is the central element of what today is understood as sustainable development (Isla 2015).

The political economy of "greening" exposes a global model of accumulation based on extractivism. This perspective shows how the ecosystems of Central and South America are becoming increasingly destabilized, especially through an ever-growing pressure for resource extraction (Isla 2015). Particularly affected in this collusion between neoliberal global capital states and NGOs are Indigenous peoples, who collectively own large amounts of land, territories, and resources that are fundamental to their physical and cultural integrity, livelihood, and sustenance. All these violations have led to an increase in Indigenous rebellions, despite the existence of international covenants, such as the International Labour Organization (ILO) Convention No. 169, the Organization of American States (OAS) Declaration that guarantees pre-existing rights, and the United Nations Declaration on the Rights of Indigenous Peoples. In Latin America, the role of governments, regardless of political orientation, has been the surveillance and intimidation of activists, who usually become victims of selective killings.

 10.4324/9780429341427-8

This chapter addresses three questions: how the regional and local networks on the governance of nature in Latin America have been organized, why extractivism strikes women hardest, and to what extent and how movements against extractivism have become involved in contesting global trends as well as national and local policies supporting them. Two well-known regional infrastructure projects for extractivism are discussed: Plan Puebla Panama (PPP) in Central America and the South American Regional Infrastructure Integration Initiative (IIRSA) (Integracion de Infraestructura Regional in Sur America). They expose the United Nations as a key force in the permanent model of colonialism found in extractivism, based on expropriation, dispossession, exploitation, depredation, and contamination.

An ecofeminist, combined with an ecosocialist frame, is used to question the message that the "green economy" and so-called sustainable development can in any way create social and gender equality, reduce poverty, confront ecological destruction, and combat climate change. Instead, this perspective shows how extractivism can be understood as robbery of bodies and territories in a new phase of capital accumulation in which extractive industries intensify – from enclosures for wind power, mining and metals, carbon, oil, natural gas, soya, sugar cane, oil palm, corn, meat to forests, natural vistas, etc. – directly contributing to the ethnocide of Indigenous peoples, dispossessing the peasantry, expropriating the soil, and destroying ecosystems while at the same time creating conditions for human rights violations and increasing violence against women.

Organizing "greening": From neoliberal development to sustainable development

In the 1970s and particularly in 1982 (debt crisis), Latin American countries were forced to open up even more to market-centred policies, so-called neoliberal development. Since then, US banks have been using International Monetary Fund (IMF) stabilization and World Bank (WB) structural adjustment policies (SAP) to reorganize internal social production and reproduction of the indebted world to favour the penetration of transnational capital. Restructuring implies setting up a new model of accumulation, new patterns of investment and saving, new income distribution, and the creation of capital in new ways. By the end of the 1980s, under Brady's debt-restructuring plan (established in 1989) and the Washington Consensus (1995), Latin American bank loans were transformed into bonds that could be easily traded on the financial markets. The US Enterprise Americas Initiative proposed debt swaps using public funds to transfer indebted countries' public enterprises to private US corporations (Isla 2015).

The 1992 Earth Summit opened the way for the direct management of nature and human resource development by giving responsibility to states and the World Bank. The World Bank developed a key concept – natural capital – which refers to the goods and services provided by the planet's stock of water, land, air, and renewable and non-renewable resources such as plant and animal species, forests, and minerals. As a result, the debt (social) crisis was entangled with the environmental crisis into what has been called "sustainable development." This discourse mistakenly argues that the tensions between poverty and ecology will be resolved in indebted countries by reconciling global economic interests and ecological interests in the market system (UNESCO 1995). W. David Pearce and J. Jeremy Warford (1993) of the World Bank environment department stress the importance of permanent growth and development. They argue that growth in real income per capita can be achieved without major degradation, by getting the price right. This means that price calculations to allocate the use of scarce resources (e.g., water, air) should be based on the laws of supply and demand operating in the marketplace. Kirk Hamilton (2001), also from the World Bank, argues that one possible definition of sustainable development is the process of creating, maintaining, and managing a nation's portfolio of assets in national asset accounting. These assets must include built infrastructure (roads), natural capital (minerals, energy, genetics, agricultural land, forest, rivers, etc.), human capital (education, healthcare), and social capital (networks, the court system, the political regime). They propose that these elements must be embedded in the economic system as natural capital to become integrated with the sustainable development framework and thereby ensure sustainable growth. From their perspective, genuine saving measures expand the national account definitions of assets.

At the 2002 + 10 Earth Summit in Johannesburg, the United Nations argued that the state has lost its capacity to lead sustainable development (SD), understood as economic growth, to confront climate change; therefore, it transferred responsibility for SD to global corporations and their shareholders. Extractivism was boosted, and mining was defined as SD, despite its fossil fuel–centred industrial model, which greatly contributes to global warming. Since then, additional legal protections to corporations to produce alternatives to fossils fuels (solar panels, eolic energy, and geothermal) were incorporated into private sector–friendly legislation and codes regarding the rights of foreign entities in the free-trade agreements. The flawed argument is that they emit less carbon dioxide but are more mineral intensive. According to the World Bank, between 2018 and 2150, the use of graphite, cobalt, and lithium will increase by 450 percent; aluminium by 103 million tons; and copper by 23 million tons by 2050 (World Bank 2018). All this represents a strong pressure on the Southern countries.

At the 2012 + 20 Earth Summit, again in Rio de Janeiro, the "green economy" was unmistakably publicized as the process to eradicate poverty. Subsequently, nature must be selectively monetized and turned into goods (water) and services (forest) traded in financial markets. The Kyoto Protocol (1997) initiated the Payment for Environmental Services (PES) that later evolved into Reduction of Emissions from Deforestation and Forest Degradation (REDD), REDD+ (focused on the forest of indebted countries), and the European Emissions Trading Systems (ETS) programmes, using global markets to manage the forest as an environmental service. A number of dollars is paid for each hectare of forest. However, it is also relevant to emphasize that the measure of emission absorption of carbon gases is not really possible since forests are living organisms that breathe and are dynamic and complex systems, so measurements are always estimates. Again, this programme is problematic as Indigenous and peasant users of the land are described as the most important agents of deforestation.

Critiques by ecosocialism and ecofeminism: the "greening" by extractivism and social costs

Ecosocialism and ecofeminism joined forces in denouncing the dynamics of the capitalist mode of production. Marxist ecosocialists have concentrated on capitalism's destructive activities in the sphere of production. They argue that civilization is a heat engine, and an economy is an entropic flow of energy (oil, gas, carbon, etc.) and materials (aluminium, copper, etc.). Elmer Altvater (1994) found major sources of contradiction in the functioning of capitalism through identifying the natural processes of time and space, described as "ecological modality," as well as the attempt by capital to codify and control time and space with the aim of speeding up the process of capital accumulation, defined as "economic modality." He argues that two different modalities of space and time conflict upon a territorial-social reality as biological time and reproduction are slower than economic time or commodity production. As capitalism is a system that only understands value in terms of money capital, the perpetual drive towards short-time or "t" (economic time) accumulation is in direct conflict with the ecological limits of "T" (historical time) that allow and provide for life on this planet. By applying the principle of entropy in the use of energy in production and consumption where no transformation of energy or matter is perfectly efficient, he concludes that recycling is thermodynamically impossible; thus, energy and raw material are used only once. Therefore, the economy looks for new energy and raw materials permanently, creating social and ecosystem disturbance. Consequently, extractivism has been creating conditions for what Foster et al. posit that, with the constant process of subsuming nature under capital metabolic rifts, vast regions of the planet are being converted into dust bowls, which

> were the social-historical product of expanding capitalism, empire, and white settler colonialism, all of which cover the destruction of land cover and soil erosion. It arose out of the expropriation of indigenous lands, the indigenous people themselves, and the fertile soils.
>
> *(Foster, Clark, and Holleman 2019, 11)*

Socialist ecofeminists have adapted the Marxian framework to explore the dynamic between production and reproduction. Ecofeminists have focused on the sphere of reproduction. They locate the origins of oppression in the interconnected systems of patriarchy, colonialism, and capitalism. Ecofeminists have long considered issues of unpaid and community work, women's work, and nature's work as foundational to economic growth (Mellor 2010; Bennholdt-Thomsen and Mies 1999). They argue that extractivism declares nature as death (Merchant 1983, 2005), encloses the public commons and creates ownership (Federici 2009), and immerses our bodies and our labour as raw material (Shiva 1989).

Ecofeminist Maria Mies (1986) explains that housewifization is an ideology to define some human beings and nature as having no value. At the centre of her analysis is the knowledge that the capitalist patriarchy creates an intersecting domination against all "unwaged" labour. As a result, the expropriation of unpaid and poorly paid labour of women is through rape, harassment, and sexual assaults; the expropriation of labour, knowledge, and subsistence economies of peasants, Indigenous people labelled "unoccupied" or "unused" and thereby easily appropriated by those who can make it "productive"; and the destruction of nature. They all are called feminized or "resourced" – and to suit the purposes of capital expansion, they have to be appropriated, exploited, raped, extracted, and destroyed. Ecofeminist Ariel Salleh suggests that as women's relation to capital and nature has been constructed differently from that of men, women's involvement across cultures in life-affirming activities has resulted in the development of gender-specific knowledge grounded in a material base and reality. Her concept "embodied materialism" is an ecological embodied knowledge that reproduces labour (women), sustains metabolic relations with nature (peasants), and creates lay knowledge (Indigenous people) and biological infrastructure (nature). She uses the term "meta-industrial class" to describe all invisible reproductive labourers (e.g., mothers, peasants, Indigenous peoples, nature) (Salleh 2004).

Consequently, "greening" disputes the message that sustainable development (SD) can in any way create social and gender equality, reduce poverty, confront ecological destruction, or combat climate change. Instead, it shows that extractivism as SD can be understood as a new phase of capital accumulation of global capital (Isla 2015). "Greening" denotes the massive expropriation of territories, depredation and contamination of the soil, and dispossession of the workers through the monetization of nature that requires the devaluing of all other forms of social existence, transforming skills into deficiencies, commons into resources, knowledge into ignorance, autonomy into dependency, and men and women into commodified labour power whose needs require the mediation of the markets. Therefore, viewed through the lens of ecofeminism, these aspects of "greening" come together to wage war on women, subsistence producers, and nature by formulating a new kind of domination based on poverty and unsustainability.

In the "greening" paradigm, two conditions exacerbate poverty and unsustainability: first, bodies are united to the territories that people inhabit. In extractive contexts, the local economy is reoriented towards the presence of the corporations (mining, oil, etc.); as a result, people lose access to natural assets in their communities that have historically allowed them to reproduce ancestral and solidarity forms of economy. The inadequate minimum wage offered by the company becomes an instrument of dependency. The feminine is considered a submissive and subordinate nature that has to sustain the reproduction of life even when ecosystems are destroyed. Furthermore, the increase in diseases caused by water pollution as a consequence of extractive activities increases the care tasks that sick people require and that fall exclusively on women. Nature, like women's bodies, is considered by corporations as a territory that has to be sacrificed to allow the reproduction of capital; nature can be exploited, violated, extracted (Colectivo Miradas Criticas 2017). The political economy of development is the second condition that exacerbates poverty. Economic growth, aligned with military dictatorships, the way the United States protects its corporation by bribing or running over governments, and the operation of its notorious School of the Americas, also known as School of Assassins, has created the violence that runs through the continent (SOA Watch n.d.). Extractivism, expanded after the 2002 Earth Summit, has been increasing militarization with the consequence that the bodies are controlled, objectified, appropriated, and violated, and sometimes, those who resist dispossession are killed. Militarism is combined with violence, force disappearance, femicides, and rapes. Paramilitaries that accompany extractive projects, as in Colombia and Mexico, are encouraged (Zibechi 2021). As a result, the extractive projects deepen the power relations of gender, causing gender inequalities towards women and children and violence towards women's bodies.

The mark of extractivism, as SD, on women's bodies is expressed in the number of women assassinated or disappeared. For instance, the latest report from the Gender Equality Observatory of the UN Economic Commission for Latin America and the Caribbean (ECLAC 2020), reveals that in 2019 at least 4,640 femicides took place in the region, noting that it is in the state where the high levels of impunity, repression, and rape by state agents occur. As mining expanded by 2005, Indigenous, Black, and rural women, who saw their children with various health ailments, declared themselves in rebellion and in open opposition to the robbery and predation committed by transnational mining companies. They began to organize "*Encuentros* (Gatherings) Against Mining," gather information, and act together in the defense of their territories and their human rights by challenging their states and justice systems that do not work for the common good or for the right to life (Ecuador 2007). As a result, violence is directed at these women, and they are harassed, stigmatized, disappeared, and murdered.

The region is raged with anti-systemic movements, including the anti-femicide movement of which I am part. In Argentina, due to the lack of judicial response to femicide (2015) women created a movement called "Ni una Menos" (No one [woman] less), which means that another woman does not die; that cannot be tolerated. It paved the way for "Un violador en tu camino" (A rapist on your path) by the Chilean group Collective Las Tesis (2019), which began as a street act during the commemoration of a new International Day against Gender Violence and the social outbreak in Chile. The message of the lyrics says that "patriarchy is a judge who criticizes us for being born, and our punishment is the violence that you already see." It claims, "The repressive state is a male rapist" before concluding that "the rapist is you." The movement "Ni una Menos" and the line "A rapist on your way" have crossed borders and in some cases have been modified according to the context of each country or city where it has been replicated.

In Mexico, 3,752 women were killed in 2020 (Amnesty International 2021). Mexican women denounce femicide in *Song Without Fear* (Quintana 2020). Here are some verses:

Let the state, the skies, the streets tremble
Let the judges and the judiciary tremble
Women's calm was rekindled
They scared us, but we grew wings

For all the compas (companeras) marching in Reforma [where the government house is located]
For all the girls fighting in Sonora
For the Comandantas battling for Chiapas
For all the mothers looking [for their disappeared children] in Tijuana

We sing without fear, we ask for justice
We scream for every missing
Let it resound loudly "we want each other alive!" [Nos Queremos Vivas!!!]
Let the femicide fall with force.

In brief, these women have become a global symbol of women's anger and their repudiation of the misogynist violence and gendered discrimination that permeates the world and the destruction of the planet. Consequently, ecofeminists are committed not only to resisting the worst consequences of this system and its spread but also to working towards totally different equal, cooperative, life-sustaining, communal forms of social and economic organization.

The next section illustrates the political economy of extractivism, also called corridors, expressed in class/gender/ethnicity struggles, taking place in Central and South America. The first moment illustrates rebellions to massive dispossession and the depredation of women, peasants, and Indigenous people in Central America by Plan Puebla Panama, and the second moment shows struggles against expropriation and exploitation in South America by IIRSA.

Moment 1: Rebellions against expropriation and dispossession in Central America under Plan Puebla Panama (PPP)

Dire social and ecological consequences result from the generalization of the commodity form on nature and human beings. In Central America, the PPP emerged from Agenda 21 and was considered the main initiative for SD, with the objective of identifying and quantifying the biodiversity of the area, organizing infrastructure, and contributing to poverty reduction and environmental reparations (World Bank 2003). Eight governments – Mexico, Guatemala, Belize, El Salvador, Honduras, Nicaragua, Costa Rica, and Panama – accepted the PPP, which involves infrastructure projects in an area of 1,026,117 square kilometres and affects almost 63 million inhabitants. PPP, combined with the North American Free Trade Agreement (NAFTA), signed in 1994 by the US, Canada, and Mexico, assembled capitalist globalization promoting free capital movements and free trade as well as institutional framework for property and contracts, setting the conditions for capital accumulation.

Minerals have been the basic inputs of industrial production. In the mineral economy, subsoil resources and water are the primary means of accumulation. But mining is a fundamentally unsustainable activity that destroys ecosystems through massive deforestation, poisoning water and lagoons, turning land into deserts, increasing traffic, polluting rivers, and draining chemical sludge and heavy metals into inhabited valleys. Moreover, mining corporations pay little in income tax or for the massive amount of water and energy they use. Furthermore, to expand mining, capital needs to destroy the self-reproducing capacity of individuals and communities; therefore, mining has generated the most conflicts because access to water for cultivation and livestock is distorted by extractive activities and made even more destructive by the effects of climate change.

Communities resist violent and corrupt incursions of government and military-backed companies and investors operating in many sectors. Pressured by Indigenous activists, since 1989, the International Labor Organization has recognized the ancestral rights of Indigenous peoples. But in a deregulated framework, there is no community right to reject mining investments. Tragic outcomes of mining projects occur when rural communities refuse to become stakeholders in what they perceive to be the plunder and contamination of their lands and resources. Consequently, civil unrest is the only option left to those who do not want mining in their areas or territories (Coumans 2010). Additionally,

> [G]overnments, including many in Latin America, are being increasingly targeted with multi-million-dollar claims from corporations to undermine efforts of mining-affected communities, courts, governments and even international human rights bodies to protect people and the environment from the harms of mineral extraction.
>
> *(MiningWatch Canada 2017)*

Changing the energy matrix and technological patterns: Nature and social disorder

Within "modern environmentalism," wind turbines are a key factor in the change of the energy matrix and technological pattern. But wind turbines are made of steel (an alloy of iron with varying amounts of carbon content which represent 71 to 79 percent of total turbine mass), fiberglass, resin or plastic (11 to 16 percent), iron or cast iron (5 to 17 percent), copper (1 percent), and aluminium (0 to 2 percent) (World Bank 2018). In Mexico, their use expanded under the Federal Electricity Commission of Mexico and the National Renewable Energy Laboratory (NREL), which is dependent on the United States Department of Energy (DOE). These groups organized the exploitation of the Isthmus of Tehuantepec for wind power, which is the narrowest area of the Mexican territory that lies between the two oceans – Pacific and Atlantic. At its narrowest point, only 200 kilometres separate the Gulf of Tehuantepec and the Gulf of Mexico. The isthmus region has one of the highest potential for wind-power generation in the world, calculated between 5,000 and 7,000 MW of annual capacity. As a result, using public transmission lines, private corporations – the French EDF, the Italian ENEL, the Australian MacQuarie Infrastructure Fund, the Dutch PGGM, the Japanese Mitsubishi, the Spanish Iberdrola, Gamesa, Acciona, Renovalia, Gas Natural Fenosa, Preneal, EYRA-ACS, and the Mexican companies Peñoles, Grupomar, Cemex, and Grupo Salinas, among others – have taken over the Isthmus of Tehuantepec for wind farms (GRIETA n.d.).

The wind projects were entered into without authorization from the community or environmental impact assessments. In 2014, there were 28 wind farms for electricity production, whose beneficiaries are mining, intensive agriculture, gas production, maquiladora industrial parks, and eco-tourism, among others (OMAL n.d.). This infrastructure required the mapping of the 59 towns that inhabit the isthmus, the militarization and para-militarization of the area, and the creation of immigration controls between the borders of Mexico and Guatemala. These projects have produced dispossession among the Olmeca, Zapoteca and other Indigenous peoples living in the territory. They have lost food sovereignty due to the displacement of agriculture (e.g., rice, corn, sorghum, and cane) and fishing, and they have also lost socio-cultural knowledge. There has been a rupture of the collective being due to the confrontations between communities, organized crime to scare people, the generation of precarious jobs, prostitution, and forced migration (Cruz 2020). To combat this, popular and Indigenous resistance has been organized – the Regional Coordinator of Community Authorities (CRAC) controls the territory and administers security and justice. Further, Zapatista women declared in February 2019. "We are going to fight with everything and with all our strength against these mega-projects. If they conquer these lands, it will be on the blood of us Zapatistas."

Max Binks-Collier (2020) examines the ongoing landmark Hudbay Minerals lawsuits in Canada. In 2004, Skye Resources, a Vancouver-based mining company, was granted permission by the government to begin work in a large area in northeastern Guatemala that was home to at least 20 Maya Q'eqchi' communities, including Lote Ocho. Earlier that year, Skye had bought the rights to the open-pit Fenix nickel mine, located near the majority-Maya town of El Estor, from the Canadian mining company INCO. Skye also bought INCO's 70 percent share of its subsidiary, EXMIBAL, which Skye then renamed CGN. But the deal also saw Skye acquire the long-festering, unresolved disputes over land left by INCO and EXMIBAL's violent past (as an example, between 2006 and 2008, CGN dispatched helicopters to terrorize those living on the land). The CGN campaign culminated in two waves of evictions targeting several Indigenous villages on 8, 9, and 17 January 2007. Eleven women from Lote Ocho were allegedly gang-raped by police officers, soldiers, and CGN's security during the last eviction. Consequently, these women have been suing Hudbay Minerals Inc., a Toronto-based mining company that bought Skye in 2008, acquiring Skye's legal liability. Five were pregnant at the time; four miscarried, and one, three days from her due date when she was allegedly gang-raped, said in a deposition that she gave birth to a stillborn who "was all blue or green." Then, men from all three groups (soldiers, police, and CGN security) splashed gasoline over the makeshift huts and the women's tattered clothing and set them ablaze. Marriages were irreparably ruined. The impoverished community eventually split and drifted apart as some members accepted jobs at CGN. Supported by Rights Action (n.d., Archives) the litigation seeking justice continues in Canada and Guatemala.

The 1998 mining law in Honduras erased any distinction between exploration and exploitation concessions, legalized open-pit cyanide mining, permitted forced expropriation in the case of conflict, reduced taxes on mining, and established no limits on water use and no closure requirements for abandoned mines. However, by 2004, an oppositional coalition forced the president to reject more than 60 mining concessions solicited and the suspension of new concessions by executive decree pending the passage of a new mining law (Bebbington, Fash, and Rogan 2018). In addition, the Supreme Court unanimous ruled in 2006 that 13 articles of the 1998 law violated the "fundamental right to harmonic conviviality with the environment and to sustainable development" (Bebbington, Fash, and Rogan 2018, 18–19). By 2007, Comite Ambientalista del Valle de Siria, organized by communities around San Andrés (concessioned to the Canadian Greenstone Resources Limited) and San Martín (concessioned to the Canadian Goldcorp), enlisted scientific studies on problems such as cyanide usage and spills and associated fish kills. They denounced health problems, such as skin infections, respiratory illness, sexual and psychological violence, and feminicide (Honduras 2007). With devastating results, the Honduran government passed a new general water law in 2007, promoting private hydroelectric dams. Between mining and dams, hundreds of defenders of freshwater resources for livestock and crops have been killed, and many others silenced. Among the dead in 2016 was Berta Caceres, murdered for leading a grassroots campaign to prevent a private energy company,

Desarrollos Energeticos Sociedad Anónima, from building a hydroelectric dam in Agua Zarca. Reports of death squad killings among the Garifuna, due to mining and palm oil activities sold to the US and Europe for biofuel, have also occurred (Spring and Russell 2011). The threats and violence continue.

In 2004, after two years of searching for gold in El Salvador, the Canadian Pacific Rim Mining Corporation requested permits to begin mining close to the Lempa River. After several years of negotiations, political manoeuvring, and conflicts with the local communities that tragically cost the lives of four environmental activists – one was eight months pregnant – the request was declined on the basis that the company had not met the necessary regulatory requirements, and a nationwide moratorium on all new mining projects was put in place. In 2009, Pacific Rim sued the government of El Salvador for USD 250 million at a World Bank International Centre for Settlement of Investment Dispute (ICSID) for alleged losses of potential profits because of not being granted permits for its project (MiningWatch 2011). Several local organizations participated in this struggle: the Economic and Social Development Association of Santa Marta (ADES), the Association for the Development of El Salvador, (CRIPDES), and the Foundation for the Study of the Application of the Law (FESPAD). All three were member organizations of the National Roundtable against Metallic Mining in El Salvador. Working with international allies, such as the Institute for Policy Studies in Washington, DC; Oxfam in the US; and MiningWatch Canada, the campaign was on solid ground. They coordinated international protests in Canada and Australia and at the World Bank offices. The case was covered by the international press. By 2016, El Salvador had scored a massive international legal and moral victory, and Pacific was ordered to pay $8 million towards El Salvador's more than $12 million in legal fees. In March 2017, legislators introduced a law that banned metal mining activities in the country (MiningWatch 2017).

In brief, PPP expanded disorder by transforming commons into resources by privatizing public resources, such as wind power (for aeolian electricity) and hydroelectric production (for mining), coupled with the expansion of physical infrastructure, highways, commercial markets, and other transnational businesses. For women and men who depend on their local commons for their livelihoods, the PPP amounted to a full-blown assault on their surroundings and meant a loss of dignity, independence, security, livelihood, health, and sometimes life. Furthermore, as life becomes unbearable under PPP, thousands of Central Americans have fled their countries for the United States, where they suffer a devastating family separation (Deutsch 2020).

The next section examines Iniciativa para la Integración de la Infraestructura Regional Sudamericana (IIRSA). This initiative allowed multinational corporations to operate without impact evaluations, with successive addendum, with preferential arbitration, and without real competition. It also examines the People's Conference on Climate Change (CCC), where forests were re-evaluated in terms of the carbon they sequestered. A key concept in the carbon credit market is the Payment for Environmental Services (PES), which is a voluntary transaction in which a buyer from the industrial world pays a supplier for a well-defined environmental service, such as a patch of forest or a form of land use.

Moment 2: Struggles against dispossession, depredation, and exploitation in South America – IIRSA

IIRSA, another corridor, is a plan to link resources of South America with Northern markets. In 2000, under the initiative of Banco Interamerican de Desarrollo, the presidents of South America countries met in Brasilia to discuss credits for governments interested in the construction of large infrastructures for ports, airports, roads, hydroelectric dams, railways, gas pipelines, and telecommunications. It was not private investment that assumed the risks in the construction of the infrastructure, but rather the state's external debt, as well as the user's taxes, paid for the infrastructure works. As a negative result, states are now involved in centennial foreign debt bonds, indebting generations until 2121. For example, Mexico has a debt of €1.5 billion, Argentina's debt is USD 2.75 billion, and that of Peru is USD 4.0 billion (Bartenstein 2020).

Infrastructure has been centralized in a new structure of the international business community, represented by Brazilian banks, such as BNDES, and corporations, such as Odebreacht and others, though with state partners that connected the main political representatives and international organizations in bribery; kickbacks to senior officials, presidents, and ministers; rigged bids; inflated contracts; and social and environmental harm (Proetica 2020). IIRSA has turned into an infrastructure fiasco, leaving in its wake the destruction of the forest-generating transformations resulting in more than 17,000 hot spots burning in the Amazonia.

Entropy in the Amazon rainforest: Destruction of ecosystems

IIRSA infrastructure increased the expropriation of the local commons. According to Centro Peruano de Estudios Sociales (CEPES), 25 percent of Peru's territory is conceded to mining companies, conferring 48,000 mining rights. As a consequence, struggles for environmental justice, such as movement against the Yanacocha open-pit mining of Newmont Mining Corporation, Buenaventura, and the World Bank emerged in Cajamarca. These corporations hoped to move on to Minas Conga (nine times bigger) as their next project but encountered the resistance of subsistence farmers, Indigenous people, and civil society (Isla 2019). The resistance movement expanded when, on 2 June, 2000, approximately 151 kilograms of elemental mercury spilled, despoiling the towns of Magdalena, San Sebastián de Choropampa, and San Juan, affecting an area of 50 square kilometers; producing high levels of poison in some 1,200 people; contaminating rivers, streams, flora, and fauna; and causing the alteration of the natural cycles of the area. The oppositional movement has been subjected to police and military presence, repression, harassment, fiscal persecution, illegal detention, prison, and death.

After several setbacks, the anti-mining movement saw victory with Maxima Acuna and her rural family's resistance. In 2012, Yanacocha alleged that Maxima and her family occupied its land, and the Peruvian court ruled in favour of the corporation. The judge gave four members of the family suspended prison sentences, which were then revoked in 2014. Since then, the family has suffered more repression at the hands of the police and the private security of Yanacocha. In February 2015, their house was demolished; in 2016, the family's potato harvest was destroyed; and in 2019, their fish pond was poisoned. The resistance movement supported the family by rebuilding their house and replanting their fields.

In the 1980s, Brazil's marginalized rubber tapper communities were losing land to timber companies and cattle ranchers. Chico Mendes proposed the creation of extractive reserves to allow forest peoples to live sustainably. His assassination in 1988 sparked worldwide attention to the plight of the Amazon's forest communities. In 2007, the government laid the groundwork for Acre's programme, the State System of Incentives for Environmental Services (SISA), by dividing all land into zones that developed programmes pricing forests, biodiversity, water, soil, climate, and traditional/cultural knowledge. Two REDD+ programmes under international guidelines outlined by the World Wildlife Fund (WWF), the International Union for the Conservation of Nature (IUCN), the Federal University of Acre (UFAC), the Amazon Environmental Research Institute (IPAM), the Woods Hole Research Center, the Brazilian Agriculture Research Corporation (EMBRAPA), and the German Agency for International Cooperation (GTZ), in collaboration with local governments, have disenfranchised

Indigenous people. For example, the programme did not help people live with or obtain their livelihood from the forests. On the contrary, restrictive measures were imposed on *seringueiros* (rubber tappers).

With respect to the Indigenous peoples of Acre, despite written promises, REDD+ money was never utilized for titling of Indigenous lands, creating incentives for outsiders to further dispossess Indigenous people by large-scale livestock, agribusiness, and logging industry activities; land grabbers; and miners (WRM 2020). Instead, the commodification of the processes of natural and social reproduction (water, carbon sequestration, and biodiversity knowledge) leads to the further alienation of people from the rights of their natural surroundings. Indigenous organizations and social groups of Acre have denounced REDD+ for violations: (a) of land and collective territory rights as Indigenous people have no land title, deepening territorial conflicts and (b) of the rights of the peoples in REDD+ occupied territories to subsistence and traditional activities, such as traditional agriculture and fishing, which have been reduced or eliminated, depriving communities of their livelihood. As a result, these programmes that place a "price" on nature have displaced entire families to the periphery of the cities, forcing some of their children into prostitution (Faustino and Furtado 2014).

Another example of the devastation of local people is lithium mining in Bolivia. Lithium is another key factor in "sustainable energy and technology," leading to geopolitical disputes over natural resources and consolidation of the current hegemonic shift. Lithium in Chile and Argentina is exploited by large multinationals, but in Bolivia, the government controls the resource and exploits lithium through the national company Yacimientos Litiferos Bolivianos. This allows Bolivia to enter the world lithium market without depending on the resources, machinery, investment, technique, or conditions of the large multinationals. To industrialize lithium, Bolivia made agreements with ACI Systems Alemania (ACISA) and Xinjiang TBEA Group Co Ltd of China. These deals in the United States' backyard displeased some US-based mining financiers and corporations; as a result, a regime change was called for (Sanchez 2019).

In 2019, the Organization of American States (OAS), under the command of its president, made a false accusation of electoral fraud against the government of Evo Morales, which gave rise to military treason and activated the fiercest repression against the Movimiento al Socialismo (MAS) militants (a network of unions, civil society, Indigenous people, and peasants). For instance, the mayor of Vinto-Cochabamba, Patricia Arce, who had her hair cut, was covered in red paint and put out on the street. The OAS report of electoral fraud was denied by prestigious institutions (Long 2020). In this circus, the "UK embassy acted as 'strategic partner' to coup regime and organised international mining event in Bolivia four months after democracy overthrown." Also "[it] brought cybersecurity company with close links to the CIA to Bolivia in March 2019, eight months before the military takeover" (Kennard 2021). Yet in November 2020, Luis Arce from MAS overwhelmingly won the election. In this way, in Bolivia, the military attempt to seize political power was repudiated.

In all these ways, by converting the local commons into global commons, IIRSA is subjugating and annihilating women, Indigenous peoples, peasant communities, and the Amazonia rainforest.

Conclusion

The term "greening" includes all the present intensification of extractivism that prioritizes economic growth over local people through militarization. In detailing the gendered processes of extractivism in Central and South America, this chapter has exposed and examined some of the fallacies of "green" capitalism and corporate-defined sustainable development. The two related capitalist projects share a commonality: the expropriation and destruction of Indigenous bodies, territories, and labour in Central and South America, reinforced by colonial/racialized policy instruments and enforced by the United Nations and its institutions and free trade agreements. Most of the projects are paid for by bank loans given to the indebted state to benefit corrupt local officials and international mafias. Natural commons (oil, mining, wind, hydroelectric, forest, scenery, etc.) are expropriated by multinational corporations. Signed international consultations (Convention 169, OAS, and other United Nations resolutions) become meaningless, and when the Latin American citizens protest, they are assassinated or disappeared by the police and their undercover agents and, in some countries, also by paramilitaries. In light of the human and planetary crises, ecosocialism that is ecofeminist is the most appropriate theoretical concept to interrupt economic growth.

References

Altvater, Elmer. 1994. "Ecological and Economic Modalities of Time and Space." In *Is Capitalism Sustainable?: Political Economy and the Politics of Ecology*, edited by martin O'Connor, 76–89. New York: Guilford Press.

Amnesty International. 2021. "Violence Against Women and Girls." www.amnesty.org/en/countries/americas/mexico/report-mexico/ (accessed 14 April 2021).

Bartenstein, Ben. 2020. "Peru's 100-Year Bond Sale Shows Markets Can Defy Political Chaos." www.bloomberg.com/news/articles/2020-11-23/peru-is-selling-century-bonds-just-days-after-political-crisis (accessed 23 November 2021).

Bebbington, Anthony, Benjamin Fash, and Jhon Rogan. 2018. "Socio-Environmental Conflict, Political Settlements, and Mining Governance: A Cross-Border Comparison, El Salvador and Honduras." *Latin America Perspectives* XX (3): 1–23.

Bennholdt-Thomsen, Veronika, and Maria Mies. 1999. *The Subsistence Perspective: Beyond the Globalized Economy*. New York: Zed Books.

Binks-Collier, Max. 2020. "Evicting Lote Ocho: How a Canadian Mining Company Infiltrate the Guatemala State." *The Intercept*, 26 September. https://theintercept.com/2020/09/26/hudbay-skye-canada-mining-guatemala/ (accessed 18 October 2020).

Colectivo Miradas Criticas del Territorio desde el Feminismo. 2017. "Mapeando el Cuerpo-Territorio. Guia Metodológica para Mujeres que Defienden su Territorios." *Ecofeminism, Publicaciones y material multimedia*. www.accionecologica.org/mapeando-el-cuerpo-territorio-guia-metodologica-para-mujeres-que-defienden-sus-territorios/ (accessed 23 April 2021).

Coumans, Catherine. 2010. "Alternative Accountability Mechanisms and Mining: The Problems of Effective Impunity, Human Rights and Agency." *Canadian Journal of Development Studies* 30 (1–2): 27–48.

Cruz, Lucilla. 2020. "La Disputa del territorio en el Istmo de Tehuantec. No al Corredor Interoceanico." Zoom presentation, 3 October.

Deutsch, Judith. 2020. "Child Emergency: Souls in Ice." *Socialist Project*. https://socialistproject.ca/2020/12/child-emergency-souls-in-ice/?fbclid=IwAR3w_qwyVfh0c-QOIZ4LwyNmMU08PcGBRwjOWK8Gi3U2hd2CtKMW9nETaV0 (accessed 20 March 2021).

ECLAC. 2020. "Gender Equality Observatory." https://oig.cepal.org/en/indicators/femicide-or-feminicide (accessed 28 February 2021).

Ecuador. 2007. "Declaracion del Encuentro de los Pueblos por la Vida." www.ocmal.org/3665/#more-106 (accessed 20 November 2020).

Faustino, Cristiane, and Fabrina Furtado. 2014. *Economía verde, pueblos de los bosques y territorios: Violaciones de derechos en el estado de Acre*. Informe preliminar de la Misión de Investigación e Incidencia. Rio Branco, Brazil. Dhesca: Plataforma de Dereitos Humanos, Economicos, Sociales, Culturais e Ambientas.

Federici, Silvia. 2009. *Caliban and the Witch: Women, the Body and Primitive Accumulation*. New York: Automedia.

Foster, John Bellamy, Brett Clark, and Hannah Holleman. 2019. "Capitalism and Robbery: The Expropriation of Land, Labour, and Corporate Life." *Monthly Review* 71 (7).

GRIETA. n.d. "Medio para armar: Corredor Eólico del Istmo de Tehuantepec." www.grieta.org.mx/index.php/parques-eolicos/corredor-eolico-del-istmo-detehuantepec/#:~:text=El%20Istmo%20de%20Tehuantepec%20es,%2C%20Veracruz%2C%20Chiapas%20y%20Tabasco.&text=La%20región%20del%20Istmo%20presenta,7000%20MW%20de%20capacidad%20anual (accessed 26 March 2021).

Hamilton, Kirk. 2001. "Genuine Savings, Population Growth and Sustaining Economic Welfare." Paper presented at Conference on Natural Capital, Poverty, and Development, Toronto, ON.

Honduras. 2007. "Comité Ambientalista del Valle de Siria." 30 January. www.ocmal.org/3661/ (accessed 20 November 2020).

Isla, Ana. 2015. *The "Greening" of Costa Rica. Women, Peasants, Indigenous People and the Remaking of Nature.* Toronto: University of Toronto Press.

Isla, Ana. 2019. "The Guardians of Conga Lagoon: Defending Land, Water and Freedom in Peru." In *Climate Chaos: Ecofeminism and the Land Question*, edited by Ana Isla, 94–124.Toronto: Inanna Publications & Education Inc.

Kennard, Matt. 2021. "Revealed: The UK Supported the Coup in Bolivia to Gain Access to Its 'White Gold'." *Declassified UK*, 8 March. www.dailymaverick.co.za/article/2021-03-08-revealed-the-uk-supported-the-coup-in-bolivia-to-gain-access-to-its-white-gold/ (accessed 11 March 2021).

Long, Guillaume. 2019. "How the OAS, and the Media's Lack of Scrutiny, Caused a Violent Coup in Bolivia." *Open Democracy*, 25 November. www.opendemocracy.net/en/democraciaabierta/how-the-oas-and-the-medias-lack-of-scrutiny-caused-a-violent-coup-in-bolivia/ (accessed 2 November 2020).

March Rights Action. n. d. "Archives – Hudbay Mineral (INCO, SKYE Resources, Solway Investment Group)." https://rightsaction.org/hudbay-minerals-archives (accessed 14 December 2020).

Mellor, Mary. 2010. *The Future of Money: From Financial Crisis to Public Resources.* New York: Pluto Press.

Merchant, Carolyn. 1983. *The Death of Nature: Women, Ecology and the Scientific Revolution.* San Francisco: Harper and Row.

Merchant, Carolyn. 2005. *Radical Ecology.* New York and London: Routledge.

Mies, Maria. 1986. *Patriarchy and Accumulation on a World Scale: Women in the International Division of Labour.* London: Zed Books.

MiningWatch Canada. 2011. "Like Water for Gold in El Salvador." https://miningwatch.ca/blog/2011/7/15/water-gold-el-salvador.

MiningWatch Canada. 2017. "El Salvador – When the Seeds of Resistance Bloom." https://miningwatch.ca/blog/2017/5/31/el-salvador-when-seeds-resistance-bloom (accessed 12 November 2020).

Observatorio de Multinacionales en America Latina (OMAL). n. d. "The Wind Farm Corridor on the Isthmus of Tehuantepec (Oaxaca, Mexico)." https://multinationales.org/IMG/pdf/enco_iberdrola_def.pdf (accessed 15 March 2021).

Pearce, David and Jeremy Warford. 1993. *World Without End: Economics, Environment, and Sustainable Development.* New York: Oxford University Press.

Proetica. "Caso Lava Jato." http://www.proetica.org.pe/casos-emblematicos/caso-lava-jato/ (accessed 12 December 2020).

Quintana, Vivir. 2020. "Cancion sin Miedo." https://www.youtube.com/watch?v=VLLyzqkH6cs

Salleh, Ariel. 2004. "Global Alternatives and the Meta-Industrial class." In *New Socialisms: Futures Beyond Globalization*, edited by Robert Albritton, John Bell, Shannon Bell, and Richard Westra, 201–211. London: Routledge.

Sanchez, Agustina. 2019. "Detrás del Golpe: La Industrialization del Litio en Bolivia." *OCMAL*, 29 November. www.ocmal.org/detras-del-golpe-la-industrializacion-del-litio-en-bolivia/ (accessed 23 April 2021).

Shiva, Vandana. 1989. *Staying Alive: Women, Ecology, and Development.* London: Zed Books.

SOA Watch. n.d. http://soaw.or/home/ (accessed 23 April 2021).

Spring, Karen, and Grahame Russell. 2011. "Military Coups, Mining & Canadian Involvement in Honduras." *Canadian Dimension* 45 (1): 27–29.

UNESCO. 1995. "Declaration and Integrated Framework on Action on Education for Peace and Human Rights and Democracy." General Conference. Twenty-Eighth Session, Paris. www.gcedclearinghouse.org/sites/default/files/resources/%5BENG%5D%20Declaration%20and%20Integrated%20Framework%20of%20Action%20on%20Education%20for%20Peace%2C%20Human%20Rights%20and%20Democracy.pdf

World Bank. 2003. *Briefing: Shareholders and Donors of Mesoamerican Biological Corridor Conference*. Paris: World Bank Conference Center, 12–13 December.

World Bank. 2018. "Minerals for Climate Action." http://pubdocs.worldbank.org/en/961711588875536384/minerals-for-ClimateAction-The-Mineral-Intensity-of-the-Clean-Energy-Transition.pdf (accessed 23 April 2020).

WRM (World Rainforest Movement). 2020. "REDD." https://wrm.org.uy/browse-by-subject/mercantilization-of-nature/redd/

Zibechi, Raul. 2021. "La Militarizacion, Fase Superior del Extractivismo." *La Jornada*. http://www.jornada.com.mx/notas/2021/03/26/politica/la-militarizacion-fase-superior-del-extractivismo/ (accessed 27 March 2021).

7

ROMANTICISM AND THE CRITIQUE OF PROGRESS

Crucial axes in the work of Michael Löwy

Renán Vega Cantor
Translated by Salvatore Engel-Di Mauro

> Socialism can only fulfil its revolutionary role through the inseparable union of sobriety and imagination, reason and hope, the rigour of the detective and the enthusiasm of the dreamer. According to an expression that has become famous, it is necessary to merge the cold current and the warm current of Marxism, both of which are equally indispensable. Bloch establishes a clear hierarchy between them: the cold current exists for the *warm current*, it is at the service of the warm current.
>
> *(Michael Löwy 2007, 18; emphasis in original)*

Michael Löwy has woven an extensive reflection on revolutionary romanticism and the Marxist critique of progress, two essential topics of revolutionary anti-capitalist thought taken up in this chapter. I take them up freely, trying to be as faithful as possible to the master's teachings, hoping that my analysis does not contradict the letter and, above all, the spirit of his extraordinary contributions.

Revolutionary romanticism: The example of William Morris

One of the issues that has preoccupied Michael Löwy is revolutionary romanticism, on which he has carried out various investigations that have helped explain the meaning and scope of the romantic critique of capitalist modernity. For revolutionary romanticism, capitalism has destroyed essential human values, including solidarity, while at the same time destroying nature through its intensified exploitation. This romantic vision criticizes individualism, market quantification, the cult of money, the construction of uninhabitable and polluted cities, the mechanization of workplaces and human beings, rationalist abstraction, and the destruction of social bonds (Löwy and Sayre 2008, 40).

An example of this kind of romanticism is that of William Morris, who, studying Marx and various romantic artists and writers, made a radical critique of bourgeois civilization. His critique stresses that it has meant no real progress for humanity that capitalism has destroyed peasant societies' communal ways of life, where human beings were skilled in some art or craft and those "moved by poetry and stories" were content to live (Thompson 1988, 242–243).

10.4324/9780429341427-9

William Morris anticipated some of today's mainstream environmentalist criticisms but linked them directly to capitalism. For him, the root of the problem lay in the fact that the capitalist system is based on profits, the main reason for social and economic problems (Thompson 1988, 688). This ecosocial critique remains very relevant today because it established a direct link between the dehumanization produced by capitalist society in the workplace and the destruction of the environment. William Morris did not seek a return to the past, but the construction of a new society – a communist one – in which the dehumanizing characteristics of capitalism would disappear. He advocated for the construction of a new civilizing project, radically different from capitalism, and not the construction of a *capitalist* society *without capitalists* that would reproduce its vices and its technical and cultural model.

This idea has been taken up by Michael Löwy in criticizing the cult of productive forces and the claim that these are neutral. In contrast, Löwy argues that an anti-capitalist civilizational transformation must also consider the radical modification of the predominant productive forces of capitalism. It is not, therefore, a matter of inheriting the productive forces of capitalism, but of modifying them, in the same direction as the state, as Karl Marx had called for after the Paris Commune, in order to overcome its destructive and predatory character of nature (Löwy 2011, 27, 36).

Walter Benjamin and the critique of the ideology of progress

A decisive contribution of Michael Löwy to revolutionary thought is to have rescued the criticisms of progress through the study of various authors of the revolutionary tradition, mainly Marxist, whose work had a certain heretical tinge. Among these authors rescued by Löwy are August Blanqui, José Carlos Mariátegui, the main exponents of the Frankfurt School, Ernst Bloch, and György Lukács. Among these authors, Walter Benjamin stands out. Löwy analyzed his various writings, but it is his commentaries on Benjamin's *Theses on the Concept of History* that stand out.

A heterodox Marxist, Walter Benjamin targeted the idea of progress with a most profound critique (Löwy 1990; 1992). His aim was to fuse historical materialism with a messianic and romantic conception that would allow Marxism to become the doctrine of the vanquished of all times. In his texts we find a denunciation of the consequences of militarism and of the dangers of the use of chemical weapons in the wars to come, weapons that were already being prepared in sophisticated laboratories. Such a terrible prospect, which would unfortunately become a painful reality a few years later, was evoked in his article title: "The Weapons of Tomorrow" (Löwy 2002, 272). From his crude analyses, Benjamin draws one of his most iconoclastic insights into the Marxist tradition, stating that the aim of the proletarian revolution is not to propel society into a more progressive phase, as an inevitable result of the accumulation of technological change. Rather, it is to stop the impending catastrophe. In other words, the revolution is conceived as a sudden interruption of an unstoppable technological progress leading to ruin. In 1938, shortly before the outbreak of the Second World War and as "civilized" Europe was sinking into barbarism, Benjamin stressed that it was "necessary to base the concept of progress on the idea of catastrophe. Let things continue 'as they are,' that is the catastrophe" (Benjamin 1982, 242).

After a systematic and rigorous exposition of Walter Benjamin's critical conceptualization of progress, Michael Löwy concludes that

> The result of this work is a reworking, a critical reformulation of Marxism, which incorporates messianic, romantic, Blanquist, libertarian and Fourierist "splinters" into

> the corpus of historical materialism. Or, rather, the fabrication, from the fusion of all these materials, of a new Marxism, heretical and radically different from all the variants – orthodox or dissenting – of its time. . . . Above all a *Marxism of unpredictability:* if history is open, if the "new" is possible, it is because the future is not known beforehand.
>
> *(Löwy 2002, 172; emphasis in original)*

These same words apply to Michael Löwy's own project, which has contributed to the renewal of Marxism in a number of areas, such as romanticism and the critique of progress.

Implications

The study of revolutionary romanticism and the systematization of the critique of progress has theoretical and political consequences, which are briefly presented here in the following way: the conversion of productive forces into destructive forces and the critique of the cult of these productive forces; the critique of instrumental reason and the dominant view of science and technology; the questioning of the commodification of the world; the critique of the conception that postulates human beings as masters and lords of nature; the historical reinterpretation of capitalism on the basis of the rebellion against destructive progress; and anti-capitalism and anti-imperialism as the foundation of the critique of progress.

The conversion of productive into destructive forces

The uncontrollable development of capitalism has meant the accelerated transformation and destruction of the species and mineral reserves that nature had been generating over thousands or millions of years. In the short span of 200 years – one second in cosmic history – capitalism has destroyed the natural basis of human existence.

This destructive tendency of modern production is presented in conventional terms – both by the apologists of capitalism and by some Marxist current worshipping the development of the productive forces – as a sign of the productive capacities of capitalism, which would point to its superiority over any of the other modes of production that have hitherto existed. From an energetics point of view, according to the Law of Entropy, it is relatively easy to demonstrate the destructive character of the productive forces under capitalism. Where the transformation of productive forces into destructive forces is especially clear is in the military industry, the most "dynamic" sector of the economy, which has been conducive to capitalist accumulation since the Cold War. Remarkable scientific and technological developments, sensational inventions, and sophisticated products have been linked in one way or another to the military industry.

A critique of instrumental rationality and the dominant version of science and technology

In taking stock of the twentieth century, it is doubtful whether technoscience can be separated from the barbarism that technological progress has become, insofar as science and scientists – one can hardly separate one from the other – have also become a destructive force, even if technoscience remains a productive force in a contradictory way. A critical view of progress, without abandoning reason, must criticize instrumental rationalism and, without disavowing science, must undertake its critique since science has become not only a productive/destructive force but also a commodity.

Criticizing science does not mean falling into irrationalism, to which science is also condemned today, but in the name of the vindication of a series of absolutely irrational practices, as is happening today in the United States, the country where a good part of the planet's scientists are concentrated. It should be remembered that it is in the United States where the belief in UFOs, extra-terrestrials, astrology, demonic possession, satanic control, and a thousand other such ideas flourish.

This raises the question of the contradictory nature of science in enabling the best and the worst: widespread genocide and ecocide alongside well-being and tranquility for all the inhabitants of the blue planet. In any case, given the dominance of capital over science, scientific work is predominantly done to benefit capital: i.e., for profit and capital accumulation.

The major problems in industrial society are not solved by more science and technology, but by society's actions and political decisions. This is the only way to combat the technocracy prevailing in today's world. In the same way, what must be fought is the anti-democratic character of science, which postulates that only the "wise" and the "experts" determine what is best for the rest of humanity and that humanity must resign itself to such "noble" purposes, even if this nobility of spirit programmes the death of half of humanity. Have we so soon forgotten that most of the Nobel Prize winners in the United States in the 1960s and 1970s became renowned for their contributions to the criminal war against Vietnam?

This does not mean that technology should be rejected for the sake of an absolute reunification with nature. That is not only impossible but also unrealistic in the current world situation. Although a dialogue with science is indispensable, such a dialogue must contribute to a new kind of science and technology that serves humanity and not the minority interests of international capital. As Michael Löwy has put it, environmental movements

> do not oppose the improvements brought about by technological progress: on the contrary, the demand for electricity, running water, the need for sewage and drainage, the establishment of medical clinics are high on the list of their demands. What they reject is pollution and the destruction of their natural environment in the name of the laws of the market and the imperatives of capitalist expansion.
>
> *(Löwy 2011, 39)*

Commodification of everything

A central element that demarcates the Marxist critique of progress from other such critiques lies in unravelling the dire consequences of the generalization of the commodity form on human beings and nature. Unmasking the character of commodity relations is fundamental to understanding what is happening in the world today and to proposing alternatives. A central premise can be put forward: *commodification must be seen as an unbearable form of social regression, even if it is the result of technological and scientific progress.*

As capitalist relations engulf the world, even the most unthinkable things, such as genes, eggs, and sperm, are commodified. The latest step has been taken with the commodification of nature and the human body. With regards to the former, there is an accelerated process of appropriation in which large multinational monopolies are participating as beachheads. As the biological wealth and diversity of ecosystems is to be found in the global South, a process of recolonization is underway, aimed at appropriating this diversity. This has served as an ideological justification for the "greening" of capitalism, with such conceits as "sustainable development." When big capitalist corporations have discovered the profit prospects of turning nature into a marketable commodity, everything begins to be disguised as "green." Under the pretext of protecting nature, capitalist companies embark on a mad rush to appropriate the last remaining natural "colonies." Thus, we observe an avalanche of experts, studies, and technicians with the aim of discovering new plants, animals, and genetic material in the global South, especially in tropical forests, in order to develop products and medicines that will be patented in the laboratories of the global North and of multinational companies. In this sense, capitalism is undergoing a remarkable transformation, as what was once considered an "externality" is now conceived as a *stock* for capital, opening the way to the capitalization of nature and the unbridled competition between multinational corporations.

The other issue of the abhorrent extremes that commodification has reached in today's capitalism is related to the commercialization of the human body, parts of it, or even the genes themselves. Thus, the biological reproduction of human beings is now an industry, involving multimillion-dollar investments and fabulous profits. This vile market has reached such an extent that in the USA a woman's uterus is contracted to reproduce a special commodity, hitherto known as a baby, who will become the property of the contractor. In this case, the *value form* reaches into the innermost, unfathomable recesses of the human body. It is not uncommon to find that this repugnant market in human organs proliferates semi-legally in "civilized" countries, with the aim of recomposing the bodies of some of their wealthy inhabitants. The consequences of such an abominable trade are seen most evidently in poor countries, where the practice of killing or mutilating children and young people to satisfy the demand for such coveted human organs, now turned into mere commodities, has become widespread. If this commodification of nature and human beings is conceived as progress, then we live in the best of all possible worlds, and we have no reason to be against the end points to which this technological and scientific progress is leading us!

Rebellion against destructive progress under capitalism

Until not so long ago, traditional Marxist historical studies, guided mainly by Soviet Marxism or some of its variants, shared a unilinear and progressive view of history which, while purporting to criticize the Western model of colonial and imperialist expansion and the apologetic discourse that flowed from it, took on board some of its key tenets: the cult of technology; the justification of the destruction of local cultures in colonized areas on the grounds that this was an irreversible trend; the superiority of the economy and "culture" of Western capitalism; the underestimation of the capacities and proposals for resistance promoted by the poor and exploited; and, in short, a manifest cult of the *progressive* "single path," outside of which there was no escape or alternative.

Very recent studies from a Marxist perspective, or close to it, are reinterpreting the history of Western *progress* and its economic, social, cultural, and ecological consequences on the colonial and peripheral world, as well as on the plebeian order of societies in which industrial transformations took place, such as England, the cradle of the Industrial Revolution. In this light, there has been a critical reconsideration of European expansion and its destruction of diverse forms of social organization that also involved other types of relations with nature (see Wolff 1994). Studies aimed at highlighting the mechanisms of resistance developed by plebeian culture during the time of the Industrial Revolution have also stood out, which has led to showing, for example, that the agricultural revolution did not necessarily have to destroy traditional forms of cultivation. It is now known that there was a logic to the peasant economy. It achieved growth in a different way from that of the large landowners, who only sought commercial profit and ignored peasant welfare (Fontana 1994, 140).

In the same way, it has been shown that the capitalist factory did not arise out of a technical disposition but out of the need to subordinate the worker and increase the extraction of surplus value, as Marx had already indicated in *Capital*. This type of factory was opposed by the workers, those who thought that

> industrial production could be organised in a more socially equitable way, without renouncing technological advances; that machines could be put at the service of the worker, instead of enslaving him, and that capital and labour should not be separated, but "indissolubly united in the hands of the workers."
>
> *(Fontana 1994, 142)*

Then, as now, those who defended their traditional forms of existence, their customs, and their culture were labelled "reactionaries," standing against the current of history that led irremediably to what was best and most beneficial for all, even if we know today that what is announced as best and most effective is only generalized catastrophe and barbarism. In a word, they opposed nothing less than progress. From this perspective, Michael Löwy vindicates the categorical imperative of the young Karl Marx to "fight to 'overthrow all social conditions under which man is degraded, subjugated, abandoned, despised'" (Löwy 2002, 182).

Anti-capitalism and anti-imperialism: Foundations of the critique of progress

Marxism, in taking up the critique of progress, must reaffirm its anti-capitalist stance, which implies that claims of the supposed "pluralist" responsibility of all human beings for the critical human and ecological situation of the world are only intended to hide the real culprit: the capitalist mode of production. In considering this question, the more serious Marxist studies now being carried out in various parts of the world tend to point to the natural limits of this predatory mode of production.

Another essential component, though more rarely named and raised, is that of anti-imperialism. On this point, Michael Löwy is clear and refuses to renounce the use of a fundamental concept within the revolutionary tradition, that of imperialism. This is something which must be emphasized because much of the world left after the collapse of the USSR has hastily and uncritically taken on the light terminology of globalization, empire, information age, liquid society, and similar terms (Löwy 1995).

The issue of imperialism connotes a fundamental concern: to consider the forms of *domination* on the planet and not to blithely assume that globalization has unified the world in such a way that relations of inequality, oppression, and exploitation no longer exist between different states and territories. For the decline of national economies and the emergence of a new international division of labour "does not mean that economic and technological power does not continue to be geographically concentrated in one part of the globe, and poverty and dependence in the other, or that there is no connection between these two phenomena, whatever we want to call it" (Hobsbawm 1993, 74).

At the present time of imperialist expansion to the furthest corners of the planet, there is an accelerated destruction of ecosystems and a drastic reduction of biodiversity. This is a direct result of the universalization of capitalism, the elimination of state regulation mechanisms, the unconditional opening of countries to multinationals, the commodification of products of natural origin, the unbridled competition between countries to gain an advantageous position in the export market, the falling prices of raw materials from the capitalist world's peripheries, the return of primary economies: in short, the inherent logic of capitalism of accumulation at the cost of the destruction of human beings and nature. Capitalism is a profoundly unequal relationship, and the largest production and consumption capacity is concentrated in the core countries (the United States, the European Union, Japan), where millions of tonnes of waste are produced. Cars, telephones, televisions, refrigerators, batteries, etc., which, after being used, soon end up in the rubbish bin, are no different. Although materials used in the production of these artefacts come from the dependent countries of the periphery (where these materials had a use value, i.e., could be used), raw materials become useless waste after being consumed in the global North and its global South emulators. And it is at this point that these poor countries are once again thought of as the receptacle for the waste produced by the unbridled consumption of the opulent in the global North. This can be viewed as typical of ecological imperialism (in the sense of Foster 1994, 91 and fn.).

Revolution, an indispensable way to avoid the catastrophe to which capitalism is leading us

Taking up Walter Benjamin's famous statement, it is necessary to claim the need for an anti-capitalist revolution that goes beyond capitalism, but which also aims to go beyond the existing industrial civilization, rebuilding it from its very foundations: for example, with regard to the use of fossil energy, the fetishism of the automobile, and enormous urban agglomerations. From this perspective, the only real possibility is the use of solar energy (Altvater 1994, 231).

This revolution must propose the overcoming of capitalism, since its limits and the problems that its existence entails for human beings and the ecosphere have already been clearly established. But also the overcoming of industrial civilization, since it is not enough to go beyond capitalism without proposing another type of modernity which rethinks vital issues, among which would be a redistribution of wealth; a parallel attack on capitalism and patriarchy; a radical transformation of the industrialization policies implemented so far; the abandonment of technological progress as the axis of relations between the economy, society, and nature; the reorientation of science and technology so that they become knowledge that benefits the majority of people; and the abandonment of profit and the squandering of resources in the name of market efficiency. An anti-capitalist revolution is necessary to avoid the catastrophe to which world capitalism is leading us. This requires a complete overhaul of the cult of progress. The revolution can no longer be carried out to propel the train of progress but rather to stop the vertiginous race of that train towards death, towards which, even if we do not want to, we are all travelling. This implies the renewal of the concept of productive/destructive forces and the depredation of nature, as well as the participation of diverse subjects and social classes.

All this today sounds more utopian than ever, but is it not precisely this criminal absence of utopia, of mobilizing dreams, that humanity is lacking while the reactionary utopias of capital, with their false paradises of prosperity and well-being, become nightmares for more than 80 percent of the world's population and while the shameful display of technology and science becomes a new pornography against the poor world, now considered unnecessary and disposable. And, of course, within the anti-progressive perspectives of a revolution, the *ecosocial* component is indispensable and requires a new way of understanding and assuming the relations not only between human beings but also between human beings and nature.

Socialism must complement the struggle against the regressive consequences of the globalization of capital, with its technical and scientific paraphernalia that sweeps away everything in its path, with the struggle in solidarity with the peoples and cultures that resist the onslaught of capitalism. Socialism must also reaffirm the principles that have always guided that struggle, such as justice, equality, fraternity, internationalism, real democracy, and communism. These principles have guided Michael Löwy's revolutionary praxis as an example of commitment to anti-capitalist struggles around the world.

References

Altvater, Eltmar. 1994. *El precio del Bienestar: Expolio del Medio Ambiente y Nuevo (Des)Orden Mundial*. Valencia: Edicions Alfons el Magnànim.

Benjamin, Walter. 1982. *Charles Baudelaire: Un Poète Lyrique à l'Apogée du Capitalisme*. Paris: Payot.

Fontana, Josep. 1994. *Europa ante el Espejo*. Barcelona: Editorial Crítica.

Foster, John Bellamy. 1994. *The Vulnerable Planet: A Short Economic History of the Environment*. New York: Monthly Review Press.

Hobsbawm, Eric. 1993. "Addressing the Questions." *Radical History Review* 57: 73–75.

Löwy, Michael. 1990. "La Escuela de Frankfurt y la Modernidad: Benjamin y Habermas." *Revista Colombiana de Sociología* 1 (1): 23–30.

Löwy, Michael. 1992. *Redemption and Utopia: Jewish Libertarian Thought in Central Europe*. Stanford, CA: Stanford University Press.

Löwy, Michael. 1995. "Imperialism: Presentation II." *Actuel Marx* 18: 19–20.

Löwy, Michael. 2002. *Walter Benjamin: Aviso de Incendio*. Buenos Aires: Fondo de Cultura Económica.

Löwy, Michael. 2007. "Utopía y Romanticismo Revolucionario en Ernst Bloch." In *Ernst Bloch: Tendencias y Latencias de un Pensamiento*, edited by Miguel Vedda. Buenos Aires: Herramienta Ediciones.

Löwy, Michael. 2011. *Ecosocialismo: La Alternativa Radical a la Catástrofe Ecológica Capitalista*. Buenos Aires: Ediciones Herramienta.

Löwy, Michael, and Robert Sayre. 2008. *Rebelión y Melancolía: El Romanticismo como Contracorriente de la Modernidad*. Buenos Aires: Ediciones Nueva Visión.

Thompson, Edward. 1988. *William Morris, de Romántico a Revolucionario*. Valencia: Edicions Alfons el Magnanim.

Wolff, Eric R. 1994. *Europa y la Gente sin Historia*. Buenos Aires: Editorial Fondo de Cultura Económica.

8

MARX'S ECOLOGY AND METABOLIC ANALYSIS

Brett Clark and John Bellamy Foster

Introduction

A planetary emergency looms over the twenty-first century as the ecological crisis increases in scale and severity, threatening human civilization with self-extermination. This situation is primarily due to the operation of a system predicated on ever-expanding capital accumulation and exponential economic growth. This drive has resulted in the transgression and crossing of planetary boundaries that define "a safe operating space for humanity," such as climate change, ocean acidification, loss of biological diversity, diminished availability and quality of fresh water, land cover change (e.g., deforestation), the rupture of the nitrogen and phosphorus cycles, and increasing pollution from synthetic chemicals, which leads to the bioaccumulation and biomagnification of toxins within living organisms and across the food chain (Rockström et al. 2009; Steffen et al. 2015). It has manifested in an ecological moment of truth, the "Great Capitalist Climacteric," whereby humanity either proceeds further down a path of ecological destruction, undermining the conditions of life, or radically shifts its productive relations as part of an ecological revolution to forge a sustainable society (Foster 2015).

Ecological Marxism offers critical insights for understanding the emergence and development of the "anthropogenic rift in the natural history of planet earth" (Foster 2000; Hamilton and Grinevald 2015, 67). As Karl Marx stressed, reflecting on the alienation of nature,

> it is not the unity of living and active humanity with the natural, inorganic conditions of their metabolic exchange with nature, and hence their appropriation of nature, which requires explanation . . . but rather the separation between these inorganic conditions of human existence and this active existence, a separation which is completely posited only in the relation of wage labor and capital.
>
> *(Marx 1973, 489)*

It is this historical separation in material reality accompanying capitalist development that has generated the "irreparable rift" between society and nature (Marx 1981, 949).

From the late 1990s on, metabolic rift scholarship has helped establish a rich tradition within ecological Marxism. By returning to Marx to address earthly questions, it has excavated the ecological foundations of classical historical materialism. It has detailed how Marx's

 10.4324/9780429341427-10

materialist conception of history was inseparably bound to his materialist conception of nature, forming a dialectical unity. It has identified, assessed, and developed the ecological analysis found within the work of Marx, exploring how this open, dialectical, materialist approach informed and was integrated into the scientific analyses and social critiques that followed. It has demonstrated how Marx's political-economic and ecological critiques of capitalism are unified, revealing how ecological relations and concerns are part and parcel of any discussions of value analysis, accumulation, labour, alienation, history, human needs, and socialism. It has involved an extensive examination and extension of Marx's metabolic analysis of the material exchanges between human society and nature, focusing on how capitalism, given its relentless drive to accumulation, produces ecological (metabolic) rifts as it transgresses against natural limits and the workings of ecosystems (Burkett 1999; Foster 2000, 2020; Foster, Clark, and York 2010; Saito 2014).

Marx's triadic metabolic scheme

Marx actively incorporated scientific concepts and knowledge into his analysis as he embedded human society within the larger biophysical world, paying particular attention to the historical interchange of matter and energy (Foster 2000; Foster and Burkett 2016). In the early nineteenth century, physiologists introduced the concept of metabolism in reference to the biochemical processes between cells and their surrounding environment. Marx's friend Roland Daniels, a physician, extended the use of metabolism to conceive of the interactions and exchanges taking place within whole complexes of organisms, serving as the basis for examining the metabolic relations and processes at higher levels of organization and interdependency (Foster and Clark 2020, 19, 206–207; Saito 2014). The German chemist Justus von Liebig (1859, 175–183, 220) employed the concept of metabolism, exploring the exchange of nutrients between earth and humans. As part of defining the soil nutrient cycle, he explained that specific nutrients – such as, but not limited to, nitrogen, phosphorus, and potassium – are needed to produce vegetation. He detailed how plants absorbed these nutrients as they grew. Thus, maintaining the health and fertility of the soil required recycling these nutrients back to the land.

Following this work, Marx integrated the concept of metabolism into his critique of political economy. He explained that there is a necessary "metabolic interaction [*Stoffwechsel*]" between humans and the earth and that labour is "a process by which man, through his own actions, mediates, regulates and controls the metabolism between himself and nature" (Marx 1976, 283, 290; see also, Marx 1975b, 133, 209; Marx and Engels 1975, vol. 24, 553). He put forward a groundbreaking metabolic analysis as part of his dialectical approach, which involves a triadic scheme consisting of "the universal metabolism of nature," the "social metabolism," and the "metabolic rift" (Foster 2013). Within this framework, he studied the historical mediation of the relations between humans and nature as part of an open, dialectical totality.

"The universal metabolism of nature" represents the broader biophysical world, including specific cycles and processes within the earth system that produce and regenerate ecological conditions (Foster 2013; Marx and Engels 1975, vol. 30, 54–66). All life depends on and interacts with this earthly metabolism, which constitutes specific boundaries. The social metabolism, which operates within and in relation to the universal metabolism, comprises the productive activities of human societies. For "all forms of society," Marx argued (1976, 133), "labor, then, as the creator of use-values, as useful labor, is . . . an eternal natural necessity which mediates the metabolism between man and nature, and therefore human life itself." This relationship of interchange is shaped by the historically specific political-economic organization of labour and production.

Under the capital system, "the social metabolic process" is directed towards "the real exchange of commodities," whereby the productive exchange with nature is organized to expand profits, resulting in "a transformation in which the dual nature of the commodity – commodity as use-value and as exchange-value – manifests itself" (Marx 1970, 51–52, 86). The social metabolism takes on an alienated form, given that the quantitative expansion of capital is supreme, progressively transgressing and violating the boundaries associated with the universal metabolism, resulting in ecological rifts in the metabolism of society and nature. Importantly, Marx and Frederick Engels eschewed reductively subsuming society into nature (considering it an emergent form of nature), as well as vice versa. They thus avoided the dangers "of both absolute idealism and mechanistic science" (Foster 2000, 250–251, 2013, 8; Napoletano et al. 2018). Marx's historical, metabolic analysis recognized the constant interactions and emergent consequences. He employed this integrative approach in his analysis of capitalist agricultural production and the ecological problem associated with the soil nutrient cycle.

Expropriation and the metabolic rift

To understand the crisis associated with soil fertility in the 1850s and 1860s, Marx extensively studied the work of Liebig, as well as Henry Carey, James F. W. Johnston, Carl Fraas, and Wilhelm George Friedrich Roscher (Foster 2000, 149–163; Saito 2014). Liebig (1859, 175–177) argued that British high-farming was systematically "alienating the crops" as the essential soil nutrients were being stripped away to supply food and fiber for distant markets. The failure to recycle lost nutrients, violating the "law of compensation," was compromising the mineral composition of cultivated fields (Liebig 1859, 254–255, 1863, 233). He explained that this situation was the primary factor leading to the "exhaustion" of the soil (Liebig 1859, 175–177, 220, 230, 1863, 180, 210). To maintain agricultural production and enrich the fields, capitalist farmers in Great Britain and other leading European countries imported massive quantities of bones, seeds for oil cakes, guano, and nitrates from abroad, without changing the social conditions that were draining soil nutrients. Liebig described the rise of industrial capitalist agriculture, its intensive techniques to increase production, and its plundering of distant lands as a "robbery system" that was despoiling the earth (Clark and Foster 2009; Foster and Clark 2020; Liebig 2018). "Great Britain," he argued, "robs all countries of the conditions of their fertility. . . . She hangs like a vampire on the neck of Europe, and seeks out its hearts-blood, without any necessity and without permanent benefit to herself" (Liebig 1862, 85). This alienated metabolic relation, which treated "the Earth" as if it was "inexhaustible in its gifts," was counter to the "rational husbandry" that was necessary for maintaining the long-term productivity of the land (Liebig 1862, 96, 101, 1863, 233).

Marx integrated these insights into his more comprehensive and systematic analysis regarding world history, the historical development of capital, the alienation of nature, and the ecological contradictions of capitalism. As part of identifying and analyzing the "irreparable rift" between society and nature under capitalism, Marx was determined to understand the historic changes in the social metabolism throughout human history, deepening and expanding his ecological studies (Foster, Clark, and Holleman 2020; Saito 2017). This led him to conduct extensive studies of natural and communal economies, both past and present, in Europe, pre-Columbian America, Asia, Africa, and Australia, using ethnological works by Hubert Howe Bancroft, John Lubbock, John Budd Phaer, Maxim Kovalevsky, Henry Sumner Maine, Georg Ludwig von Mauer, Lewis Henry Morgan, and numerous others. In his notes, letters, and *Ethnological Notebooks*, which contain extensive excerpts and interpolations, Marx detailed the diversity of common property relations and the communal organization of these societies, as well as the tensions and transformations associated with European colonialism (Marx 1974b, 1975a, 1983). Marx was especially interested in the "natural economy" of Indigenous peoples in the Americas, particularly the Incas. In the *Grundrisse* and the first volume of *Capital*, he emphasized that the land had not been alienated in these societies and communal exchange, rather than commodities, was the basis of the natural economy (Foster, Clark, and Holleman 2020).

While these natural economies varied from locality to locality, as well as historically, they were rooted in collective property and established social relations with the land, organized to meet the immediate needs of the communities. In England, the rights to the commons – which included common fields, meadows, marshes, streams, and woodlands – were enshrined in the thirteenth century in the *Magna Carta* and the *Charter of the Forest*. These rights were predicated on maintaining relations and activities that contributed to the collective reproduction of communities, which created a diversified, makeshift, subsistence economy, based on use values. Historian Peter Linebaugh (2008, 44–45, 59, 289) indicates that "common rights are embedded" within the specific ecologies of various locations, whereby commoners have a collective responsibility to organize their social metabolic relations in ways to take care of the land and people. As Engels (1978, 77–83) noted in his study of the Mark, in the early German system of communal association, this entailed manuring the fields with dung from cattle and sheep to enrich the soil.

In the section "So-Called Primitive Accumulation" in the first volume of *Capital*, Marx emphasized that the expropriation of the land and people throughout the world took place through the enclosure of the commons and colonialism, serving as the preconditions for the agricultural and Industrial Revolutions in the seventeenth and eighteenth centuries. He detailed the five phases of the long enclosure moment, from the late fifteenth century to the early nineteenth century, which progressively privatized the land, dismantled common rights, annihilated the peasantry, and severed the social relations that connected commoners to the land (Foster, Clark, and Holleman 2021; Marx 1976, 873–940). Coinciding with the enclosures, from the long sixteenth century and after, European colonialism decimated many natural economies via extreme forms of expropriation, such as "the extirpation, enslavement and entombment in mines of the indigenous population" in the Americas, the transatlantic slave trade, and the "conquest and plunder of India" (Marx 1976, 915). Marx detailed how the genesis of both the capitalist farmer and industrial capitalist is rooted in expropriation of nature and people. These processes helped give rise to a distinct social metabolism, associated with the capital system, in which private riches subordinated public wealth (Foster, Holleman, and Clark 2020; Mészáros 1995).

This new alienated social metabolism played a major role creating a soil crisis in the nineteenth century. Marx understood that soil fertility was influenced by the productive relations people had with the land. The enclosure movement, the changes in property rights, the division between town and country, the proletarianization of the peasants, the new industrial system, the application of novel agricultural techniques, and the drive to maximize profits reorganized the metabolic interchange. Capitalist farmers cleared the land of trees to create more uniformity, making it easier to incorporate machinery to cultivate the land and to increase the scale of operations (Marx 1976, 908; Morton 1859). These intensive efforts increased the rate at which nutrients were being extracted from the soil. Food and fiber were shipped to distant markets, transferring the nutrients to cities to feed and clothe the concentrated population. In the third volume of *Capital*, Marx highlighted the ecological contradictions that were emerging in relation to the social metabolism of capitalist agriculture:

> We have both the excrement produced by man's natural metabolism and the form in which useful articles survive after use has been made of them. . . . The natural human

> waste products, remains of clothing in the form of rags, etc. are the refuse of consumption. The latter are of the greatest importance for agriculture. But there is a colossal wastage in the capitalist economy in proportion to their actual use.
>
> (Marx 1981, 195)

The nutrients necessary to maintain soil fertility accumulated as waste within cities, leading to concerns regarding sanitation and disease. While detailed proposals were made to capture and return the nutrients to the countryside, such endeavours were not profitable. Thus, the nutrients ended up creating a pollution problem (Angus 2018; Clark and Longo 2018).

By failing to recycle soil nutrients, capitalist agriculture was violating the "law of compensation." As a result, Marx (1976, 637) indicated that it was progressively disturbing "the metabolic interaction between man and the earth," preventing "the return to the soil of its constituent elements consumed by man in the form of food and clothing; hence it hinders the operation of the eternal natural condition for the lasting fertility of the soil." In other words, these conditions were creating "an irreparable rift in the interdependent process of the social metabolism, a metabolism prescribed by the natural laws of life itself. The result of this is a squandering of the vitality of the soil" (Marx 1981, 949).

This ecological rift, Marx (1973, 527) stressed, was a consequence of the destruction of "self-sustaining agriculture" as profit-driven industrial agriculture arose, transgressing against the earthly metabolism of the soil nutrient cycle. As a result, "agriculture no longer finds the natural conditions of its own production within itself, naturally, arisen, spontaneous, and ready to hand, but these exist as an independent industry separate from it," thus requiring "machinery, chemical fertilizer acquired through exchange, seeds from distant countries." In an effort to replenish the land with needed nutrients, bones from battlefields across Europe and from the catacombs in Sicily were ground up and spread across agricultural land (Mårald 2002, 74). Between 1840 and 1880, millions of tons of Peruvian guano, extracted by Chinese labourers as *de facto* slaves as part of ecological imperialism, were shipped to Great Britain and other countries in the global North to enrich exhausted fields (Clark, Auerbach, and Zhang 2018; Clark and Foster 2009). Even the mass production of synthetic fertilizer in the twentieth century has not been able resolve the metabolic rift in the soil nutrient cycle, given the growth imperative of capital, the failure to recycle nutrients, and the intensification of agricultural practices (Magdoff 2011; Mancus 2007). This problem persists, given the very logic of capital and its social metabolism. As Marx emphasized,

> All progress in capitalist agriculture is a progress in the art, not only of robbing the worker, but of robbing the soil; all progress in increasing the fertility of the soil for a given time is progress towards ruining the more long-lasting sources of that fertility Capitalist production, therefore, only develops the technique and the degree of combination of the social process of production by simultaneously undermining the original sources of all wealth – the soil and the worker.
>
> (Marx 1976, 638)

Deepening and expanding Marx's ecology

Ecological Marxism continues to burgeon on multiple fronts, deepening and expanding Marx's ecological analyses, extending metabolic analysis, and addressing many of the most pressing ecological challenges today. Andreas Malm (2018, 177) contends that the scholarship on Marx's ecology and metabolic rift is one of immense "creativity and productivity." A critical part of this work involves the ongoing exploration, recovery, and development of Marx's historical-materialist, dialectical approach.

Paul Burkett (1999) has demonstrated how Marx's critique of capital incorporates rich ecological insights, which are integrated throughout his analysis of political economy. He illuminates Marx's understanding of natural limits and the expropriation of nature and the sophistication of Marx's value analysis for revealing the ecological contradictions of capital (for further discussion of value, see Foster and Burkett 2018). Additional work offers an extensive investigation of Marx and Engels's assessments of the science and debates regarding energy and thermodynamics, especially as these related to questions regarding the economy and labour (Foster and Burkett 2016).

Ecological Marxism is benefitting from integrating and building on important scholarship on imperialism, racial capitalism, settler colonialism, and social reproduction (Bhattacharya 2017; Beckert 2014; Coulthard 2014; Cox 1964; Du Bois 1992; Dunbar-Ortiz 2014; Estes 2019; Federici 2004; Fraser 2014, 2016; Salleh 2010). This work is revealing the historical and ongoing role of the expropriation of nature and people in relation to interlocking forms of oppression and inequalities that have been figurative in the emergence and development of capitalism. It has led to extensive studies of Marx's analysis of the historic economic and ecological robbery of Ireland by Britain, which culminated in widespread environmental disruption and extermination in the nineteenth century (Flaherty 2013; Foster and Clark 2020; Slater 2018a, 2018b; Slater and McDonough 2008). It has generated further exploration of Marx's analysis of natural economies, his critique of the genocide of Indigenous peoples throughout the world, his examination of the political economy of slaveowner capitalism, his analysis of how capital expropriates reproductive work, and his critique of alienated speciesism (Foster and Clark 2020; Foster, Clark, and Holleman 2020; Foster, Holleman, and Clark 2020; Fraser 2014, 2016). This exploration of the hidden abodes of capitalism is offering a more comprehensive understanding of the relationships between expropriation and exploitation, which define the system.

This body of research offers rich analyses of how classical historical materialism's open, dialectical, materialist ecological approach continued to develop, though primarily in the "second foundation" represented by natural science, throughout the nineteenth and twentieth centuries (Foster 2020, 7; see also Angus 2017; Empson 2013; Flaherty 2019; Saito 2017). Through this work, it is clear that Marx's ecology and metabolic analysis has a long lineage within both the natural and social sciences. In fact, socialist engagement and contributions have been central to the historical development of ecological science and ecology as a field of study (Foster 2020). Burkett (2006) illuminates how the integration of Marxist political economy into ecological economics can greatly enrich the field, positioning it to address the most pressing environmental problems. The significance of Marx's conception of labour as a metabolic relation, an integrated dialectic of nature and society, was recognized by Erich Fromm (1970, 153–154), Georg Lukács (2003, 96, 106, 113–114, 130–131), and Herbert Marcuse (1978, 16). French Marxist Henri Lefebvre (2016) drew directly on Marx's theory of metabolic rift in describing how the capital system was leading to total alienation and ecological destruction (see also Foster, Clark, and Holleman 2020; Napoletano et al. 2020).

The significance of metabolic rift research has become increasing apparent. As Del Weston contends,

> the metabolic rift permeates our dominant and pervasive acquisitive and technocratic thinking whereby human beings are conceptualised as being apart from nature, exempt from nature's laws, where it is believed that science and technology and human activities can dominate, manipulate and control nature, and that we find expression and

> satisfaction through external, material goods. . . . [I]t denotes the disjuncture between social systems and the rest of nature.
>
> *(Weston 2014, 66)*

She stresses that "the rift between humans and nature (and humans as nature) is manifested in many different ways" (ibid., 67). Thus, metabolic rift research continues to evolve, studying how the social metabolism of capitalism is producing ecological ruptures and shifts on numerous fronts, culminating in the planetary emergency in the twenty-first century (Clark and York 2008).

In *Fossil Capital*, Andreas Malm (2016) provides an extensive historical-materialist analysis of the rise of steam power and how the burning of fossil fuels has come to reside at the heart of capital. Through this development, capitalism is generating a metabolic rift in the carbon cycle via the constant burning of fossil fuels, the degradation of the carbon sinks (e.g., deforestation), and the flooding of the atmosphere with greenhouse gas emissions (Angus 2016; Clark and York 2005; Foster 2019; Klein 2014; Weston 2014). This rift is only amplified as the social metabolic order of capital demands more energy and raw materials for generalized commodity production (Burkett 2006; Dickens 2014; Foster, Clark, and York 2010).

Metabolic rift research has been extended to overlooked realms, such as marine systems, examining how the social metabolism of capitalism is altering ecosystem dynamics and life cycles (Longo and Clark 2016). For instance, the capital accumulation process plays a primary role in the structure and function of the fishing industry as prized fish are commodified for the global market (Longo, Clausen, and Clark 2015). Fish are being harvested at a rate faster than they can reproduce, which is contributing to the collapse of fisheries (Clausen and Clark 2005; Longo 2012). Recent analysis has focused on the social metabolism associated with capitalist food production. Concentrated animal feeding operations separate animals from pasture, as well as fish from marine systems. Feed is grown on distant land or captured at sea and transferred to animal production sites. Animal wastes, including important soil nutrients, accumulate in cesspools, polluting water systems (Carolan 2012; Clausen and Clark 2005; Clow and McLauchlin 2007; Longo, Clausen, and Clark 2014; Weis 2007). These operations enhance the ability of corporate enterprises to control the entire life cycle of animals in an attempt to decrease the time between birth and slaughter. At the same time, these enterprises are able to increase commodity production but, more importantly, increase value. Factory farms require massive amounts of animal feed, growth hormones, and antibiotics. They also generate enormous quantities of waste not readily reincorporated into ecosystems, creating ecological rifts on numerous fronts (Gunderson 2011).

Another area of investigation within this tradition has been forest ecology, including the ecological and temporal contradictions associated with capitalist logging practices, whereby the focus on short-term profits violates the regenerative requirements associated with the earthly metabolism of sustaining forests (Auerbach and Clark 2018). It has analyzed how white settler colonialism, racial capitalism, and capitalist agricultural development produced dust bowls that devasted the United States in the 1930s (Holleman 2018). Ecological Marxists, in pursuing metabolic rift analysis, have recovered Marx's food regime theory (Foster and Clark 2020, 113–125; Wallace 2020, 109). They have detailed how the capitalist global food system is destroying ecosystems and decimating wildlife in ways that are facilitating the spread of diseases from other animals to humans, which manifest in novel viruses and contributed to the COVID pandemic (Wallace 2016, 2020; Wallace et al. 2020). One line of inquiry presents how these social and ecological disruptions are creating corporeal rifts, including within the human microbiome (Foster and Clark 2020; Friedman 2018).

As a dynamic system constantly seeking to expand accumulation, capital inevitability confronts natural limits and transgresses against environmental obstacles associated with the earthly metabolism. Metabolic rift scholars detail how capital responds by geographically shifting extraction and production, employing technological fixes, and/or incorporating new resources as substitutes for other raw materials. Through these actions, capital avoids attending to the metabolic rifts it generates, often shifting them around while expanding the scale and scope of environmental degradation (Clark and York 2008; Foster, Clark, and York 2010; Napoletano, Paneque-Gálvez, and Vieyra 2015; Napoletano et al. 2019). Weston (2014, 67) stresses that ecological rifts are continuing to grow "in dimensions and complexity, to the point where economic activities of human society are causing unprecedented change in the Earth's biosphere, its lands, forests, water and air." The hyper-disruptive social metabolism of capital represents an "alienated mediation" of nature and society, one with potentially irreversible, catastrophic impacts, undermining the very conditions of life itself (Marx 1974a, 260–261).

Capitalism is antithetical to sustainability, as its logic leads to "exterminism" (Angus 2016, 179–180; Foster 2020, 525–526; Thompson 1982, 64). Ecological ruin or ecological revolution is the choice associated with the Great Capitalist Climacteric. The future of humanity depends on creating a long social and ecological revolution, in which associated producers are able to create a new social metabolism, one in accord with the requirements of the universal metabolism of nature. Ecological sustainability, substantive equality, and the restoration of the associated producers as the self-mediating beings of nature regulating the human metabolism with nature constitute the only possible bases of future social advance: *eco*socialism.

References

Angus, Ian. 2016. *Facing the Anthropocene*. New York: Monthly Review Press.

Angus, Ian. 2017. *A Redder Shade of Green*. New York: Monthly Review Press.

Angus, Ian. 2018. "Cesspools, Sewage, and Social Murder." *Monthly Review* 70 (3): 32–68.

Auerbach, Daniel, and Brett Clark. 2018. "Metabolic Rifts, Temporal Imperatives, and Geographical Shifts." *International Critical Thought* 8 (3): 468–486.

Beckert, Sven. 2014. *Empire of Cotton*. New York: Vintage.

Bhattacharya, Tithi, ed. 2017. *Social Reproduction Theory*. London: Pluto.

Burkett, Paul. 1999. *Marx and Nature*. New York: St. Martin's Press.

Burkett, Paul. 2006. *Marxism and Ecological Economics*. Leiden: Brill.

Carolan, Michael. 2012. *The Sociology of Food and Agriculture*. New York: Routledge.

Clark, Brett, Daniel Auerbach, and Karen Xuan Zhang. 2018. "The Du Bois Nexus." *Environmental Sociology* 4 (1): 54–66.

Clark, Brett, and John Bellamy Foster. 2009. "Ecological Imperialism and the Global Metabolic Rift." *International Journal of Comparative Sociology* 50 (3–4): 311–334.

Clark, Brett, and Richard York. 2005. "Carbon Metabolism." *Theory and Society* 34 (4): 391–428.

Clark, Brett, and Richard York. 2008. "Rifts and Shifts." *Monthly Review* 60 (6): 13–24.

Clark, Brett, and Stefano B. Longo. 2018. "Land-Sea Ecological Rifts." *Monthly Review* 70 (3): 106–121.

Clausen, Rebecca, and Brett Clark. 2005. "The Metabolic Rift and Marine Ecology." *Organization & Environment* 18 (4): 422–444.

Clow, Michael and D. McLauchlin. 2007. "Healing the Metabolic Rift Between Farming and the Ecosystem." *Socialist Studies/Études Socialistes* 5–25, Spring.

Coulthard, Glen Sean. 2014. *Red Skin, White Masks*. Minneapolis: University of Minnesota Press.

Cox, Oliver Cromwell. 1964. *Capitalism as a System*. New York: Monthly Review Press.

Dickens, Peter. 2014. *Nature and Society*. Cambridge: Polity.

Du Bois, W.E.B. 1992. *Black Reconstruction in America*. New York: Atheneum.

Dunbar-Ortiz, Roxanne. 2014. *An Indigenous Peoples' History of the United States*. Boston: Beacon Press.

Empson, Martin. 2013. *Land and Labour*. London: Bookmarks.

Engels, Frederick. 1978. *Socialism: Utopian and Scientific*. New York: International Publishers.

Estes, Nick. 2019. *Our History Is the Future*. London: Verso.

Federici, Silvia. 2004. *Caliban and the Witch*. Brooklyn, NY: Autonomedia.

Flaherty, Eoin. 2013. "Geographies of Communality, Colonialism, and Capitalism." *Historical Geography* 41: 39–79.

Flaherty, Eoin. 2019. *Complexity and Resilience in the Social and Ecological Sciences*. London: Palgrave Macmillan.

Foster, John Bellamy. 2000. *Marx's Ecology*. New York: Monthly Review Press.

Foster, John Bellamy. 2013. "Marx and the Rift in the Universal Metabolism of Nature." *Monthly Review* 65 (7): 1–19.

Foster, John Bellamy. 2015. "The Great Capitalist Climacteric." *Monthly Review* 67 (6): 1–18.

Foster, John Bellamy. 2019. "On Fire This Time." *Monthly Review* 71 (6): 1–17.

Foster, John Bellamy. 2020. *The Return of Nature*. New York: Monthly Review Press.

Foster, John Bellamy, and Paul Burkett. 2016. *Marx and the Earth*. Leiden: Brill.

Foster, John Bellamy, and Paul Burkett. 2018. "Value Isn't Everything." *Monthly Review* 70 (6): 1–17.

Foster, John Bellamy, and Brett Clark. 2020. *The Robbery of Nature*. New York: Monthly Review Press.

Foster, John Bellamy, Brett Clark, and Hannah Holleman. 2020. "Marx and the Indigenous." *Monthly Review* 71 (9): 1–19.

Foster, John Bellamy, Brett Clark, and Hannah Holleman. 2021 [forthcoming]. "Marx and the Commons." *Social Research: An International Quarterly*.

Foster, John Bellamy, Brett Clark, and Richard York. 2010. *The Ecological Rift*. New York: Monthly Review Press.

Foster, John Bellamy, Hannah Holleman, and Brett Clark. 2020. "Marx and Slavery." *Monthly Review* 72 (3): 96–117.

Fraser, Nancy. 2014. "Behind Marx's Hidden Abode." *New Left Review* 86: 55–72.

Fraser, Nancy. 2016. "Expropriation and Exploitation in Racialized Capitalism." *Critical Historical Studies* 3 (1): 163–178.

Friedman, Michael. 2018. "Metabolic Rift and the Human Microbiome." *Monthly Review* 70 (3): 70–105.

Fromm, Erich. 1970. *The Crisis of Psychoanalysis*. Greenwich, CT: Fawcett.

Gunderson, Ryan. 2011. "The Metabolic Rifts of Livestock Production." *Organization & Environment* 24 (4): 404–422.

Hamilton, Clive, and Jacques Grinevald. 2015. "Was the Anthropocene Anticipated?" *The Anthropocene Review* 2 (1): 59–72.

Holleman, Hannah. 2018. *Dust Bowls of Empire*. New Haven, CT: Yale University Press.

Klein, Naomi. 2014. *This Changes Everything*. New York: Simon & Schuster.

Lefebvre, Henri. 2016. *Marxist Thought and the City*. Minneapolis: University of Minnesota.

Liebig, Justus von. 1859. *Letters on Modern Agriculture*. London: Walton and Maberly.

Liebig, Justus von. 1862. "Introduction." In *Agricultural Chemistry*, translated by Lady Gilbert, seventh ed. Hertfordshire: Archives at Rothamsted Research.

Liebig, Justus von. 1863. *The Natural Laws of Husbandry*. New York: Appleton.

Liebig, Justus von. 2018. "1862 Preface to *Agricultural Chemistry*." *Monthly Review* 70 (3): 146–150.

Linebaugh, Peter. 2008. *The Magna Carta Manifesto*. Berkeley: University of California Press.

Longo, Stefano B. 2012. "Mediterranean Rift." *Critical Sociology* 38 (3): 417–436.

Longo, Stefano B., and Brett Clark. 2016. "An Ocean of Troubles." *Social Problems* 63 (4): 463–479.

Longo, Stefano B., Rebecca Clausen, and Brett Clark. 2014. "Capitalism and the Commodification of Salmon." *Monthly Review* 66 (7): 35–55.

Longo, Stefano B., Rebecca Clausen, and Brett Clark. 2015. *The Tragedy of the Commodity*. New Brunswick, NJ: Rutgers University Press.

Lukács, Georg. 2003. *A Defence of 'History and Class Consciousness': Totalitarianism and the Dialectic*. London: Verso.

Magdoff, Fred. 2011. "Ecological Civilization." *Monthly Review* 62 (8): 1–25.

Malm, Andreas. 2016. *Fossil Capital*. London: Verso.

Malm, Andreas. 2018. *The Progress of This Storm*. New York: Verso.

Mancus, Phillip. 2007. "Nitrogen Fertilizer Dependency and its Contradictions." *Rural Sociology* 72 (2): 269–288.

Mårald, Erland. 2002. "Everything Circulates." *Environment and History* 8: 65–84.

Marcuse, Herbert. 1978. *The Aesthetic Dimension*. Boston: Beacon.

Marx, Karl. 1970. *A Contribution to the Critique of Political Economy*. Moscow: Progress Publishers.

Marx, Karl. 1973. *Grundrisse*. London: Penguin Books.

Marx, Karl. 1974a. *Early Writings*. London: Penguin Books.

Marx, Karl. 1974b. *The Ethnological Notebooks*. Assen, Netherlands: Van Gorcum and Co.

Marx, Karl. 1975a. "Excerpts from M.M. Kovalevskii (Kovalevsky)." In *The Asiatic Mode of Production*, edited by Lawrence Krader, 343–412. Assen, Netherlands: Van Gorcum.

Marx, Karl. 1975b. *Texts on Method*. Oxford: Blackwell.

Marx, Karl. 1976. *Capital*. Vol. 1. London: Penguin Books.

Marx, Karl. 1981. *Capital*. Vol. 3. London: Penguin Books.

Marx, Karl. 1983. "Marx-Zasulich Correspondence: Letters and Draft." In *Late Marx and the Russian Road*, edited by Teodor Shanin, 97–26. New York: Monthly Review Press.

Marx, Karl, and Frederick Engels. 1975. *Collected Works*. New York: International Publishers.

Mészáros, István. 1995. *Beyond Capital*. New York: Monthly Review Press.

Morton, John Chalmers. 1859. "On the Forces Used in Agriculture." *Journal of the Society of Arts* 53–68, 9 December.

Napoletano, Brian M., Brett Clark, John Bellamy Foster, and Pedro S. Urquijo. 2020. "Sustainability and Metabolic Revolution in the Works of Henri Lefebvre." *World* 1 (3): 300–317.

Napoletano, Brian M., Jaime Paneque-Gálvez, Yadira Méndez-Lemus, and Antonio Vieyra. 2019. "Geographical Rift in the Urban Periphery, and Its Concrete Manifestations in Morelia, Mexico." *Journal of Latin American Geography* 18 (1): 38–64.

Napoletano, Brian M., Jaime Paneque-Gálvez, and Antonio Vieyra. 2015. "Spatial Fix and Metabolic Rift as Conceptual Tools in Land-Change Science." *Capitalism Nature Socialism* 26 (4): 198–214.

Napoletano, Brian M., Pedro S. Urquijo, Jaime Paneque-Gálvez, Brett Clark, Richard York, Iván Franch-Pardo, Yadira Méndez-Lemus, and Antonio Vieyra. 2018. "Has (Even Marxist) Political Ecology Really Transcended Metabolic Rift?" *Geoforum* 92: 92–95.

Rockström, Johan, Will Steffen, Kevin Noone, Åsa Persson, F. Stuart Chapin, III, Eric F. Lambin, Timothy M. Lenton, Marten Scheffer, Carl Folke, Hans Joachim Schellnhuber, Björn Nykvist, Cynthia A. de Wit, Terry Hughes, Sander van der Leeuw, Henning Rodhe, Sverker Sörlin, Peter K. Snyder, Robert Constanza, Uno Svedin, Malin Falkenmark, Louise Karlberg, Robert W. Corell, Victoria J. Fabry, James Hansen, Brian Walker, Diana Liverman, Katherine Richardson, Paul Crutzen, and Jonathan A. Foley. 2009. "A Safe Operating Space for Humanity." *Nature* 461 (24): 472–475.

Saito, Kohei. 2014. "The Emergence of Marx's Critique of Modern Agriculture: Ecological Insights from His Excerpt Notebooks." *Monthly Review* 66 (5): 25–46.

Saito, Kohei. 2017. *Karl Marx's Ecosocialism*. New York: Monthly Review Press.

Salleh, Ariel. 2010. "From Metabolic Rift to 'Metabolic Value'." *Organization & Environment* 23 (2): 205–219.

Slater, Eamonn. 2018a. "Engels on Ireland's Dialectics of Nature." *Capitalism Nature Socialism* 29 (4): 31–50.

Slater, Eamonn. 2018b. "Marx on the Colonization of Irish Soil." MUSSI Working Paper Series, http://mural.maynoothuniversity.ie/9342/1/ES-Marx-2018.pdf (accessed 12 March 2021).

Slater, Eamonn, and Terrence McDonough. 2008. "Marx on Nineteenth-Century Colonial Ireland." *Irish Historical Studies* 36 (142): 153–172.

Steffen, Will, Katherine Richardson, Johan Rockström, Sarah E. Cornell, Ingo Fetzer, Elena M. Bennett, Reinette Biggs, Stephen R. Carpenter, Wim de Vries, Cynthia A. de Wit, Carl Folke, Dieter Gerten, Jens Heinke, Georgina M. Mace, Linn M. Persson, Veerabhadran Ramanathan, Belinda Reyers, and Sverker Sörlin. 2015. "Planetary Boundaries." *Science* 347 (6223): 1259855.

Thompson, Edward P. 1982. *Beyond the Cold War*. New York: Pantheon.

Wallace, Rob. 2016. *Big Farms Make Flu*. New York: Monthly Review Press.

Wallace, Rob. 2020. *Dead Epidemiologists*. New York: Monthly Review Press.

Wallace, Rob, Alex Liebman, Luis Fernando Chaves, and Rodrick Wallace. 2020. "COVID-19 and Circuits of Capital." *Monthly Review* 72 (1): 1–15.

Weis, Tony. 2007. *The Global Food Economy*. New York: Zed Books.

Weston, Del. 2014. *The Political Economy of Global Warming*. New York: Routledge.

PART II

Extending Marxist roots

9

ECOLOGICAL AND ECONOMIC MODALITIES OF TIME AND SPACE[1]

Elmar Altvater

Introduction

If physical activity could be reduced to an infinitesimally small period of time, the concept of physical space would become meaningless. Since real world activity "takes time," space coordinates constitute the frame of reference of all social and material activity. So an economy without space and time exists only in neo-classical models of "pure economics"; its theoretical relevance remains limited precisely because of this heroic feat of abstraction. Even if the period of a particular activity cannot be reduced to "zero," this is, in fact, the aim of capital. To shorten the circulation time of capital is a principle inherent in capitalist development. Thanks to modern transport and communications technologies, the meaning of space defined quantitatively (i.e., distance) and qualitatively (i.e., physical relief and concrete characteristics of social structures) has been diminished. Today, it is possible to travel between Berlin and New York in under ten hours and "500 million dollars can be telexed from Singapore to the Bahamas via London by pressing a button, just as if no physical distance existed between these places." With the separation of money from its material form (metal, paper) and its transformation into energetic, or electronic, money, the spatiality of money circulation tends to vanish. Space is overcome with the speed of light.

So time is abridged, making space meaningless. And, conversely, physical space is conditioned in a way which compresses the time of activity. For the acceleration of material transport, time is nothing but an ensemble of impediments. Natural, cultural, and social impediments to the circulation of capital must be removed. This "usurpation of space" has as its purpose the removal of impediments to the acceleration of productive and transportation activity. This usurpation is simultaneously the "production of space" and the construction of a "second nature."

The contradiction between economy and ecology

Physical space must be adjusted to shorten the time period of economic activity. The logic of shortening the time of economic activity and the removal of qualitative and quantitative impediments in space is precisely the imperative of capitalist valorization, or (in Weber's categories) the "rationality of occidental 'world domination.'" This means to tailor space and time coordinates of activity to the principles of means-ends optimization. The creation of a

10.4324/9780429341427-12

form-specific spatial and temporal social system of coordinates through the production of material structures and immaterial norms abstracts from traditional, pristine, and "natural" spatial and temporal coordinates. This is the production of "sociality": "A unified perception of time, space, cause, number, etc. . . . Society cannot relinquish these categories to the arbitrariness of individuals without surrendering itself. To live, society needs not only sufficient moral conformity, a minimum of logical conformity must exist, too." Or, in Norbert Elias' words, time has "the character of a social institution, a regulator of social events, a mode of human experience; and chronometers are an integral part of a social order which cannot function without them." Chronometers, or instruments for the measurement of time, were invented only recently. Previously, activity was a measure of time. Now, time itself would be the measure of activity. Space and time thus are social categories. Moreover, societies are defined by their specific normative commitments. This explains the resurgent interest in the linkages between "social relations and spatial structures," which has resulted in a closer connection between social scientists and human geographers at a time when, with few exceptions (Lefebvre, Poulantzas), neither group has theoretically studied territorial space as a social matrix.

However, this is not the decisive problem to be addressed in this article. It is not a question of the "sociality" of space and time as such, but rather of a social principle that endeavours to reduce time intervals by submitting the quantity and quality of space to the principle of acceleration. Of course, it is not possible to abstract entirely from space and time since all activity as well as all production and consumption presupposes the transformation of matter and energy from forms with which we are endowed to different forms which we need as human beings. In production and consumption, it is not possible to disregard the quantitative and qualitative properties of space and time nor the use-value aspects of particular products and production processes. Even the reduction of economic processes to money payments by systems theory cannot escape this fact without invalidating its own premises. Even communicative information has a material substrate – paper currency, for example – which is the result of deliberate and intelligent material and energetic transformation processes. Its consequence in economic communication systems could be abstracted only if communication could be arranged without information, which is clearly an absurdity.

Production means production of space and production of nature. The results of production (and consumption) manifest themselves spatially as cultural landscapes, buildings, cities, streets, ruins of nuclear power plants, canals, sewers and smog, deserts, garbage dumps, and so on. As Marx observed, what makes a place a hunting ground is the fact that specific tribes hunt in it. What makes a region a mining region is the fact that metal ore is mined there by mining companies. What makes a region an industrial area is the spatial realization of entrepreneurial decisions or state planning with the establishment of an industrial region as a goal. What makes a particular space a recreation area is the destruction of other areas and the translocation of possibilities for the satisfaction of the human need for vacations in a territory defined by its use as a vacation spot.

In short, it is human beings' own socially organized, normatively imprinted, and politically influenced process of production and consumption which forms their regional environments. This has a double character. The production of space is at the same time its valorization. Its production and consumption has both material and value-oriented facets. Their spatial and temporal coordinates thus have ecological and economic dimensions, respectively. The logics of different functions are therefore at work in the same territorial area. This is already evident in the fact that the transformation of matter and energy in the process of production and consumption is a particular physical process, hence bound spatially and temporally in particular ways. But the circulation process, with its logic of time compression and destruction of qualitative and quantitative spatial obstructions, transforms production into a moment of global capitalist reproduction: i.e., the world market. At this point, it is no longer a question of the transformation of resources into an exchangeable use value – from iron ore to raw iron. Now the staple commodity, processed under specific spatial conditions, becomes one element of the total amount of raw iron on the world market and "compares itself" to the same commodity extracted and produced under completely different spatial conditions.

Competition on the world market forces capital to approximate the average spatial conditions. The specificities, singularities, and particularities of spatially bound production and consumption are in this way "equalized." Countries and landscapes lose their unmistakable characteristics and transform themselves into segments of world market circulation and global communication. Local, regional, and national particularities of communication (eating, legal forms, traffic regulations, language, and so on) are more likely to be regarded as obstructive. Patterns of consumption, transported by way of the circulation process and competition in the world market, express a specific spatial reality, as do patterns of production. The conditions of commodity production and the conditions of capital valorization must conform to one another for reasons of "competitiveness." The world market manifests itself as pure objectivity. In this process, natural milieu and social relations alike are transformed according to a "plan" which adheres to the conditions of the world market and which abstracts from specific regions and natural, cultural, and social conditions of reproduction within particular regions.

Competitiveness would not exist if profit was not the aim of production and marketing. But profit is a surplus produced in the production process of specific commodities while factor inputs are money valued. Economic activity, specifically production, is thus determined in two ways. On the one hand, production is nothing but the transformation of matter and energy. On the other hand, it is the creation of surpluses measured in money units (thus abstracted from the quality of use value). Transformation of material and energy qualitatively follows certain laws of nature with coordinates defined in terms of physical time and physical space. Georgescu-Roegen has differentiated between two times in the concept of time (a differentiation which could be made analogously with the concept of space): time "T," which can be described as a "stream of consciousness" or as a "continuous sequence of moments" and time "t," which denotes the time intervals between two activities using a mechanical chronometer. During time defined as a "stream of consciousness," it is irrelevant when a specific physical process (e.g., the oscillation of a pendulum) takes place. One can simply measure the intervals. However, an interlinkage between "historical time" (T) and "dynamic time" (t) is not possible. "In other words, mechanical phenomena are timeless ('Zeitlos') but not timeless ('zeitlos')." In this mechanistic-temporal sense, processes can be unambiguously forecasted, which requires the elimination of all elements of time "T." In a similar way, the measurement of time with continuously more perfected chronometers defines time as nothing more than the interval between two activities, regardless of the historical arrangement of the latter.

In social life as well there are tendencies towards the dehistorization of time: e.g., the measurement of a sprint or ski race, which, of course, can never be fully realized. Regardless of where these activities take place, the only measure of interest is the interval (distance and time) between the starting and finishing lines. Time "t" is independent of human activities; it is not part of consciousness and is therefore irrelevant in terms of fixing the coordinates of activities. However, physical processes in space and time have an important *result* for human consciousness and activity. In the interval between two activities – i.e., during the process itself – *entropy* has increased, and something *irreversible* has occurred. The temporal aftermath of the activity has left the world changed. Thus, a particular logic develops in the space and time coordinates (as in the social and economic coordinates): economic surplus production is guided by the quantitative imperative of growth by way of reducing the time spans of human activities (especially those of production and consumption). It does this by accelerating and transcending the quantitative and qualitative impediments in space in order to compress time, thus setting "T" into "t." There are thus two coordinating systems of space and time, which, in the form of two patterns of "functional spaces," are fixed upon a territorialsocial reality. This is what is meant by the "contradiction of economy and ecology." Thus, Durkheim's understanding of society as a part – indeed, the highest form – of nature is incorrect He depicts the category of time merely as giving rhythm to social life. In fact, the space and time of a society, especially capitalist society, and the physical time and space of nature are in no way identical. The logic of their respective functional spaces collides.

Entropy and scarcity

Energy cannot be produced but only transformed from one form to another. The two laws of thermodynamics espoused by Claudius in 1865 are first, that the universe's energy is constant, and material and energetic inputs are always equivalent to outputs, and second, the entropy of the world strives to maximize itself. Used energy and matter are transformed from the order of unequal distribution to the disorder of equal distribution and thus are no longer very useful. "We can use a specific amount of low entropy only once." From the standpoint of the criteria of human use, no transformation of energy or matter is perfectly efficient; a portion is always lost in the form of heat. And if heat is spread evenly, the flow of heat from which energy derives completely stops. Georgescu-Roegen provides us with a convincing example: in comparison with the amount of heat in the ocean, the heat of a ship's boiler is infinitesimally small. Yet the ocean's heat cannot be used, or can be used only with considerable difficulty; the heat from the boiler can be transformed to propel the ship. As processes of material and energetic transformation, production and consumption as well are subject to the law of increasing entropy. This means that the economic system and the tendencies inherent in it cannot be conceptually grasped without a reference to their conditionality: i.e., the modes of action of natural laws.

An increase in entropy is inevitable in closed systems. In open systems, however, entropy can remain constant or decrease through entropy migration. This explains the rampant growth of certain biosystems which are capable of assimilating nutrients and energy from other systems. Even in closed systems, the efficiency of energy and matter transformation is variable. Variability is measurable in terms of inputs of energy and matter compared with that part of the output which is useful to human beings. (In passing, we note that, formulated this way, the law of thermodynamics is obviously anthropocentric.) Thus, one can make use of energy and matter more or less efficiently, sparingly or wastefully, sensibly or senselessly. The speed of the inevitable entropy increase can accelerate or decelerate.

In biological systems, for example, it appears that the "rate of entropy production" depends on the degree of complexity and diversity of the system. These determine the scale of nutrient recycling and the necessity for external energy and matter inputs, as well as the susceptibility and responsiveness to external shocks. In tropical rainforests, for example, it is possible to show that the transition to monocultural forms severely increases the injurability of the ecological system as a result of external shocks, with the possibility of complete collapse of the system. In sociological systems, however, the "rate of entropy production" depends on what one might call "system intelligence." This intelligence decides how great the rate of exploitation of renewable resources is, if and how processes of substitution for nonrenewable resources are carried out, and how much it is possible to develop the use of time and space in ways that optimize resource supplies and their reproduction cycles. Social conditions and mechanisms of regulation of society and nature are an immaterial and therefore, in principle, renewable resource – and obviously decisive for the rate of entropy production. Indeed, ecological economic approaches (e.g., waste recycling and electronically steered savings of chemically bound fossil energy) rest upon social and political options which either increase entropy growth or slow down the pace of this irreversible process and limit its spatial dimensions. The question is, does the potential for system intelligence have restrictions which are embedded in the structure and function of the socio-economic system itself?

Even if the rate of entropy production can be slowed down, it can never be reduced to zero in a closed system, which brings up the problem of alternative use since resources are scarce. At this point, economics as the science of the rational use of scarce resources is needed. Without scarcity, there is no need for economics. If entropy increase was equal to zero or even negative, there would be no scarcity, and economics would lose its purpose. In Georgescu-Roegen's thermodynamics-oriented "bio-economics," scarcity is assumed on the basis of the Second Law of Thermodynamics. This is an economics of irreversible processes while mainstream economics is premised on the reversibility of economic cycles. Accordingly, mainstream economics is unaware of the thermodynamic basis of its own central category – scarcity.

Economics and time

At this point, the modality of time and the connectedness of past, present, and future material and social processes again confront us. In economics, the modality of time and the connectedness of these processes have been largely eliminated by the introduction of the concept of "interest," or the discounting of future economic values. In contrast, "bio-economics" stresses that "we must emphasize that every Cadillac let alone every instrument of war means fewer ploughshares for future generations and future human beings, too." The ore of the world's largest ore mine in Carajas in the eastern Amazon region (about 18 billion tons of iron ore with an iron content of 66 percent) will last about 500 years at an annual rate of extraction of 35 million tons, according to present estimates. Five hundred years is a long but finite time. Compared to the millions of years that the deposits have existed, it is a tiny time span. Earth time, resource time, and human time use different intervals to measure the span between past, present, and future. In this connection, the calculations of the Club of Rome pertaining to resource supplies and resource consumption justifiably make sense. They give us a strong impression of the finiteness of resource supplies and their exhaustibility in time, even if their depletion rate (in terms of present generations) is far in the future. Resources are mobilized by the economic process in a comparatively short time and then are available only in quantitatively reduced and qualitatively degraded forms. Or they are completely – i.e., irreversibly – consumed. They then are remembered in the form of their heritage as radioactive wastes, seas of red mud from the production of aluminium, and so on.

On this thermodynamic dimension of scarcity is superimposed an economic mask, so to speak, because resources can be economically scarce when it is not economically worthwhile to prospect for, develop, and exploit them because of negative cost/gain ratios. Scarcity is defined not only by the finiteness of resources and the irreversibility of their consumption, but also economically by the principle of rationality which the functional space of economics (i.e., world market) provides. Paradoxically, scarcity in economic functional space can even lead to the profusion of resource supply. This occurs when "scarce resources" become expensive, and high prices encourage the increase in resource exploitation. Recent examples are the exploitation of North Sea oil deposits, the opening of new fields in Texas, and the development of oil surrogates (e.g., produced from sugar cane in the Proalcool programme in Brazil). Conversely, a resource field or a programme of resource substitution can become uneconomical when resource prices fall. Prices in the functional space of economics indicate changing scarcities. This is the simple explanation provided by economics for the mobilization of resource supplies (and their suspension) as profitability dictates. Economics, using this scarcity concept and the calculative rationality based on it, thus claims to solve all problems or to conjure them away.

However, the problems are more complicated, not just because prices (and interest rates) are determined according to the erratic reactions of the world market. More important, the temporal range of economic calculations and the price movements resulting from these calculations diverge sharply from resource times. The planning horizon of nuclear power companies, for example, is at most several decades. The half-life period of radioactive waste, however, is some 100,000 years. Economics is, in effect, the science of the "*avant le deluge*." On its banner could be written, "*apres moi le deluge*."

Thermodynamics and economic surpluses

According to the laws of thermodynamics, production is nothing but the transformation of matter and energy – a process in which an available input is transformed into needed output. A "throughput" is thereby produced – one whose external effects are effective beyond the temporal and spatial horizons of economic agents. But these economic agents (capitalists, entrepreneurs) are not content with merely transforming matter and energy. In fact, they are indifferent to such transformations so long as a capital surplus is achieved which sets into the motion the transformation process in the first place. In the concept of surplus, it is possible to see the circularity of the economic process: i.e., the reflexivity of the results in relation to their precondition – namely, input and output equality in thermodynamics (ecology) and surplus production in economics. This contradiction structures the relationship between economics and ecology in the capitalist mode of production.

This contradiction has a social dynamic. As Marx showed, the production process is both a labour process in which the transformation of matter and energy is carried out according to the laws of nature and a valorization process in the course of which an increase in labour value is added to the money capital which the capitalist has advanced. This double character of production and reproduction is possible because of the form in which it occurs. Its precondition is that labour has been transformed into wage labour and that wage labourers perform surplus labour: i.e., that they are exploited. In the absence of this specific social form of labour, there would be no difference between the ecological transformation of matter and energy and economic surplus production. Only within the wage form of labour is it possible that matter and energy can be transformed according to an intelligent plan and, at the same time, that a qualitative redistribution of matter and energy flows between social classes – i.e., from labour to capital – can occur. The processes in the functional space of economics thus undergo a change of measurement in which units of energy and matter are valued. Materials that have been transformed into use values simultaneously have exchange values. They are subsumed under the value form and thus under the money form. This is the precondition for the realization of the "logic" of economic functional space, which, in principle, is twofold. First, all energy and matter conversions are brought into a qualitatively equalized unit as flows of money and thus can be differentiated only quantitatively. This process is obviously very different from the perception of energy and matter transformations in accordance with their qualitative differentiation, on which the real perception of change is based, retreats from, or falls back behind, quantitative differentiation, which makes historically possible an orientation towards spatial expansion of quantitative accumulation.

It is this aspect of the particular form of economic processes that is neglected in most economic analysis, even when the analysis does take into account spatiality and temporarily (hence is able to grasp the contradiction between economics and ecology). However, the contradiction is totally ignored when value is attributed to nature without noting the form of the value of nature. Labour only produces value as wage labour. Which form must nature then take to produce value? Georgescu-Roegen has capitulated, faced with this question. For him, the origin of value devolves to the "enjoyment of life itself'; thus, he neglects the specific form of value in capitalism. (Value remains a subjective category.) In an anthology on the subject of "social relations and spatial structures," none of the authors included attempt to tackle the problem of the value form. Space is conceived only as a real substrate of "*Vergesellschaftung*" and not in its relation to nature, where processes are initiated through social action based on thermodynamic laws. After Chernobyl, this is clearly an unjustifiable form of reductionism; the social sciences quite obviously need to enlarge the scope of their analysis.

The social form (value form) thus makes two things possible: first, the quantitative logic of capital valorization abstracts from the qualitative limitation of use value. In fact, capital finds its fulfilment in technologically transcending, economically externalizing, socially marginalizing, and politically reprimanding all obstacles to the growth of quantitative value (or profit production and accumulation). The possibility of surplus production and the accumulation of capital create a social tendency to detach the economic process from all qualitative limitations. The reduction of all qualitative peculiarities to a common denominator which can be expressed in monetary form has made possible the enormous advance of Eurocentric civilization during the last two centuries, at the same time destroying whole social formations and modes of production. Also, in this process, the natural environment has been aggressively and powerfully changed and often degraded or destroyed. Whole mountains have been leveled; oceans have been fished out; species have been exterminated, rainforests destroyed, and huge areas have been transformed into refuge dumps and poisonous seas, lakes, and rivers. All this has occurred in the name of valorization and growth.

Second, the expansionist pressure inherent in the economic logic of surplus production has a territorial dimension (as production is necessarily always spatial). Surplus production is thus identical to the economic conquest – exploration, development, penetration, and exploitation – of space: i.e., the "production of space." At first, space is conquered extensively; subsequently, it is capitalized intensively. The "propagandistic tendency of the world market" (Marx) follows from the logic of capital valorization. The effect is the globalization of the principle of the contradictory functional spaces of economy and ecology – leaving nothing untouched on the entire globe.

Frontiers and borders

The process of capitalist growth and spatial expansion has no inherent borders, but it is in fact limited by external factors. When "the last tree has been cut, one will realize that one cannot eat money," as the saying goes among West German ecologists. Ecological borders facing capital in its push to widen its frontiers do exist, but they are far away. Up to now, destroyed landscapes in the industrial centres have been transformed into artificial parks by the recreation industry or bypassed by the offer of trips to the "intact world" of undamaged nature. The precondition, of course, is the monetization of damages, a mechanism which (at least in the rich industrial countries) still functions.

However, sooner or later, these sanctuaries as well will be destroyed. Then what happens? This means that borders must be set up ahead of time – not ecological borders but social ones.

As this chapter has shown, it is the social (value) form which not only produces and enshrouds but also brings to a head the contradiction between economy and ecology. Thus, immanent borders can be created only if the forms of social reproduction are transformed. The contradiction between physical and social modalities of the historical time regime and historical spatiality of capital can only be diminished (never nullified because of the law of entropy) by an increase in system intelligence and a removal of the obstacles to a conscious and considerate intercourse with nature which are today inherent in the social (value) form: the principle of surplus production (profit) and expansion (accumulation). We must create social and political border lines before the frontier of capitalist expansion reaches the last ecological border, which would be fatal to the conditions of survival of the human race. At this point, the discussion about ecological reform could begin, which would bring us back to more familiar social science and reformist debates.

Note

1 Editors' Note: This chapter appeared originally as Elmar Altvater. 1989. "Ecological and Economic Modalities of Time and Space." *Capitalism Nature Socialism* 1 (3): 59–70. It was an abridged translation by Michael Schatzschneider of Elmar Altvater. 1987. "Ökologische und Ökonomische Modalitäten von Zeit und Raum." *Prokla* 17 (61): 35–54.

10
THE ECOLOGY OF MISOGYNY

Leigh Brownhill

Misogyny is a matter of ecology

Ecofeminism is the recognition of and struggle against capitalists' racist colonization and exploitation of (that is, extraction of profits from) nature and women (and all non-binary gender identities). Ecofeminism, insofar as it is characterized by efforts to unite the exploited across historic social divisions (e.g., waged and unwaged), is *the revolutionary way* to an ecosocialist, post-capitalist future (Barca 2019; Brownhill and Turner 2018; Feder 2019; Giacomini et al. 2018; Kovel 2005).

Indigenous women's stewardship of the commons and reinvention of subsistence livelihoods across the world are at the core of "ecosocialist ecofeminism" (Brownhill, Kaara, and Turner 2016; Gómez-Barris 2017, 2018; Hess 2018; LaDuke 2020; Red Nation 2019, 2020; Lewis 2015; see also Terisa E. Turner, Ana Isla, Mary Mellor, and Terran Giacomini in this volume). This ecofeminism is notable for its radical grassroots and working-class origins, as well as its propensity for building the kind of "unity in diversity" that wins peoples' struggles, whether in women-led ecological movements (Brownhill, Kaara, and Turner 1997) or in struggles of women, youth, and people of colour in urban and built environments (Turner and Brownhill 2007; Blumberg et al. 2018). And it is this kind of ecofeminist discourse, rather than a discourse of gender equality and inclusion within the bounds of neoliberal capitalism, that can reveal the depth of the crisis faced by humanity today and its resolution in the "re-enchantment of the commons" (Federici 2018a).

Misogyny is, as a whole, a matter of ecology. Rape culture significantly changes or defines the relationship of almost every woman, trans, and non-binary person to the built and natural environment (Smith 2019). A term in use since the 1970s, rape culture is

> a set of general cultural beliefs supporting men's violence against women, including the idea that this violence is a fact of life, that there is an association between violence and sexuality, that men are active while women are passive, and that men have a right to sexual intercourse.
>
> *(Phipps et al. 2018, 1)*

10.4324/9780429341427-13

The Chilean feminist anti-rape anthem, "Un Violador en Tu Camino" ("A Rapist in Your Path"), performed on 25 November 2019 in recognition of the International Day for the Elimination of Violence Against Women, within weeks had been performed in protest by large groups of women in dozens of countries (McGowan 2019):

> *It's femicide, impunity for my murderer, it's disappearance, it's rape.*
> *And it wasn't my fault,*
> *Not where I was, nor how I was dressed.*
> *You are the rapist, you are the rapist.*
> *It's the police, the judges, the state, the president.*
> *The oppressive state is a macho rapist.*

The lyrics, the accompanying dance, and their viral popularity are testimonies to the ubiquity of rape as a systemic delimitation of women's ability to move freely and stay alive. The persistence of sexist and racist violence in the culture and political economy of late capitalism is tied, in the first place, to peoples' alienation from nature and from their diverse means of independent subsistence.

Enclosures pre-exist wages. The capitalist system of waged slavery began with historical dispossessions from commons and otherwise available independent means of life. In Europe and in the New World, historic enclosures entailed the systematic torture and brutalization of women in the witch hunts (Federici 2004, 2018b), tied in, there and elsewhere, with centuries of imperialism, colonization, and slavery. Capitalist social relations require the perpetuation of the alienation of people from their daily means of life to keep their lands and territories available for private and state profiteers. This has also entailed alienation from the body and its corporeal means of production.

To ensure that the majority find it too difficult to unite and resist their own exploitation and the destruction of our earthly commons, capital relies on the perpetual division of the 99 percent through racism, sexism, xenophobia, and other divisions, including the calculated persistence of racist and misogynistic violence and clinically stratified wage gaps.

Elsewhere, we have argued that a global hierarchy of labour power is cemented together with "male deals" between capitalists and those dispossessed men who are enticed, impressed, acculturated, or employed to channel use and intrinsic values from (waged and unwaged) labour and nature into corporate commodity value chains (Turner 1994; Turner and Brownhill 2001; Brownhill 2009; see also Mies and Shiva 1993; Mies and Bennholdt-Thomsen 1999; see also chapter by Terisa E. Turner in this volume).

When we write of "male deals," we must clarify that these are not always or only made by people identified as "men" and that no one should conflate the word "deal" with the idea of a "fair deal" or a deal between equals. For the male deal is always deeply unequal. Male deals cannot be done better. Institutions structured by male deals can coopt and absorb "equity, diversity, and inclusion," for instance, and still remain intact as vehicles of profit making. They have from the earliest imperial and slavery days involved cross-ethnic collaborations.

Lord Lugard made male deals when he applied his doctrine of indirect rule to entrench British imperialism around the world. The doctrine was to find the local men who would, by force or compromise, administer colonial policy; give those men enough compensation and power (and brutal oversight); and let them organize the collection of taxes for the colonial administration, recruit labour for the white settlers, and enforce colonial by-laws that undermined the livelihoods and land rights of their own people (Lugard 1922; Matthews 1937).

Male deals are never fair. They are deeply unfair for everyone, though unequally so, and insidiously rely on racism, cooptation, and indifference to block the potential unity of the majority. Male deals are bad for the earth. As profit channels, they are a social mechanism that capitalists deploy in nearly every place and culture, to extract whichever profit potential, resource, or widget possible.[1] These relations of exploitation are maintained by the perpetuation of intergenerational inequality and the imposition of commodified culture that reinforces and normalizes gross disparities in wealth and access to basic subsistence.

While male deals are initiated by capitalists for their own benefit, it could be argued that, if we consider the spiritual, emotional, and moral aspects of human existence, the male deal is not even good for individual capitalists themselves. Male deals, it turns out, are only good for the perpetuation of capital, which is itself not human (see Neocleous 2003). Capital holds power even over the stockholders and CEOs. Marx intimated as much in *Capital, Volume One*:

> As capitalist, he is only capital personified. His soul is the soul of capital. But capital has one single life impulse, the tendency to create value and surplus-value, to make its constant factor, the means of production, absorb the greatest possible amount of surplus-labour. Capital is dead labour, that, vampire-like, only lives by sucking living labour, and lives the more, the more labour it sucks.
>
> *(1867 [1967], 233)*

If capital is the accumulation of alienated and expropriated values, then the male deal is the valve that keeps this exploitation going and sustains an undead planetary overlord.

In the early colonial days in Kenya, Uganda, and Zambia, anti-colonial militancy was in part expressed in stories of vampires and "terrifying rumors that Africans who worked for white colonists captured unwary residents and took their blood" (White 2000). Across Turtle Island, a similar story was told. In the liner notes to her ballad, "The Priests of the Golden Bull," Buffy St. Marie explains that "the Windego monster to Cree people is like the Vampire is in Europe; it's a metaphor for mindless greed that cannibalizes indiscriminately for the satisfaction of only itself" (http://buffysainte-marie.com/?page_id=759#tpotgb).

Winona LaDuke (2020) also references "Wiindigoo" in the greed of the extractive capitalist system, against which Indigenous water protectors and global alliances of resistance with Indigenous people have flourished, exemplified in the Standing Rock struggle. One of the ways that allegories like this serve as a strength in the struggle against capitalism is that they crystallize the problem not as an individual character flaw in capitalists, but as a rapacious social relation of domination that creates both capitalists and the working class, cemented, as argued earlier, by the male deals that bind them. Marx stated in *The Grundrisse*,

> The *production of capitalists and wage labourers is thus a chief product of capital's realization process.* Ordinary economics, which looks only at the things produced, forgets this completely. When objectified labour is, in this process, at the same time posited as the worker's non-objectivity, as the objectivity of a subjectivity antithetical to the worker, as property of a will alien to him, then capital is necessarily at the same time the capitalist, and the idea held by some socialists that we need capital but not the capitalists is altogether wrong. It is posited within the concept of capital that the objective conditions of labour – and these are its own product – take on a personality towards it, or, what is the same, that they are posited as the property of a personality alien to the worker. The concept of capital contains the capitalist.
>
> *(Marx 1857 [1978], 275–276)*

The task of ecofeminist ecosocialism, then, is to rid humanity of the capital relation, starting with resisting and undoing male deals and (re)building political economies based on alliance, cooperation, and reinvention of the commons.

Overcoming accumulation

Rape culture, as a chronic affliction of the capitalist body politic, is not just a horrific, episodic *side-effect* of the colonization, enclosure, and alienation processes of capitalism; it is a constantly present *mechanism* of capital accumulation and the preservation of private property. Rape culture enforces capitalists' profit-making exploitation. Its constancy speaks to rape culture's connection to the political economy of capitalism within which it is situated, a political economy that happens to rely completely on the value and surplus value created by women's sexual reproduction and their daily work of producing (their own and others') labour power.

The pervasive violence of misogyny and its imminent threats serve to discipline women, on a global scale, into producing labour power for capitalist exploitation, cannon fodder, and other surplus value. Without misogyny and its violence, women (and men) would not continue succumbing to the enclosure of their commons, be these land and territories or city blocks and taxpayer-financed government coffers. Violence against women, whether through child abuse, pornography, prostitution, domestic violence, or rape, constrains and limits the freedom, mobility, health, and safety of its victims and survivors. It also limits and constrains all others, who, to varying degrees, calculate and adjust their lives to reduce the daily risk of sexual violence. Rape culture serves the purpose of capitalist profit extraction insofar as it maintains and entrenches division, isolation, and the alienation of gendered and ethnicized members of the dispossessed class. From above, this is a strategy of divide and rule.

Rape culture pushes binary gender roles and stereotypes of "men," their genitalia, and muscularity as inherently dominant and dominating. As a result, rape culture also dehumanizes and alienates men, especially those who resist or do not fit those stereotypes. Rape culture also involves the vicious disciplining of gender non-binary, "two-spirit," and trans women and men for daring to transgress rigid notions of gender and personhood. And their resistance to marginalization and victimization is often connected to land, housing, and ecological struggle so conceived (McNeil-Seymour 2017).

Racist misogyny in North America is strong evidence of the genocidal tendencies that accompany corporate-led eco-imperialism, significantly in communities on the frontlines of extractive industries (*Democracy Now* 2018; Faber 1993; Martens 2019; Ray 2018). Indigenous legal scholar Beverly Jacobs (2017, 50) argues that

> As a result of generations of abuse and control, Indigenous peoples have become victims in a long-standing abusive relationship and have been silenced through their lack of control over lands and resources, the genocidal policies of the residential school and child welfare systems, and the disrespect and violence against Indigenous women.

Aboriginal women in Canada "are seven times more likely to be victims of homicide than non-Aboriginal women" (Watson 2018, 204). Systemic violence against Indigenous women is core to the ongoing process of settler colonization and genocide. The old First Nations tenet is eerily resonant: "A people is not conquered until the hearts of their women are on the ground." This principle, expressive of Indigenous women's power in their communities, also helps explain the cold, extractive logic behind the scourge of sexual violence against Indigenous women and girls.

Anti-colonial ecofeminist activists are confronting these realities (Campbell 2003), including through mobilizing on behalf of the murdered and missing (Tolley, Martin, and Gilchrist 2012). In October 2019 Kanahus Manuel and Isha Jules of the Tiny House Warriors in British Columbia, Canada, were arrested for blocking work on the construction of a new Trans Mountain pipeline workers' accommodation complex, or "man camp." They were arrested for their protests "over fears of violence against Indigenous women and girls" (Martens 2019; see also http://tinyhousewarriors.com/; Manuel 2017). This is the logic of colonization, the ecology of the misogyny of the fossil fuel goliaths, that ecofeminism stands against.

We can trace a centuries-long thread from the rapes at the ports of Christopher Columbus's Niña, Pinta, and Santa Maria, to the anti-rape struggles of women in unceded Secwepemc territory in British Columbia, to North Dakota's Bakken oil fields, to the workers in hotel laundries and on Hollywood sets, and the Indigenous peoples living in the Amazon and Congo rainforests, where women continue to face harassment, violation, rape, trafficking, disappearance, and death, as the captains of industry undertake ongoing enclosures.

Sexism and racism are the blunt instruments that maintain the specific divisions and hierarchies that endless cycles of capital's exploitation of labour and nature require. The same capitalist hierarchy has generated existential threats against humanity in the form of rapidly unfolding climate crises and pandemics. Overcoming climate chaos and coronavirus means overcoming capitalism, and overcoming capitalism means uniting the exploited class to undo the male deals and to realize and universalize the life-centred, Indigenous-informed political economies and ways of life that are prefigured all around us as the ecofeminist and ecosocialist alternative to capitalism.

At every enclosure and industrial fence line, people are struggling to have agency over their lives, their health, and their environments; consequently, theirs is a struggle to overcome the capital relation. The way to win in these struggles is to unite across gender, racial, and other divides that have been cemented by male deals to instead build alliances that can hold the line on inclusive demands best expressed by peasants, the Indigenous, and women of color, often, but not always, unwaged. Social movements and union campaigns of women at the bottom of the hierarchy of labour power are uniquely situated to enact the unity of the exploited class across genders and ethnicities and to thereby encompass and stand for the experiences, needs, demands, and rights of all: in other words, to express the general class interest.

The unique power of grassroots ecofeminist knowledges and social movements is not, per se, the sex and race or the "bare feet" of the activists, but rather their brilliant use of the capacity of people at the bottom of the hierarchy of labour power *to unite every segment of the global working class above them*, both waged and unwaged (see Terisa E. Turner's chapter in this volume). Being at the "bottom" of the hierarchy of labour power does not automatically make one a revolutionary. But on the principle that people must lead themselves, and not be led by those "from outside" or "from above," it is only in solidarity with the claims, demands, and movements of women of color that anyone else above them on the hierarchy of labour power can be sure that their activism is inclusive of the general class interest. So, when grassroots and Indigenous movements do actualize their unique capacity to catalyze the unity of all sections of the exploited class (for instance, in a coordinated producer-consumer strike), powerful global shifts can follow (Tobocman et al. 2004). In their life-oriented principles, knowledge, already-existing commons, and "use value chains," and in the global class unity that they are uniquely able to foment, lie the tools and horizontal social relations indispensable to the building of an ecofeminist ecosocialist future for all.

Waged and unwaged Indigenous women and women of color are struggling to maintain, defend, and elaborate new and already-existing solutions to the economic, social, and ecological crises wrought by capitalism. They are uniquely capable of forging class unity. Peasant-, Indigenous-, and women of color–led movements have never been limited to protecting their own rights and interests. Their struggles and demands are also about the human rights and natural and civil commons that impact us all. These struggles concern the air we breathe and the well-being of the planet's atmosphere. They include Nigerian women's 1998 "gift to humanity" in their struggle to close down natural gas flares to reduce global emissions and the Indigenous "Sky Protectors," who oppose carbon colonialism (Turner and Brownhill 2004, 2007; Indigenous Environmental Network n.d.). These elder-led peaceful protest movements have set the stage for the children-led climate strikes that are now sweeping the world. From Standing Rock, to teachers' and airline workers' struggles for better working conditions, to the youth-led climate and anti-gun-violence movements, gendered-ethnicized class alliances are being formed in struggles to protect not only the rights of one group of people, but the waters, lands, and skies for all people, all life on earth.

Note

1 McMurtry (2001) describes a "corporate male gang" enforcing members' compliance to the law of the market.

References

Barca, Stefania. 2019. "The Labor(s) of Degrowth." *Capitalism Nature Socialism* 30 (2): 207–216.

Blumberg, Renata, Rosa Huitzitzilin, Claudia Urdanivia, and Brian C. Lorio. 2018. "Raíces Del Sur: Cultivating Ecofeminist Visions in Urban New Jersey." *Capitalism Nature Socialism* 29 (1): 58–68.

Brownhill, Leigh. 2009. *Land Food Freedom: Struggles for the Gendered Commons in Kenya, 1870–2007.* Trenton, NJ: Africa World Press.

Brownhill, Leigh, Wahu M. Kaara, and Terisa E. Turner. 1997. "Gender Relations and Sustainable Agriculture: Rural Women's Resistance to Structural Adjustment in Kenya." *Canadian Woman Studies/Les Cahiers de la Femme* 17 (2): 40–44.

Brownhill, Leigh, Wahu M. Kaara, and Terisa E. Turner. 2016. "Building Food Sovereignty Through Ecofeminism in Kenya: From Export to Local Agricultural Value Chains." *Canadian Woman Studies* 31 (1–2): 106–112.

Brownhill, Leigh, and Terisa E. Turner. 2018. "Ecofeminism at the Heart of Ecosocialism." *Capitalism Nature Socialism* 30 (1): 1–10.

Campbell, Horace. 2003. *Reclaiming Zimbabwe: The Exhaustion of the Patriarchal Model of Liberation.* Trenton, NJ: Africa World Press.

Democracy Now. 2018. "Body of Olivia Lone Bear Found in N. Dakota as Native Women Face Crisis of Murders, Disappearances." 3 August. www.democracynow.org/2018/8/3/body_of_olivia_lone_bear_found

Faber, Daniel. 1993. *Environment under Fire: Imperialism and the Ecological Crisis in Central America.* New York: Monthly Review Press.

Feder, Helena. 2019. "'Never Waste a Good Crisis': An Interview with Mary Mellor." *Capitalism Nature Socialism* 30 (4): 44–54.

Federici, Silvia. 2004. *Caliban and the Witch: Women, the Body and Primitive Accumulation.* New York: Autonomedia.

Federici, Silvia. 2018a. *Re-Enchanting the World: Feminism and the Politics of the Commons.* Oakland, CA: PM Press.

Federici, Silvia. 2018b. *Witches, Witch-Hunting, and Women.* Oakland, CA: PM Press.

Giacomini, Terran, Terisa E. Turner, Ana Isla, and Leigh Brownhill. 2018. "Ecofeminism Against Capitalism and for the Commons." *Capitalism Nature Socialism* 29 (1): 1–6.

Gómez-Barris, Macarena. 2017. *The Extractive Zone: Social Ecologies and Decolonial Perspectives.* Durham, NC: Duke University Press.

Gómez-Barris, Macarena. 2018. *Beyond the Pink Tide: Art and Political Undercurrents in the Americas.* Berkeley: University of California Press.

Hess, Shirin. 2018. "Zapatista Women Inspire the Fight Against Patriarchy." *Waging Nonviolence*, 3 April. https://wagingnonviolence.org/2018/04/zapatista-women-inspire-fight-against-patriarchy/

Indigenous Environmental Network. n.d. "Sky Protectors' Against California's Carbon Colonialism." www.ienearth.org/?s=sky+defenders

Jacobs, Beverly. 2017. "Decolonizing the Violence Against Indigenous Women." In *Whose Land Is It Anyway?*, edited by Peter McFarlane and Nicole Schabus, 47–51. Vancouver: Federation of Post-Secondary Educators of BC. https://fpse.ca/decolonization_manual_whose_land_is_it_anyway

Kovel, Joel. 2005. "The Ecofeminist Ground of Ecosocialism." *Capitalism Nature Socialism* 16 (2): 1–8.

LaDuke, Winona. 2020. *To Be A Water Protector: The Rise of the Wiindigoo Slayers*. Halifax: Fernwood.

Lewis, Avi, dir. 2015. *This Changes Everything*. Ottawa: National Film Board of Canada.

Lugard, Frederick. J.D. 1922. *Dual Mandate in British Tropical Africa*. London: Frank Cass.

Manuel, Kanahus. 2017. "Decolonization: The Frontline Struggle." In *Whose Land Is It Anyway?*, edited by Peter McFarlane and Nicole Schabus, 42–46. Vancouver: Federation of Post-Secondary Educators of BC.

Martens, Kathleen. 2019. "Kanahus Manuel Arrested, Injured During Pipeline Protest, Say Tiny House Warriors." *APTN National News*, 21 October. https://aptnnews.ca/2019/10/21/arrest/

Marx, Karl. 1857 [1978]. "The Grundrisse." In *The Marx-Engels Reader*, edited by Robert C. Tucker, 221–293. New York: W.W. Norton & Company. www.marxists.org/archive/marx/works/1857/grundrisse/ch09.htm

Marx, Karl. 1867 [1967]. *Capital: A Critical Analysis of Capitalist Production, Volume One: The Process of Capitalist Production*, edited by Frederick Engels. New York: International Publishers. See also www.marxists.org/archive/marx/works/1867-c1/ch10.htm#3

Matthews, Z.K. 1937. "An African View of Indirect Rule in Africa." *Journal of the Royal African Society* 36 (145): 433–437.

McGowan, Charis. 2019. "Chilean Anti-rape Anthem Becomes International Feminist Phenomenon." *The Guardian*, 6 December. www.theguardian.com/world/2019/dec/06/chilean-anti-rape-anthem-becomes-international-feminist-phenomenon

McMurtry, John. 2001. "The Life-Ground, the Civil Commons and the Corporate Male Gang." *Canadian Journal of Development Studies* XXII: 819–854.

McNeil-Seymour, Jeffrey. 2017. "Two-Spirit Resistance." In *Whose Land Is It Anyway?*, edited by Peter McFarlane and Nicole Schabus, 52–56. Vancouver: Federation of Post-Secondary Educators of BC.

Mies, Maria, and Veronica Bennholdt-Thomsen. 1999. *The Subsistence Perspective: Beyond the Globalised Economy*. London: Zed Books.

Mies, Maria, and Vandana Shiva. 1993. *Ecofeminism*. London: Zed Books.

Neocleous, Mark. 2003. "The Political Economy of the Dead: Marx's Vampires." *History of Political Thought* 24 (4): 668–684.

Phipps, Alison, Jessica Ringrose, Emma Renold, and Carolyn Patricia Jackson. 2018. "Rape Culture, Lad Culture and Everyday Sexism: Researching, Conceptualizing and Politicizing New Mediations of Gender and Sexual Violence." *Journal of Gender Studies* 27 (1): 1–8, January.

Ray, Lana. 2018. "Pipelines, Prostitution and Indigenous Women: A Critical Analysis of Contemporary Discourse." *Canadian Woman Studies* 33 (1–2): 107–116.

Red Nation. 2019. "Indigenous Feminism Does Not Discriminate." September. https://therednation.org/2019/09/07/indigenous-feminism-does-not-discriminate/

Red Nation. 2020. "Communism Is the Horizon: Queer Indigenous Feminism is the Way." September. http://therednation.org/wp-content/uploads/2020/09/TRN-pamphlet-final.pdf

Smith, Allison. 2019. "The 'Feminist City' Should Be Built for and by Women." *Rabble*, 6 December. https://rabble.ca/books/reviews/2019/12/feminist-city-should-be-built-and-women

Tobocman, Seth, Terisa E. Turner, Leigh Brownhill, and Laird Ogden. 2004. "Nakedness and Power." *World War 3 Illustrated* 35: 55–63.

Tolley, Bridget, Sue Martin, and Kristen Gilchrist. 2012. "Families of Sisters in Spirit: Solidarity and Relationship-Building in the Family-Led Movement on Behalf of Missing and Murdered Indigenous Women." *Our Schools Our Selves* 21 (3): 133–140.

Turner, Terisa E. 1994. *Arise Ye Mighty People: Gender, Race and Class in Popular Struggles*. Trenton, NJ: Africa World Press.

Turner, Terisa E., and Leigh Brownhill. 2001. "African Jubilee: Mau Mau Resurgence and the Fight for Fertility in Kenya, 1986–2002." *Canadian Journal of Development Studies* XXII: 1037–1108.

Turner, Terisa E., and Leigh Brownhill. 2004. "'Why Women Are at War with Chevron': Nigerian Subsistence Struggles Against the International Oil Industry." *Journal of Asian and African Studies* 39 (2–3): 63–93, March.

Turner, Terisa E., and Leigh Brownhill. 2007. "Climate Justice and Nigerian Women's Gift to Humanity." *Women and Environments International, Special Issue on Women and Global Climate Change* 74–75: 47–48.

Watson, Kaitlin. 2018. "Missing and Murdered Indigenous Women: The Role of Grassroots Organizations and Social Media in Education." *Canadian Woman Studies* 33 (1–2): 204–210.

White, Louise. 2000. *Speaking with Vampires: Rumor and History in Colonial Africa.* Oakland: University of California Press. www.ucpress.edu/book/9780520217041/speaking-with-vampires

11

FROM MARX TO ECOSOCIALISM[1]

Michael Löwy

Since the Industrial Revolution, capitalist societies (and more recently, the late bureaucratic societies of Eastern Europe) have been characterized by an ever-growing rationalization. Following Max Weber, we can distinguish three closely related aspects of this:

1. *Zweckrationalität*, or the rationality of ends: that is, the utilization of rational means to attain objectives that are not at all rational themselves. Bureaucracy is the ideal-typical institutional expression of this pattern. This is what the Frankfurt School referred to as instrumental rationality, a type of ratio compatible with the most monstrous substantive irrationalities – the rational-bureaucratic administration of genocide, for instance, to take the limiting case. But apart from such extremes, as Ernest Mandel has pointed out, the combination of partial rationality with overall irrationality is intrinsic to the "normal" functioning of the capitalist economy and its bureaucratic institutions (Mandel 1992, 182).
2. The differentiation and autonomization of domains, resulting in the separation of the economic, social, political, and cultural spheres. The market economy becomes a self-regulating system that is no longer "embedded" in the society (to use Polanyi's famous expression), thereby escaping social, moral, and political control.
3. *Rechenhaftigkeit*, or the spirit of rational calculation and the general tendency to quantification. This tendency finds its most direct expression in the monetarization of social relations and the unchecked dominion of the exchange value of commodities; through its action, qualitative, ethical, social, and ecological values are inexorably eroded, jettisoned, or destroyed.

Today, under the auspices of such institutions as the International Monetary Fund, the World Bank, the World Trade Organization, and the G-7, the rational "pursuit of maximum profit," along with the globalization process, have achieved a truly planetary scale for the working out of these tendencies. Unfortunately, the neoliberal European Union of Maastricht has not escaped this logic.

The first critics of the capitalist-industrial model of civilization were the romantics. From Rousseau in the second half of the eighteenth century to contemporary critics (such as the English historian E.P. Thompson), romanticism, through affirmation of precapitalist cultural, social, or ethical values, has protested the quantification, mechanization, and disenchantment of the world of capital.

10.4324/9780429341427-14

In the course of the history of romanticism, the nostalgia for a lost paradise and for pre-modern organic communities has taken forms that have sometimes been backward looking and retrograde, sometimes revolutionary and utopian. In the latter case, it is no longer a question of a return to the past, but of a detour through the past towards the future. For Pierre Leroux, William Morris, and Herbert Marcuse, to name a few, the future utopia allows recovery of lost community but in a new guise, incorporating as well the achievements of modernity: liberty, equality, fraternity, and democracy.

Socialism and political ecology (or at least, certain tendencies they contain) are, each in its own way, heirs to the romantic critique. They share the goal of going beyond instrumental rationality, the reign of quantification, and production as an end in itself. They insist on a society beyond the autonomization of the economy, the dominion of money, and the reduction of the social universe to the calculation of profit and the accumulation of capital. Instead, they posit qualitative values: use value and the satisfaction of needs. One may emphasize social equality, the other, the preservation of nature and ecological equilibrium. But both conceive of the economy as "embedded" in the social and natural environment.

That said, fundamental differences have so far divided the "reds" and the "greens," the Marxists and the ecologists. One important question concerns a frequent accusation made by ecologists: that Marx and Engels were infatuated with capitalist production. This charge of "productivism" admits to a number of interpretations.

From one aspect, it is entirely unjustified. For no one condemned the capitalist logic of production for the sake of production, or the accumulation of capital, wealth, and goods as ends in themselves, more than Marx. The very notion of socialism (never realized by its miserable bureaucratic avatars) is founded on the production of use values, of goods required for the satisfaction of human needs. For Marx, the final purpose of technical progress is not "having" (the infinite growth of possessions) but "being" (beginning with the reduction of the work day and an expansion of free time).

Nevertheless, one often finds in Marx and Engels (and even more in later Marxism) an uncritical attitude towards those aspects of industrial civilization that have contributed to its destructive relationship to the environment. This has chiefly appeared as a tendency to make the "development of the productive forces" the principal vector of progress. The "canonical" text for this point of view is the famous *Preface to the Contribution to the Critique of Political Economy* (1859), one of Marx's writings most permeated by a certain evolutionism, by the philosophy of progress, by scientism (that is, a valorization of the natural science model), and by an utterly unproblematic vision of the productive forces.

In reality, there is evidence for both interpretations in the writings of Marx and Engels. The following passage from the *Grundrisse* is a good example of Marx's uncritical admiration for the "civilizing" effect of capitalist production, including its brutal instrumentalization of nature:

> Thus, production based on capital creates, on the one hand, [universal] industry . . . surplus labor . . . creation of value; on the other, general exploitation of . . . nature and man. . . . Creation of bourgeois society, universal appropriation of nature . . . incorporating all members of the society: *such is the great civilizing effect of capital.*
>
> *(emphasis added)*

> It rises to such a [high] social level that all previous societies appear as purely *local developments* of man-kind and as *nature-worshippers*. . . . Nature becomes a . . . useful

> object . . . no longer a power. . . . Knowledge of natural laws . . . submitting nature to human needs, whether as object of consumption or as means of production.
>
> *(Marx 1859 [1967], 366–367; emphasis added)*

However, there are also a certain number of passages by Marx and Engels that show a more critical vision of the "productive forces." For example, in *The German Ideology*, one finds the following statement:

> In the development of the productive forces, it comes to a stage where productive forces . . . emerge that in the context of the existing [social] relations can only be negative, no longer productive but destructive. . . (mechanization and money).
>
> *(Marx 1846 [1967], 67–68)*

This idea is not developed by Marx, and it is not clear if the destruction in question includes the destruction of nature. One example in which this is the case is the well-known passage on capitalist agriculture in *Capital*, where we find a dialectical vision of the inherent contradictions of the "progress" created by the productive forces, along with explicit reference to the havoc wreaked by capital on the natural environment:

> Thus it destroys both the physical health of the urban worker and the spiritual life of the rural worker. Every stage in the development of capitalist agriculture, each short-term advance in fertility, represents at the same time a stage in the destruction of the long-term basis of this fertility. The more a country develops . . . large-scale industry (US for example), the more rapidly this destructive process moves forward. Capitalist production thus only develops . . . by at the same time exhausting the two springs from which flow all wealth: the land and the laborer.
>
> *(Marx 1867 [1967], 360–361)*

Even with Engels, who so often applauded the human "mastery" and "domination" of nature, one finds writings that call very explicit attention to the dangers of such a stance. Consider, for example, the following passage from the essay "The role of work in the transformation of monkey to man" (1876):

> We should not boast overmuch of our human victories over nature. For each of these victories, nature takes its revenge. . . . The peoples of Mesopotamia, Greece, Asia Minor . . . who destroyed the forests to create arable land, never imagined that . . . they were creating the condition for the present desolate state of this land. The Italians of the Alps cut the forests. . . . [T]hey had no idea . . . they destroyed the basis for the dairy industry . . . even . . . depriving their mountain springs of water. . . . We must always remember that we can never rule nature . . . as a ruler over conquered people, as if outside of nature. . . . [W]e belong to it body and soul. . . . [A]ll our domination rests on the advantage we have over other creatures in knowing its laws and being able to use them wisely.
>
> *(Engels 1876 [1968], 180–181)*

It would not be difficult to find other examples. It remains the case, however, that Marx and Engels do not possess an overall ecological perspective. Indeed, their optimistic conception of the unlimited development of the productive forces which is to take place once these are freed from the trammels of capitalist production relations is no longer defensible today. Not only from the strictly economic point of view (given the risk of resource depletion), but more fundamentally, in view of the threat posed to the earth's ecological equilibrium by the productivist logic of capital (or of that represented by its poor imitation, the late "socialist" bureaucracies).

One might provisionally conclude this discussion with the very pertinent suggestion advanced in the recent – and remarkable – study of Marx by Daniel Bensäid. As Bensäid observes, it would be as unjustified to exonerate Marx from the "progressivist" or "Promethean" illusions of his time as it would be to make him the evangelist of an unfettered industrialism. Instead, he suggests a much more promising tack: to move right in with Marx's contradictions and take them fully to heart – in particular, those between the productivist credo announced in certain texts and the intuition that progress may cause irreversible environmental harm (Bensäid 1995, 347).

At the beginning of the twenty-first century, it is the ecological question, in my view, that poses the major challenge to a renewal of Marxist thought. It demands a thorough and critical revision of the traditional Marxist conception of the "productive forces" and implied by this a radical break with the technological and economic paradigm of modern industrial civilization and with the ideology of progress.

Walter Benjamin was one of the first Marxists of the twentieth century to raise this kind of question. As early as 1928, in his book *Sens Unique*, he denounced the notion of dominating nature as an "imperialist idea," putting forward instead a new conception of technology as "mastery of the relationship between nature and humanity" (Benjamin 1978, 243). Several years later, in his *Thèses sur le Concept d'Histoire*, he suggests enriching historical materialism with the ideas of Fourier, that utopian visionary who had dreamed of a new form of labour that "far from exploiting nature, can bring life to the creations that lie dormant in her womb" (Benjamin 1971, 190).[2]

Even today, Marxism is far from having corrected its shortcomings in this respect. But certain lines of analysis have begun to tackle the problem, notably the "Marxist-Polanyist" James O'Connor, with his fruitful suggestion that we add to the first contradiction of capitalism – the contradiction between the forces and relations of production examined by Marx – a second contradiction, between the forces of production and the conditions of production (nature, workers, urban space). Capital, by virtue of its expansionist dynamic, endangers or destroys its own conditions of existence, beginning with the natural environment – a possibility that Marx had not taken sufficiently into consideration (O'Connor 1992).

A recent work by the Italian "eco-Marxist" Tiziano Bagarolo builds on the passage in *The German Ideology* cited earlier to suggest another interesting approach:

> The formula that posits a transformation of potentially productive forces into effectively destructive forces, particularly with respect to the environment, seems to us more appropriate and significant than the well-known schema of the contradiction between (dynamic) productive forces and production relations (that hold them back). Besides, this formula opens the way to a critical and non-apologetic foundation for economic, technological, and scientific development, and thus to the elaboration of a "differentiated" concept of progress (E. Bloch).
>
> *(Bagarolo 1992, 25)*

Whether Marxist or not, the traditional labour movement in Europe – unions, social-democratic parties, and communists – remains profoundly imprinted by productivism and the ideology of "progress." In some cases, labour even goes so far as to defend nuclear energy or the automobile industry without asking the necessary questions as to their effects on the global ecology. Nonetheless, the beginnings of an ecological consciousness are taking shape (particularly in the Nordic countries, Spain, and Germany) in the unions and parties of the Left.

The great contribution of ecology has been to make us aware of the dangers that threaten the planet as a result of the current mode of production and consumption: the exponential growth of pollution of the water, earth, and air; the massive extinction of living species; the desertification of fertile lands; the build-up of dangerous nuclear wastes; the constant threat of new Chernobyls; the destruction of forests at a dizzying pace; the greenhouse effect; and the possible rupture of the ozone layer, with catastrophic effects on all organic life. Together, these create a doomsday scenario that puts in question the very survival of humanity. We are confronted, in fact, with a civilizational crisis that requires radical change.

If socialist thought fails to address this through its residual productivism, the proposals advanced by political ecology tend to exemplify another dimension of inadequacy. Here, the major weakness has been to ignore the essential connection between productivism and capitalism. This approach leads to the illusion of a "clean capitalism" or of the possibility of controlling its "excesses" with "eco-taxes" or other reforms. Alternatively, on the grounds that bureaucratic command economies simply imitated Western productivism, most political ecologists lump capitalism and socialism together as variants of the same model – an argument that, with the disintegration of the supposed "real existing socialism," has lost much of its power.

The ecologists deceive themselves if they imagine they can do without a Marxist critique of capitalism. An ecology that does not take account of the relationship between "productivism" and the profit motive is doomed to failure – or worse, to cooptation by the system. There is no lack of examples.

Considering the workers irremediably given over to productivism, certain ecologists simply ignore the labour movement, raising the banner of "neither Left nor Right." Ex-Marxists converted to ecology hastily bid "farewell to the working class" while others insist that one must abandon the "red" (Marxism and socialism, that is) to join the "green," the new paradigm that should resolve all economic and social problems.

Finally, in what are called "fundamentalist" circles or deep ecology, a certain rejection of humanism – supposedly to combat "anthropocentrism" – is taking shape. This leads to a relativist position in which all living species are put on the same level. But must we really believe that the Koch bacillus or the anopheles mosquito has the same right to life as a child afflicted with tuberculosis or malaria?

Clearly, an ecosocialist alternative needs to be developed in order to avoid these pitfalls. Incorporating the fundamental insights of Marxism – while discarding its productivist dross – ecosocialism understands that the logic of profit and the market (like that of the techno-bureaucratic authoritarianism of the defunct "popular democracies") is incompatible with ecological needs. While criticizing the prevailing ideological tendencies of the labour movement, it knows that the workers and their organizations are essential partners in any radical systemic change.

Building on the work of several late-nineteenth- and early-twentieth-century Russian pioneers like Serge Podolinsky and Vladimir Vernadsky, such an ecosocialism has begun to emerge in the course of the last 25 years. It is indebted to thinkers of such stature as Manuel Sacristan, Raymond Williams, and Rudolf Bahro and André Gorz (in their early writings), as well as to the valuable recent contributions of James O'Connor, Barry Commoner, Ted Benton, Juan Martínez Alier, Francisco Frenandez Buey, Jorge Riechman, Jean-Paul Deléage, Jutta Dittfurth, Thomas Ebermann, Ranier Trampert, Erhard Eppler, Elmar Altvater, Frieder Otto Wolf, and many others, featured in various journals such as *Capitalism Nature Socialism*, *Écologie Politique*, and others.

The ecosocialist tendency is found in the green parties and the "green-red" movements on the far Left, and even in the bosom of the "classic" Left. Although hardly homogeneous politically, most of its spokespeople share certain common themes. Breaking with the productivist ideology of progress (in its capitalist and/or "real socialist" bureaucratic forms), and opposing the infinite expansion of a mode of production and consumption that destroys the environment, ecosocialism represents the most advanced wing of the ecological camp. This is reflected in its sensitivity to the interests of the workers and the peoples of the South, as well as through its radical rejection of the notion of "sustainable development" within a capitalist market framework.

The ecosocialist rationale rests on two essential arguments:

1 The current mode of production and consumption, based on a logic of unlimited accumulation (of goods, profits, and capital), ostentatious consumption, waste of resources, and accelerated destruction of the environment, can never be extended to the planet as a whole without causing a major ecological crisis. According to recent estimates, for instance, if the average energy consumption of the United States was generalized to the whole population of the world, known petroleum reserves would be exhausted in *19 days* (Mies 1992). The economic system is thus predicated on the maintenance and exacerbation of a blatant inequality between North and South.

Moreover, due to a deliberate policy of "exportation of pollution" by the imperialist countries, neoliberal globalization is intensifying the ecological problems of Asia, Africa, and Latin America. This strategy has even received a penultimate economic "legitimation" (from the capitalist point of view), recently formulated by the eminent World Bank expert (later to become the US Secretary of the Treasury and president of Harvard) Lawrence Summers: the poor cost less! To cite his own words:

> [T]he measure of the costs of pollution harmful to health depends on the output lost because of increased morbidity and mortality. From this point of view a given quantity of pollution harmful to health should be concentrated in countries with the lowest costs, that is the countries with the lowest salaries.
>
> *(Summers 1992)*[3]

This cynical formulation reveals much more about the logic of global capital than all the soothing speeches about "development" produced by the international financial institutions.

2 In any case, the ongoing reproduction of capitalist "progress" and market society – even in this brutally inegalitarian form – directly threatens, however hazardous the prediction, the very survival of the human species in the short or medium term. The safeguard of the natural environment is thus a humanist imperative as well.

The narrow rationality of the market, with its short-term calculations of profit and loss, and ecological rationality, which takes into account the full duration of natural cycles, are inherently contradictory. Neoliberalism lends its support to the prevailing fetishism of commodities and the reified autonomization of the economy. For ecosocialists, on the other hand, what is essential is the inauguration of a "moral economy," in E.P. Thompson's meaning of the term: that is, a political economy founded on extra-economic, nonmonetary criteria. To put it another way, the economic is to be re-embedded into the ecological, the social, and the political (Bensäid 1995, 385–386).

Partial reforms are totally inadequate. A social and ecological macro-rationality must replace the micro-rationality of profit – something that demands a veritable civilizational change (Reichmann 1991).[4] This is impossible without a profound technological reorientation, whose foundation is the replacement of current sources of energy with non-polluting and renewable alternatives, such as solar energy (Schwartzman 1996). Control over the means of production, especially over decisions concerning investments and technological change, must be thoroughly reorganized according to nonmarket criteria encompassing the real needs of the population (for which there may be inadequate "effective demand") and the protection of the environment. The only possible ground for this can be a democratic choice of priorities and investments by the population itself – rather than the "laws of the market" or an omniscient politburo.

In sum, we envision an economy of transition to ecosocialism, re-embedded, as Polanyi would put it, in the social and natural environment. The transition is to an alternative way of life, to a new civilization, beyond the rule of money and the artificial consumption habits created by advertising, beyond the unlimited production of commodities, such as the automobile, that harm the environment.

A utopia? In the etymological sense of "no place," then surely. But – unless one agrees with Hegel that "all that is real is rational, and all that is rational is real" – how can we even conceive of a more substantive rationality without resorting to utopias? On condition that it be based on real contradictions and real social movements, utopia is an indispensable element of social change. This is the case with ecosocialism, which proposes a strategy of alliance between the "reds" and the "greens," the labour and ecological movements, and of solidarity with the oppressed and exploited of the South.

This alliance could find its first strategic site in Europe, where the two movements, red and green, are active, and the barriers that separate them are beginning to fall. But this assumes that ecology can give up its naturalist and anti-humanist proclivities, along with its pretentions to do without, or even replace, a critique of political economy. Marxism, on its side, must give up its productivism, its mechanistic schema of opposition between the (developing) forces of production and the (restrictive) relations of production. The convergence with ecology requires, instead, the much more fruitful conception of forces that are potentially productive being transformed into forces that are effectively destructive, and vice versa (Bensäid 1995).

The fact that green socialism or a solar communism must be posited as a revolutionary utopia does not mean that we should not act and starting now. Recognizing the impossibility of "ecologizing" capitalism does not mean that we should not engage in the struggle for immediate reforms. Certain types of eco-taxes, for example, might be useful, on condition that they be informed by an egalitarian social logic (that it is the polluters, not the consumers, who should pay) and that one give up the myth that the "market price" of ecological damage can be calculated in economic terms. (These are incommensurable variables from a monetary point of view.) We have a desperate need to make up for lost time, to struggle immediately for a ban on the CFCs that destroy the ozone layer, for severe restrictions on the gas emissions that cause the greenhouse effect, and for the promotion of public transportation as opposed to the polluting and anti-social individual passenger car (Reichmann 1995).

The struggle for ecosocial reforms could spark a new dynamic of change, a transition from minimal to maximal programmes and demands – but only on condition that one eschew the arguments and pressures deployed by the dominant interests in the name of "competitiveness," "modernization," and "market laws."

Certain immediate demands are already, or may rapidly become, points of convergence between "reds" and "greens." These include the following:

- The promotion of free or inexpensive public transportation – trains, subways, buses, trams – as an alternative to the asphyxiation and pollution of the cities and countrysides by the road transport system and passenger cars
- The struggle against the system of ultra-liberal "adjustment" and debt imposed by the IMF and the World Bank on the countries of the South, and with such dramatic social and ecological consequences: massive unemployment and destruction of social safety nets, abandonment of food crops, and destruction of natural resources in favor of exports
- The protection of public health from the pollution of the air, the ground water, and the food supply (by-products of the greed of large-scale capitalist enterprise)
- The reduction of work time, both as a solution to unemployment and as an expression of a larger social vision, one placing greater value on free time than on the accumulation of goods (Rousett 1996, 8–9)

In the struggle for a new civilization, however, it is the entire array of emancipatory social movements that need to be interlinked. As Reichmann puts it so well:

> This project must not reject any of the colors of the rainbow: neither the red of the anticapitalist and egalitarian labor movement, the pink of the women's liberation struggle, the white of the non-violent peace movements, nor the black anti-authoritarianism of the libertarians and anarchists, still less the green of the struggle for a just and free humanity on a habitable planet.
>
> *(Reichman 1996, 57)*

This is a truly planetary cause. But the European Union, if it can escape the neoliberal straightjacket of Maastricht and reunite around new goals, can become a major crucible of change, working to recast the future in a different mold.

Notes

1 Editors' Note: This chapter appeared originally as Michael Löwy. 2002. "From Marx to Ecosocialism." *Capitalism Nature Socialism* 13 (1): 121–133. The original was translated by K.P Mosely, including quoted text, but the citations are in the original. The original publication has been modified to conform to the style adopted for *The Routledge Handbook on Ecosocialism* and the cited references have been amended by completing missing information and removing any redundancies.

2 One might also mention the Austrian socialist Julius Dickmann, author of a pioneering essay (Dickmann 1933). In his view, socialism would not result from a "dramatic take-off of the productive forces," but rather, would be a necessity imposed by the "shrinking of the reservoir of natural resources" being depleted by capital. The "heedless" development of the productive forces by capitalism undermines the very conditions of existence of the human species.

3 Another striking example: a Working Group of the Intergovernmental Workshop on Climate Change, at a meeting in Geneva in July 1995, discussed a report which asked if it was "cost-effective" to take measures against the greenhouse effect, considering that the effects will be felt above all in the poor countries. According to the experts, the cost of a life in a rich country is $1.5 million, in a poor country, only $100,000 (see Lovejoy 1996, 274).

4 On this point, see also the remarkable essay by Jorge Reichmann (1996).

References

Bagarolo, Tiziano. 1992. "Encore sur Marxisme et Écologie." *Quatrième Internationale* 44: 7–32.

Benjamin, Walter. 1971. "Thèses sur la Philosophie de l'Histoire." In *L'homme, le langage et la culture*. Paris: Denoel.

Benjamin, Walter. 1978. *Sens Unique*. Paris: Lettres Nouvelles, Maurice Nadeau.

Bensäid, Daniel. 1995. *Marx l'Intempestif: Grandeurs et Misères d'une Aventure Critique*. Paris: Fayard.

Dickmann, Julius. 1933. "La Véritable Limite de la Production Capitaliste." *La Critique Sociale* 9: 108–113.

Engels, Friedrich. 1876 [1968]. *La Dialectique de la Nature*. Paris: Éditions Sociales.

Lovejoy, Derek. 1996. "Limits to Growth?" *Science & Society* 60 (3): 266–278.

Mandel, Ernest. 1992. *Power and Money: A Marxist Theory of Bureaucracy*. London: Verso.

Marx, Karl. 1846 [1967]. *L'Idéologie Allemande*. Paris: Editions Sociales.

Marx, Karl. 1859 [1967]. *Fondements de la Critique de l'Économie Politique*. Paris: Anthropos.

Marx, Karl. 1867 [1967]. *Le Capital*. Translated by Joseph Roy. Tome 1. Paris: Editions Sociales.

Mies, M. 1992. "Liberación del Consumo o Politización de la Vida Cotidiana." *Mentras Tanto (Barcelona)* 48: 73.

O'Connor, James. 1992. "La Seconde Contradiction du Capitalisme: Causes et Conséquences." *Actuel Marx* 12: 30, 36.

Reichmann, Jorge. 1991. *Problemas con los Frenos de Emergencia?* Madrid: Editorial Revolución.

Reichmann, Jorge. 1995. *De la Economía a la Ecología*. Madrid: Editorial Revolución.

Reichmann, Jorge. 1996. "El Socialismo Puede Llegar Solo en Bicicleta." *Papales de la Fondation de Investigaciones Marxistas (Madrid)* 6.

Rousett, Pierre. 1996. "Convergence de Combats. L'Écologique et le Social." *Rouge* 8–9, 16 Mai.

Schwartzman, David. 1996. "Solar Communism." *Science & Society*, 60 (3): 307–331.

Summers, Larry. 1992. "Let Them Eat Pollution." *The Economist*, 82, 8–14 February.

12

DIALECTICAL ECOLOGY

The origins of dialectical ecology

John Clark

The basic principles, or anti-principles, of dialectical thought were developed in the "Axial Period," about two and half millennia ago, in India (Shakyamuni Buddha), China (Laozi, Zhuangzi), and Southwest Asia (Heraclitus). Heraclitus's dialectic introduced such fundamental themes as the universality of change and flux, the illusory nature of sameness or identity, and the mutual determination of opposites. In addition, his concept of the logos expressed the principle of dialectical reason, the concept that there is a truth of the whole (or becoming whole) within the context of non-identity and all-pervasive change.

In this same period, early Buddhism developed such dialectical concepts as *anitya* or impermanence, *anātman* or non-self-identity, *śūnyatā* or emptiness (non-essentialism or non-substantiality), and *pratītyasamutpāda* or dependent origination (internal relations between all beings, which have no "self-being").

It can be argued that early Daoist philosophy was the first developed dialectical ecology in history. Laozi taught that opposites are complementary and mutually determining within a developing whole. He undertook an immanent critique of existing ideologies and values, including the hierarchical and dualistic images of male and female, adult and child, human and natural. Zhuangzi, the other major early Daoist thinker, developed a radically naturalistic perspectivism as part of his dialectical critique of anthropocentrism and egocentrism.

The radical dialectical tradition has remained strong in Asian philosophy through the Buddhist and Daoist traditions while it receded into the background in European thought until its major revival in German idealism, culminating in Hegel. Hegel's dialectical analyses of the movement of thought in the *Logic* and his immanent critique of forms of consciousness and social phenomena in the *Phenomenology of Spirit* and *Philosophy of Right* have become paradigms of dialectic. Hegelian influence is strong in dialectical ecology, especially in Marxist ecology and social ecology, which see part of their project as the naturalization of Hegelian idealist dialectic.

Marx, Engels, and the dialectics of nature

Friedrich Engels is in some ways the most important precursor of modern dialectical ecology. In his *Dialectics of Nature*, he identifies three "laws" of dialectics: "the transformation of quantity into quality and vice versa," "the interpenetration of opposites," and "the negation of the negation" (Engels 1934, 62). The interpenetration of opposites means that they are

10.4324/9780429341427-15

not "irreconcilable" realities, but rather poles that "represent the truth only in their reciprocal action" (Engels 1934, 216). Engels applies this far-ranging principle to various problems, including those of identity and difference.

One implication is a rejection of all concepts of "abstract identity" and a recognition that in nature, phenomena have no more than a relative identity that is a moment in their dialectical development. Thus, "the plant, the animal, every cell, is at every moment of its life identical with itself and yet becoming distinct from itself." Such development occurs through diverse processes including "incessant molecular changes which make up life" (Engels 1934, 214).

Engels holds that dialectic is universal and encompasses all ontological realms. It includes both an objective form that "prevails throughout nature" and a subjective form that is expressed in thought and "is only the reflection of the motion through opposites which asserts itself everywhere in nature" (Engels 1934, 211).

Marx must also be looked upon as one of the pre-eminent precursors of dialectical ecology, in view of his development of the most significant position in modern philosophy that is both deeply dialectical and deeply naturalistic. Nevertheless, his philosophy exhibits an ambiguous standpoint towards nature that has repercussions in contemporary debates in ecophilosophy.

Thus, we find his naturalistic description of labour as "a process in which both man and Nature participate" and in which "man opposes himself to Nature as one of her own forces" (Marx n.d. [1867], 127). However, in the same passage, we find the problematic anthropocentric view that "man of his own accord starts, regulates, and controls the material re-actions between himself and Nature," and does so "in order to appropriate Nature's productions in a form adapted to his own wants" (Marx n.d. [1867], 127).

In another important passage, Marx notes that capitalist production "disturbs the circulation of matter between man and the soil" and thus "violates the conditions necessary to lasting fertility of the soil," concluding that "all progress in increasing the fertility of the soil for a given time, is a progress towards ruining the lasting sources of that fertility" (Marx n.d. [1867], 330). This dimension of Marx's thought becomes the basis for the widely discussed metabolic tendency in contemporary environmental Marxism.

Finally, there are momentous ecological implications in Marx's analysis of the contradictions of capitalism. He argues that capital itself becomes the ultimate barrier to capitalist production since the overriding goal of "unconditional development of the productive forces" comes into conflict with "the self-expansion of the existing capital" (Marx 1959, 250). This concept, extended and ecologized, becomes the basis for James O'Connor's theorization of the "second contradiction of capitalism": that is, the ecological contradiction between the capitalist forces and relations of production and the often-neglected conditions of production (O'Connor 1998, 158–175).

Levins and Lewontin's dialectical biology

Among the more recent precursors of contemporary dialectical ecology, it is necessary to mention Richard Levins and Richard Lewontin, authors of *The Dialectical Biologist*. Since their analysis merges into dialectical ecology itself, they should not only be considered major precursors but also counted among its founding thinkers. Their work still remains far in advance, theoretically, of most attempts to develop a dialectical ecology.

Levins and Lewontin argue for the dialectical concept of "organism-environment interpenetration." They point out the many ways in which organisms "select their environments," "modify their environments," "transform the statistical structure of their environment," "determine what aspects of their environment are relevant," and "respond to their environments," thus undercutting any dualistic split between the two (Levins and Lewontin 1985, 53–56).

They reject both an idealistic holism that takes the community to be "the only causal reality" and an atomizing reductionism that "sees the individual species, or ultimately the individuals . . . as the only 'real' object." They argue it is necessary to reject such approaches and to investigate "the actual material relationship among entities at all levels" (Levins and Lewontin 1985, 135).

They interpret the part-whole relationship dialectically, showing that each side of the polarity can only be understood in relation to the other (Levins and Lewontin 1985, 273). Moreover, their critical-dialectical view demolishes idealizing ideologies that present wholes as "inherently balanced or harmonious" or as having some fixed identity (Levins and Lewontin 1985, 274).

Further, they adopt the radically dialectical position that "objects of laws of transformation" can become subjects that change these laws and that systems can "destroy the conditions that brought them about" and "create the possibilities of new transformations" (Levins and Lewontin 1985, 277).

Finally, they reject as undialectical and "bourgeois" the idea that development is nothing more than a "regular unfolding" of what is already present, that it can be explained by a mere description of a sequence of changes, or that it can be reduced to the passage through certain necessary stages (Levins and Lewontin 1985, 276).

Bookchin's dialectical naturalism

Bookchin's theory of dialectical naturalism is his effort to develop a philosophical basis for his social ecology. Its guiding concept is what Bookchin sees as the dialectical unfolding of potentiality within both the natural and social worlds, which he terms "first" and "second" nature.

Bookchin focuses on a kind of immanent dialectic that becomes his basic explanatory and also normative principle in both realms. He cites a crucial passage from Hegel as a kind of proof text on dialectic. In it, the philosopher describes the development of a plant, in which "that which is implicit comes into existence" and "passes into change, yet it remains one and the same" (Hegel 1955, 22). The fully developed being is "hidden and ideally contained" in the embryo, which is "impelled towards development, since it presents the contradiction of being only implicit and yet not desiring so to be" (Hegel 1955, 22).

According to Bookchin, all social and natural beings can be understood according to this model of "what-is" versus "what-should-be" (Bookchin 1996, 21–24). He rejects the idea that such a philosophy of nature is teleological since, in his view, teleology implies that development is "inevitable" or "predetermined" (Bookchin 1996, 16, 63). In fact, Aristotelian final causation means that a plant, an animal, or a human being aims at developing into an actualized "good of its kind," not that it is predestined to actualize this potential.

Bookchin's dialectical naturalism is firmly in the tradition of self-realization theory in ethics (those typically called "teleological") and has much in common with mainstream theories such as the human capabilities approach of Nussbaum and Sen. The extent to which it is dialectical in any strong sense is more questionable.

Bookchin's dialectical naturalism is combined with a radical critique of domination and a politics of revolutionary ecopolitical transformation that he calls "libertarian municipalism," "confederalism," or "communalism." He contends that this specific politics can be demonstrated to follow from the phenomena of social and natural history, using a process of "eduction," based on his version of immanent dialectic.

Kovel's ecological Marxism

Joel Kovel, co-author of the *Ecosocialist Manifesto*, is one of the most important figures in the development of ecosocialism and a major theorist of ecological Marxism.

Although Kovel holds a fundamentally Marxist position, he does not hesitate to point out the limitations of classical Marxism. He argues that Marx conceived of nature as "an organ subordinated to the master's mind and an instrument of labor," the earth as "an instrument and even a kind of slave," and the worker as a force that is not only "opposed to nature" but even "*outside* of nature" (Kovel 2019, 219). Moreover, he contends that pre-ecosocialist Marxism suffered from "a failure to recognize sufficient intrinsic value in nature" (Kovel 2019, 15).

Kovel's position is strongly dialectical in its analysis of all levels of being. Dialectic, he says, "expresses . . . something about the working of the real" (Kovel 2019, 19). He holds, like Bookchin, that there is "a directedness within becoming" and a "formative process within the real" (Kovel 2019, 20). But, contrary to Bookchin, such processes are dialectical in a deep sense that involves radical negativity. "Dialectic is an interplay of absences and presences, related negatively," so there is no "grand unfolding" but rather "myriad foci of absence/presence mediating the emergence of being" (Kovel 2019, 19–20).

Kovel, unlike many other ecological thinkers, stresses the inexorable gap between the human and natural worlds. The human relationship to nature always remains "radically different from other organic processes" (Kovel 2019, 64) because of its symbolic dimension. This leads him to reject the naïve idea that there is an ideal point at which the dialectical tension between the two worlds can be resolved in a future ideal and harmonious ecological society.

Kovel is also unusual among Left ecologists in stressing the importance of developing an ecosocialist conception of intrinsic value, in which we value nature "for itself, irrespective of what we would do to it" (Kovel 2019, 275). He claims that "values do not exist in nature but in the mind," though they can be "built" on that which is "external" (Kovel 2019, 275). This recognition of intrinsic value constitutes a very important advance in dialectical ecology. However, it does not clearly address arguments for objective value in the non-human world or the possibility of a more dialectical conception of value that goes *beyond* intrinsic value in not having simple location in specific beings and their experience.

Foster's metabolic theory

Another well-known contemporary effort at dialectical ecology is the metabolic theory of John Bellamy Foster and his colleagues. Both the strength and the weakness of this position lie in its firm grounding in Marx's thought. On the one hand, it yields a very robust form of human-centred radical environmentalism, which, in environmental philosophy, would be a labeled a stewardship ethic or an environmental justice approach. On the other hand, such a radical environmentalism is, contrary to its claims, a dialectical ecology in only a weak sense.

Foster and colleagues are quite clear about their theoretical goals. For example, they cite approvingly Marx's project of defining socialism "in terms of a process of sustainable human development," which it "understood as the necessity of maintaining the earth for future generations, coupled with the greatest development of human freedom and potential" (Foster and Clark 2016). Their defense of Marx as an ecological thinker is based on the view that such a sustainable development position can be considered a strongly ecological one.

They are, of course, fully aware of the most deeply dialectical dimensions of Marx. They cite various strongly dialectical passages concerning nature, such as his assertion in his dissertation that "In hearing nature hears itself, in smelling it smells itself, in seeing it sees itself" (Foster,

Clark, and York 2010, 227) and the dialectical and naturalistic precept from *The 1844 Manuscripts* that "a being which does not have its nature outside itself is not a natural being and does not share in the being of nature" (Marx 1964, 207).

Furthermore, they recognize that Marx's "basic ontological scheme for understanding the world" involved a commitment to the dialectical doctrine of internal relations (Foster, Clark and York 2010, 235). However, their specific analyses of society, nature, and ecological crisis do not always reflect these radically dialectical precepts.

A controversial aspect of metabolic rift theory is its contention that capitalist industrialization and commodification disrupt an "eternal-natural relationship," resulting in a "systematic and intensive metabolic rift in agriculture, whereby the return of essential nutrients (e.g., nitrogen, phosphorus, and potassium) to the soil is disrupted" (Foster and Clark 2016). Such a rift, it is claimed, destroys "a metabolism prescribed by the natural laws of life itself" (Foster and Clark 2016).

The analysis of value is central to the project of metabolic theory. It follows closely Marx's value theory, which it describes as an "ecological" one. According to that theory, true wealth consists of use values, which have a "natural-material basis" and are derived from both human labour and nature, whereas capitalist value is based entirely on the exploitation of abstract social labour and has no such natural basis (Foster, Clark, and York. 2010, 63). The relationship between these forms of value and intrinsic value is not explored.

Foster finds in Marx and Engels "a *materialist ontology of emergence* – one that encompasses human beings themselves" (Foster 2000, 233). He, much like Bookchin, believes such an ontology to be the alternative to both teleology and determinism. However, the philosophical details of this ontology have yet to be developed, other than through references to the thought of the ancient atomist philosopher Epicurus.

Moore's world-ecology

Jason Moore, in *Capitalism in the Web of Life* and other extensive publications, carries out an explicit and consistent project of formulating a dialectical ecology. He does this through his interpretation of capitalism as what he calls a "world-ecology" and through key concepts such as "the web of life" and "the *oikeios*."

Moore begins his analysis by establishing his non-dualistic position through the concept of a "double internality" according to which we replace the dualistic conception of "humanity and nature" with the idea of a dialectically constituted whole of "humanity-in-nature/nature-in-humanity" (Moore 2015, 5). The web of life is this "nature as a whole": that is, "nature as us, as inside us, as around us. It is nature as a flow of flows" (Moore 2015, 2–3).

The complementary concept of capitalism being a "world-ecology" means that it synthesizes "the accumulation of capital, the pursuit of power, and the co-production of nature in dialectical unity" (Moore 2015, 4).

Moore's fundamental concept of the *oikeios* is also a dialectical one, in that it is defined in relational rather than objectifying or substantialist terms. The *oikeios* is described as "the creative, generative, and multi-layered relation of species and environment," as "the relation through which humans act – and are acted upon by the whole of nature" (Moore 2015, 4), and as "the creative, historical, and dialectical relation between, and also always within, human and extra-human natures" (Moore 2015, 35).

Moore claims that through such a framework, problems that are usually conceptualized as resource problems are analyzed primarily as "relational problems" and only secondarily as "object problems." Through such an approach, dialectical interaction can be seen on many levels. Thus, on the most general level, "the agency of nature is recognized," and, more specifically, "food, water, and oil become real historical actors" (Moore 2015, 36).

Moore's term "bundling," for the conjoining of natural and social phenomena into relative wholes, does not, perhaps, convey very well its dialectical dimensions. Indeed, he recognizes that it "inadequately grasps . . . the intimacy, porosity, and permeability of humans and human organizations within the web of life" (Moore 2015, 7). Nevertheless, the concept shifts analysis in a distinctly dialectical direction by "charting the emergence of definite historical relations through the *oikeios* that bring together (bundle) definite human and extra-human activities and movements" (Moore 2015, 46).

Moore rejects Foster's metabolic theory as inadequately dialectical, contending that its conception of metabolism "has been cleansed of its double internality" (Moore 2015, 75). He argues that despite its abstract affirmation of dialectical concepts, its analysis lapses into "a practical-analytical acceptance of the Nature/Society dualism" (Moore 2015, 75).

In his own analysis of capitalism, Moore shows that exploitation and appropriation are dialectically related and that the two moments must be unified theoretically. Thus, one "cannot think about the accumulation of capital without abstract social labor and the struggle to reduce socially necessary labor-time," and neither can one conceive of "the accumulation of capital without the symbolic praxis of abstract social nature, allowing for the appropriation of unpaid work on a scale that dwarfs the exploitation of labor-power" (Moore 2015, 302).

In short, both the rate of exploitation and the degree of ecological degradation are dependent on "the fruits of appropriation" in which nature is looked upon as a free gift. Moore conceptualizes the resources appropriated in this manner as what he calls the " 'Four Cheaps' of labor-power, food, energy, and raw materials" (Moore 2015, 17). He analyzes in some detail the vicissitudes of the inseparable social and natural history of their appropriation.

Moore also places both exploitation and appropriation within a larger dialectical context. He interprets the historical development of the law of value as the result of a dialectic between these economic factors, and other social processes, such as "the proliferation of knowledges and symbolic regimes that constructed nature as external, space as flat and geometrical, and time as linear (the field of abstract social nature)" (Moore 2015, 191).

In some ways, Moore's analysis of value echoes an enduring ecofeminist critique. Value, he says, "has been premised on valuing some nature (e.g., wage-labor) and not-valuing most nature ("women, nature, colonies")," and this value dualism has correlated with an ontological dualism that entailed "a powerfully alienating conception of Nature as external" (Moore 2015, 86). It is at this point that the implications of the analysis for political practice become most evident.

Salleh's dialectical ecofeminism

It is these implications that are directly at the centre of Ariel Salleh's dialectical ecofeminism. Salleh's classic *Ecofeminism as Politics* is perhaps the most significant contemporary work in ecofeminist theory. In this and other works, Salleh makes a major contribution to the development of a radically critical and dialectical ecology.

Central to her project is a sustained critique of the hierarchical dualism and ideologies of domination that divide all beings into those that have standing and value and those that have none (Salleh 2017, 62). On this basis, she reveals not only the fundamental contradictions and oppressions of the dominant system but also the inadequacies of most Left ideologies in their attempts to break with it.

While Salleh lists many points at which the Marxian "dialectic of internal relations" intersects with ecofeminism, she is also incisively critical of Marx and some of his latter-day apologists.

She points out Marx's anthropocentric faith in "the degree to which men can interfere with and control complex, partially understood phenomena" and his "androcentric" depiction of nature as feminine (Salleh 2017, 112).

While some contemporary eco-Marxists argue for the innocuousness of Marx's ideas concerning the mastery of nature, Salleh urges us to recognize the ways in which his theory of value "places women's reproductive and restorative activities on the unproductive resource side of the equation" (Salleh 2017, 117). Salleh, like other feminist theorists such as Dinnerstein and Chodorow, argues that the entire historical problematic of "mastery" has reflected a disordered "reaction to 'dependency' on the originary M/Other" (Salleh 2017, 117) and has deeply infected social ideology and practice.

Salleh suggests that in considering the human metabolism with nature, rather than limiting ourselves to material exchanges, we should look at the "more dialectical logic" of "reciprocity and nurture" that is reflected in "the sensuous praxis of women workers, from Third World subsistence farmers to urban mothers" (Salleh 2017, 129).

She argues that the traditional consignment of women to a social, moral, and, indeed, ontological place "between men and nature" is the fundamental contradiction not only of capitalism but also of world history since the beginnings of patriarchal domination. This ideology of domination has been accompanied by a social practice in which women's bodies have been imagined and controlled as if they were a natural resource. Thus, "the uterus as organ of birthing labor" is "the material origin of 'formal labor' as such" (Salleh 2017, 143).

As a materialist feminist, Salleh holds that social transformation will not be effected merely through abstract critique but demands embodied practice. Society, and men in particular, must open themselves to the perspective of ordinary women and "to share holding labors" (Salleh 2017, 248). Social transformation that moves caring labour to the centre of culture, politics, and economics is "more emancipatory" than reforms that merely integrate the oppressed into "gender-dysfunctional and unsustainable" institutions (Salleh 2017, 248).

Salleh's position is deeply dialectical in its approach to the question of perspective and the constitution of subjectivity. "A voice is privileged," she argues (quoting J. Cheney), "to the extent that it is constructed from a position that enables it to spot distortions, mystifications, and colonizing and totalizing tendencies within other discourses" (Salleh 2017, 260–261). Thus, it has the concrete universality sought by dialectic. Rather than making "an abstract claim to universal objectivity; it is relational, telling the world from a particular sense of place at a particular time in history" (Salleh 2017, 261).

Engel-Di Mauro's dialectical eco-critique

Another of the most incisive critiques of the ways in which current Left ecologies are either inadequately ecological or dialectical has been presented by Salvatore Engel-Di Mauro. He points out that most Left positions neglect "change in environmental processes" and thus "cannot distinguish autonomous from human-induced environmental changes and instead appeal to unsupported natural equilibrium arguments" (Engel-Di Mauro 2014, 135).

According to Engel-Di Mauro, Foster's analysis, by focusing on metabolic rift, reduces Marx to a "a shallow non-dialectical thinker," showing that he "studied neither the relationship between people and environment, nor any biophysical process" (Engel-Di Mauro 2014, 136). Moreover, when the concept of metabolism is conceptualized as "material exchange between society and environment," it is inadequately dialectical or ecological since it "systematically excludes the importance of material exchanges not involving people" (Engel-Di Mauro 2014, 141).

Engel-Di Mauro also claims that there are certain inadequacies in Moore's analysis. He argues that despite Moore's intended focus on the entire "web of life," he sometimes presents capital as "the sole protagonist" (Engel-Di Mauro 2014, 146). He finds this shortcoming in Moore's commodity frontier thesis, in which, Engel-Di Mauro claims, "the social (the part)" is prioritized over "the ecological (the whole)"; consequently, "social struggles and nonhuman processes" become "mere residuals of capitalists' active shaping of the world" (Engel-Di Mauro 2014, 146).

One of the many distinctive contributions of Engel-Di Mauro's analysis is its recognition of the role of Indigenous and other traditional peoples in geohistory. He points out the fact that soil scientists have ignored Indigenous soil classification systems and Indigenous soil knowledge in general (Engel-Di Mauro 2014, 16). Furthermore, Indigenous peoples have been treated as "homogenous and historically static," and their impacts on soils have been ignored (Engel-Di Mauro 2014, 121), including their "conservation techniques and technologies" (Engel-Di Mauro 2014, 128).

In short, Engel-Di Mauro promotes a more authentically dialectical ecology through a critique of residual dualism in putatively Left positions that fail to recognize adequately the agency of the other within the geohistorical dialectic, whether this other be women, Indigenous peoples, peasants, other species, the soil, or the earth.

The future of dialectical ecology

Dialectical ecology has a crucial role to play in world history and the crisis of the earth. To carry out this role, it must engage in a serious process of self-critique in which it roots out residual dualism and attachment to ideology and dogma, corrects failures to apply consistently the dialectical principles that it espouses, maintains a central focus on geohistory and the ecological crisis, and, above all, works to make dialectical ecological theory a direct force for transformative practice.

To achieve this, it must encompass dialectical eco-psychological or eco-therapeutic practice, in order to help facilitate the self-transformation of subjects of domination into eco-communal beings. It must undertake dialectical social-ecological and cultural-ecological practice to help pacified and alienated mass society transform itself into a global community of liberatory eco-communities. And, finally, facing the precondition for all conditions, it must play an active role in dialectical revolutionary practice aimed at world-historical revolution to change the genocidal and ecocidal course of world history.

The dialectical ecofeminist standpoint analysis discussed by Salleh does most to suggest the direction needed. This analysis addresses the position of the subject (including the collective subject) as a thinking and caring being that is at once within and outside the system of domination. It points out the creative and revolutionary potentialities that this position generates. Such an analysis suggests a form of engaged ecological dialectic in which the emerging revolutionary subject is the person-community-world in a process of liberatory self-transformation.

References

Bookchin, Murray. 1996. *The Philosophy of Social Ecology: Essays on Dialectical Naturalism*. Montréal: Black Rose Books.

Engel-Di Mauro, Salvatore. 2014. *Ecology, Soils, and the Left: An Eco-Social Approach*. New York: Palgrave Macmillan.

Engels, Fredrich. 1934. *Dialectics of Nature*. Moscow: Progress Publishers.

Foster, John Bellamy. 2000. *Marx's Ecology: Materialism and Nature*. New York: Monthly Review Press.

Foster, John Bellamy, and Brett Clark. 2016. "Marxism and the Dialectics of Ecology", *Monthly Review* 68: 1–17. https://monthlyreview.org/2016/10/01/marxism-and-the-dialectics-of-ecology/
Foster, John Bellamy, Brett Clark, and Richard York. 2010. *The Ecological Rift: Capitalism's War on the Earth.* New York: Monthly Review Press.
Hegel, Georg Wilhelm. 1955. *Lectures on the History of Philosophy*. Vol. 1. New York: Humanities Press.
Kovel, Joel. 2019. *The Emergence of Ecosocialism: Collected Essays by Joel Kovel.* New York: Leaf Press.
Levins, Richard, and Richard Lewontin. 1985. *The Dialectical Biologist.* Cambridge, MA: Harvard University Press.
Marx, Karl. 1959. *Capital: A Critique of Political Economy.* Vol. II. Moscow: Progress Publishers.
Marx, Karl. 1964. *Early Writings.* New York: McGraw-Hill.
Marx, Karl. n.d. [1867]. *Capital: A Critique of Political Economy.* Vol. I. Moscow: Progress Publishers.
Moore, Jason. 2015. *Capitalism in the Web of Life.* London: Verso.
O'Connor, James. 1998. *Natural Causes: Essays in Ecological Marxism.* New York: Guilford Press.
Salleh, Ariel. 2017. *Ecofeminism as Politics: Nature, Marx and the Postmodern.* London: Zed.

13

THE LEGACY OF KARL MARX'S ECOSOCIALISM IN THE TWENTY-FIRST CENTURY

Kohei Saito

Introduction

The "Anthropocene" denotes a new geological epoch in which humankind becomes a "major geological force" (Crutzen and Stoermer 2000, 18). Obviously enough, this unprecedented ontological situation in the history of our planet cannot be separated from the development of capitalism after WWII, which led some to employ the term "Capitalocene" (Moore 2016). After the collapse of the USSR, with neoliberal global capitalism triumphant, some declared the "end of history" (Fukuyama 1992). Finding new markets and cheap labour forces in the East, capital found a great opportunity for "buying time" (Streeck 2017) and continued to expand as if there were no limits. However, the earth is obviously finite. Capital seriously undermines the material foundation for freedom and justice by causing global ecological crises such as climate breakdown, mass extinction, and soil erosion. It is about to realize "end of (human) history."

Capital confronts a serious dilemma. On the one hand, it obviously needs to take an unprecedented scale of countermeasures against climate change. Otherwise, it will suffer from a legitimation crisis in front of the deepening ecological crisis and the corresponding degradation of people's living standards. On the other hand, such countermeasures demand a great transformation of the current system to depart from the endless cycle of mass production and consumption for the sake of eternal economic growth. Countermeasures require planning of social production, regulation of the market, and taxing of the rich, as well as abandoning neoliberal policies.

Some continue to believe that green capitalism and green growth are possible (Pollin 2018), but there is no empirical evidence of absolute decoupling (Hickel and Kallis 2020). What is left for capital is to accelerate the further "ecomodernist" or "geoconstructivist" intervention in the earth system to terraform the planet (Neyrad 2019), which is technologically uncertain and ethically questionable. In the fact of capital's dilemma, it becomes increasingly clear that a needed radical solution cannot come from inside the capitalist system; it must come from outside. A clear radicalization of environmental movements confirms this point. This is also why radial ideas and theories are more necessary than ever. In this context, Naomi Klein, though not a Marxist, has radicalized her critique of neoliberalism and argued for "ecosocialism" in *On Fire*:

> Let's acknowledge this fact [that actually existing socialism caused serious environment degradation], while also pointing out that countries with a strong democratic socialist

 10.4324/9780429341427-16

> tradition – like Denmark, Sweden, and Uruguay – have some of the most visionary environmental policies in the world. From this we can conclude that socialism isn't necessarily ecological, but that a new form of democratic eco-socialism, with the humility to learn from Indigenous teachings about the duties to future generations and the interconnection of all of life, appears to be humanity's best shot at collective survival.
>
> *(Klein 2019, 264)*

Behind Klein's endorsement of ecosocialism there exists a deepening of ecosocialist theory and practice in the last 20 years, to which the Marxist legacy of socialism makes an important contribution. This recent revival of socialist ideas 30 years after the collapse of really existing socialism is fueled by the "rediscovery" of Marx's ecology in the twenty-first century, which pivots around the concepts of "metabolic rift" (Burkett 1999; Foster 2000) and the "second contradiction of capitalism" (O'Connor 1998). To avoid the "end of history," firmly identifying this Marxist tradition is essential. It rehabilitates our imagination in order to envision a more sustainable post-capitalist society amid the global ecological crisis of the Anthropocene.

Ecosocialism and Marxism

Simply put, "ecosocialism" is the fusion of "red" (socialist ideas and working-class movement) and "green" (environmental thought and movement). This combination may sound easy to achieve, but their relationship has been antagonistic for a long time. Environmentalism, on the one hand, must recognize the limitation of green capitalism and start advocating for a much more radical transformation of the market system (Empson 2019). On the other hand, socialist movements must reject productivism, which seeks domination over nature for the sake of working-class affluence, and put more emphasis on prosperity within given natural limits.

Especially with regard to the latter, there have been copious critiques of Marx's "Prometheanism" (Giddens 1981, 60), which endorses the idea of maximization of productive forces as a necessary precondition for human emancipation. Here, Marx was understood in the modernist tradition since Bacon, who rejects the recognition of any natural limits on the free manipulation of nature. Marx's alleged naïve endorsement of Prometheanism was thus criticized for sharing "the Biblico-Cartesian ideology of the conquest of nature" (Lipietz 2000, 75). This kind of critique was strengthened even more due to serious environmental degradation in the USSR (Cole 1993, 36).

Harsh criticisms from the greens and the collapse of the USSR compelled Marxism to fundamentally rethink its earlier productivist attitude. Responding to the criticisms from environmentalists, Michael Löwy declared: "Marx does not possess an integrated ecological perspective. His optimistic, 'promethean' conception of the limitless development of the productive forces once the limits of capitalist relations of production are removed is today indefensible" (Löwy 1997). Ted Benton (1989) famously attempted to abandon Prometheanism in Marxism and fuse the red and the green. His aim is noteworthy. He intended to bring elements of the green into Marxism which do *not* exist in Marx's own thought. According to Benton, Marx was afraid of falling into Malthusianism by recognizing the limits of nature, so he ended up putting forward productivist claims about overcoming the finitude of natural resources. Admitting the validity of criticism from environmental thought, Benton sought a "greening Marxism" (Benton 1996) by paying attention to the uniqueness of the "eco-regulatory" labour processes.

This kind of approach became influential as Alain Lipietz (2000) and Hans Immler (2011) joined in criticism of Marx. Notably, they emphasized the limitation of Marxian theory to develop an ecological critique of capitalism in a systematic manner, even proclaiming: "Forget Marx!" (Immler 2011, 11). This attitude revealed their real intention. Benton, Lipietz, and Immler, despite their pretense to succeed Marx's critique of political economy by greening Marxism, actually regard Marx as a hinderance or obstacle to their attempt to subsume the labour movement under the green movement. However, its price was high. After abandoning Marx's theory of value, reification, and class, the content that a green movement can integrate from the legacy of Marxism turns out to be quite limited. In other words, the difficulty of fusing ecological and socialist projects weakens the content of "ecosocialism" as a result of excluding the central concepts of Marx's critique of capitalism.

Unfortunately, such attitudes towards Marx continue today among critical theorists. Axel Honneth (2017, 45) criticizes the limitation of Marxism in that one of the ideas inherent in Marxism is a "technological determinism" that supposes the linear progress of productive forces for the sake of "domination over nature" (*Naturbeherrschung*). Nancy Fraser (2014, 56) also joins in this: "Yet [Marx's thought] fails to reckon systematically with gender, ecology and political power as structuring principles and axes of inequality in capitalist societies – let alone as stakes and premises of social struggle." Sven-Eric Liedman also points to the "danger of presenting Marx as a champion of the environment, or at least an ecologically conscious person in the modern sense" because "[t]his was not the case" (Liedman 2018, 480).

The rediscovery of Marx's ecology

Fortunately, this is not the whole story. There were various Marxists who critically succeeded in bringing out Marx's legacy of developing an analysis of environmental destruction under capitalism and envisioning an alternative for a more sustainable society long before 1990. Karl W. Kapp (1963), Barry Commoner (1971), and István Mészáros (1972) had already attempted to integrate Marxist insights into an ecological critique of capitalism. The situation was very different in Japan, too. Even before 1990, leading economists such as Kenichi Miyamoto (1967) and Shigeto Tsuru (1976) developed critical analyses of the relationship between capitalism and industrial pollution based on Marx's *Capital*. It is not too much to say that Japanese environmental economics did not emerge without contribution from Marxian economics.

Although these theoretical contributions were marginalized under the hegemony of neoliberal globalization, some of this ecosocialist tradition was "rediscovered" after the collapse of the USSR. István Mészáros (1995) revitalized the critique of the destructive tendency of capitalist production based on Marx's concept of "metabolism." James O'Connor (1998), Joel Kovel (2007), André Gorz (2018), and Michael Löwy (2015) made important contributions. More significant contributions came from John Bellamy Foster (2000) and Paul Burkett (1999). Unlike Benton, Immler, and Lipietz, Foster and Burkett carefully analyzed Marx's and Engels' texts to demonstrate in a convincing manner that an ecological critique of capitalism exists in Marx's own theory, and it is possible to integrate it into his critique of political economy.

Obviously enough, Marx did not anticipate everything, including climate change (Engel-Di Mauro 2014; Tanuro 2003), and the imposition of a Godlike omnipotence on Marx would surely be wrong (Kovel 2007, 232). Yet Foster and Burkett's point is rather that Marx's analysis, which pivots around the concept of "metabolic rift," provides a "methodological foundation" for criticizing today's ecological crisis based on his critical comprehension of the logic of capital accumulation (Foster and Burkett 2016, 8). In other words, precisely by reappropriating and extending Marx's critique of political economy instead of rejecting it, it is possible to grasp today's global ecological crisis as a manifestation of the contradiction of capitalism. Such a critique is essential for revealing the social and material conditions for establishing a more sustainable society beyond capitalism.

In this way, Foster and Burkett elaborate on the concept of "metabolic rift" as the central concept for Marxist ecology. Justus von Liebig, a German chemist, plays a central role in this. After reading the seventh edition of Liebig's *Agricultural Chemistry* (1862), Marx harshly criticized the irrationality of modern agriculture as a robbery system. For the sake of short-term profit, capitalist agriculture violates the natural law of the soil by taking as much soil nutrients as possible into crops without returning them to the soil after harvest. This ultimately leads to soil exhaustion by disturbing the metabolic cycle between humans and nature, as stated in *Capital* volume I:

> [Capitalist production] disturbs the metabolic interaction between man and the earth, i.e. it prevents the return to the soil of its constituent elements consumed by man in the form of food and clothing; hence it hinders the operation of the eternal natural condition for the lasting fertility of the soil. Thus it destroys at the same time the physical health of the urban worker, and the intellectual life of the rural worker.
>
> *(Marx 1976, 637)*

Furthermore, Marx pointed to the "irreparable rift" of social and natural metabolism on a global scale in *Capital* volume III:

> On the other hand, large landed property reduces the agricultural population to an ever decreasing minimum and confronts it with an ever growing industrial population crammed together in large towns; in this way it produces conditions that provoke an irreparable rift in the interdependent process between social metabolism and natural metabolism prescribed by the natural laws of the soil. The result of this is a squandering of the vitality of the soil, and *trade carries this devastation far beyond the bounds of a single country* (Liebig).
>
> *(Marx 1992, 752–753)*

These became key passages for the "metabolic rift" approach (Foster, York, and Clark 2011). Thanks to this formulation by Foster and Burkett, while many fields of Marxian studies suffer from decline, Marxist ecology has gained more influence since the collapse of the USSR and with the deepening of the global ecological crisis, as attested in various sociological studies on marine ecology (Longo, Clausen, and Clark 2015), agribusiness (Gunderson 2011), nitrogen cycle disruption (Mancus 2007), and climate change (Klein 2014; Weston 2014).

Again, Liebig's theoretical contribution in *Capital* will not be new to Japanese readers. Shigeaki Shiina (1976), Fumikazu Yoshida (1980), and Masami Fukutomi (1989) have already carefully analyzed the intellectual relationship between Marx and Liebig. This brought about an unfortunate result because the recent revival of Marxian ecology in the Anglophone world was largely underestimated in Japan. Combined with the rapid decline of Marxism in Japan after the 1990s, the theory of metabolic rift was not introduced until quite recently (Saito 2019a).

O'Connor's "second contradiction of capitalism" as well as Foster's "metabolic rift" intend to analyze the historical dynamics of capital accumulation and the ongoing degradation of the natural environment on a global scale. This does not mean that capitalism will simply collapse one day. Through the constant process of subsuming nature under capital (Boyd, Prudham, and Schurman 2001) and shifting the ecological contradiction through science and technology (York and Clark 2010), there is actually no compelling reason to believe so. Capital's attempt to ruthlessly appropriate cheap nature as well as to terraform the entire planet – fracking and geoengineering being two examples – allows capital to continue its self-valorization, though accompanied by more serious ecological disasters. Capital can profit even from natural degradation by finding new opportunities for investments in such disasters, too (Burkett 2006, 136). As far as the logic of capital's accumulation being estranged from human life and ecological sustainability, the capitalist system might continue to exist, even if all planetary boundaries are fully exceeded, and most of the earth becomes unsuitable for living beings. At the same time, capital "shifts" its negative consequences mainly to the global South. By shifting rifts to the global South, the shifts become largely "invisible" to people in the global North. The "imperial mode of living" (Brand and Wissen 2021) and the "externalization society" (Lessenich 2019) are necessary conditions for affluence in the global North but at the cost of others somewhere else.

MEGA and Marx's ecology

However, there are still a series of works that do not fully recognize the possibility of developing an ecological critique of capitalism following Marx's *Capital*. Those Marxists believe that it is not possible to systematically elaborate on the Marxist critique of capitalism by gathering sporadic remarks of Marx and Engels. Daniel Bensaid bemoans that "It is a matter for regret that Marx did not extend this understanding to raw materials, energy and the environment" (Bensaïd 2009, 317). In this vein, some criticize Foster for overemphasizing the systematic character of Marx's ecology and its contemporary scientific validity as he "extrapolates the ecological in Marx from brief and vague excursions in texts addressing subjects other than ecological dynamics" (Engel-Di Mauro 2014, 137). Ecology must rest on what actual ecologists and environmental scientists have been finding out since Marx's time, not on what Marx found out about the environment through the scientists of his age.

The publication of the new complete works of Marx and Engels, *Marx-Engels-Gesamtausgabe* (MEGA), changes this situation. In the fourth section of the MEGA, Marx's notebooks are published for the first time. They are important considering the fact that *Capital* is an unfinished work. Marx strove very hard in his late years to study the natural sciences while preparing Volumes II and III of *Capital*. The topics are quite diverse, ranging from agricultural chemistry and botany to mineralogy and geology, and the amount of writing in the notebooks is enormous. Nevertheless, he was not able to integrate his new findings into *Capital*, and his ecological insights remained in his personal notebooks (Saito 2017).

In other words, while Marx's ecological critique of capitalism in his *Capital* and other published writings may look unsystematic and sporadic, his notebooks document that he was aware of this theoretical hole and attempted to elaborate more on ecological contradiction of capitalism based on his critique of political economy. These notebooks document that Marx did not absolutize Liebig's agricultural chemistry but constantly updated his knowledge, inspired by the rapid development of natural science in the late nineteenth century. In this way, the unfinished character of Marx's political economy makes ecosocialism much more open to recent findings of ecological science.

Unfortunately, even serious scholars ignored the existence of Marx's notebooks because they remained unpublished for more than a century after his death. One reason for this neglect was that "traditional Marxism" in the USSR treated Marx's historical materialism as a closed dialectical system. The problem here is that the more they emphasized the closedness of the system in *Capital*, the harder it became for traditional Marxists to pay attention to Marx's economic manuscripts and notebooks. This is because these documents confirm the incomplete character of *Capital*.[1] In fact, Engels, when he attempted to establish Marxism as a theoretical foundation for a social and political movement, did not mention the existence of numerous notebooks, even though he knew about them (Saito 2019b). As a consequence, the following generations of Marxists simply took for granted that Marx had little to say about nature, but Engels was an expert on the developments in natural sciences, as expressed in *Dialectics of Nature* and *Anti-Dühring*.

On the other hand, there are Marxists who rejected a view of dialectics as all encompassing. "Western Marxism" was especially determined to save Marx's philosophy by blaming Engels for establishing such a problematic worldview of traditional Marxism. Lukács famously argued in *History and Class Consciousness*:

> It is of the first importance to realize that the method is limited here to the realms of history and society [*historisch-soziale Wirklichkeit*]. The misunderstandings that arise from Engels' account of dialectics can in the main be put down to the fact that Engels – following Hegel's mistaken lead – extended the method to apply also to knowledge of nature. However, the crucial determinants of dialectics – the interaction of subject and object, the unity of theory and practice, the historical changes in the reality underlying the categories as the root cause of changes in thought, etc. – are absent from our knowledge of nature.
>
> *(Lukács 1971, 24)*

According to this view, it was Engels's mistake to expand dialectics to nature, even though it was supposed to be a method of social analysis. This misapplication is the cause of economic determinism and mechanistic positivism. This was an endeavour to save Marx, but the price was high. Western Marxism ended up excluding nature and natural science from its analysis, so Western Marxism was unable to treat ecological issues. In short, Marx's engagement with natural science was largely neglected by both traditional Marxism and Western Marxism.

However, this situation has clearly changed with the publication of Marx's economic manuscripts and notebooks in the MEGA. It is no longer justifiable to exclude the sphere of nature from Marx's critique of political economy. Especially in his late years, Marx clearly aimed at integrating concrete description about the environmental destruction under capitalism into his critique of capitalism. Unlike Liedman's view, Marx was an "ecologically conscious person."

There were various discussions not only on soil exhaustion but also on excessive deforestation, maltreatment of animals, and species extinction (Saito 2017). In this vein, various socialists of Marx's time also recognized the importance of the ecological issue in their struggle for social justice, so it was discussed within the International Workingmen's Association (IWA). For example, the resolution of IWA's Brussels Congress in 1868 says:

> Considering that the abandonment of forests to private individuals causes the destruction of woods necessary for the conservation of springs, and, as a matter of course, of the good qualities the soil as well as the health and lives of the population, the Congress thinks that the forests ought to remain the property of society.
>
> *(Marx and Engels 2009, 1955)*

Marx commented on this resolution in the General Council meeting:

> The small peasantry is not at the Congresses but their idealistic representatives are there. The Proudhonists are very strong upon the point & they were at Brussels. The

> Council is not responsible for the resolutions they were shaped by the Brussels committee, by men who well knew the opposition they had to deal with. I am not against recasting them.
>
> (*ibid.* 672)[2]

Marx was not against the socialization of forests and water. Rather, he was criticizing the insufficiency of the resolution of the Brussels Congress. Marx was demanding a more comprehensive social control that would include arable lands, but Proudhonists were trying to protect private land ownership from the standpoint of small-scale peasants.

Marx's more comprehensive view was underlain by his theory on metabolism. Through that theory, he aimed to analyze how the world is materially transformed through the formal and real subsumption of nature under capital. The material world is elastic, so it is modifiable in favor of capital accumulation. However, its elasticity is not infinite (Akashi 2016). The discrepancy between nature and capital expands and ultimately results in the qualitative degradation of the material conditions of production. Marx was trying to analyze its effects on capital accumulation in relation to the turnover of capital and profit rates.

By paying attention to the notebooks, it becomes clear that Marx was becoming increasingly critical of environmental destruction under capitalism as he deepened his critique of political economy. Critiques of Marx overgeneralize his apparently Promethean remarks in the *Communist Manifesto*. In his late years, it is not possible to find overtly optimistic remarks about the progressive tendency of capitalist development. It is important to recognize this change in Marx's ecological consciousness (Löwy 2017, 11). According to the late Marx, the development of productive forces under capitalism are organized as the "productive forces of capital," which are in favor of the valorization of capital. Such a development of the productive forces does not realize the emancipation of workers from labour and does not make for a sustainable production system in society. In contrast, the "productive forces of capital" dominate humans and aid in carrying out a much more systematized robbery from nature. Thus, the development of productive forces must include the dimension of sustainability. If it is not sustainable, the increase of productive forces is mere *robbery*. Productive forces basically express a human ability to consciously regulate metabolism with nature in a sustainable manner.[3]

As Marx came to abandon a productivist view of history and to recognize natural limits, he started to emphasize their incompatibility with the limitless desire of capital for its valorization. If red seriously reflects upon the flaws of its earlier productivism, ecosocialism requires that prosperity be achieved within natural limits, which indicates the need for more active engagements with proponents of steady-state economy as well as degrowth. To overcome the alienated relationship with nature is a central task for both red and green. In this context, Marx's critique of the metabolic rift provides a methodological foundation for a critical analysis of the current global ecological crisis; it is our task today to substantiate and update Marx's idea of ecosocialism for the twenty-first century by developing a synthetic analysis of political economy and natural science.

Acknowledgements

This work was supported by JSPS Kakenhi Grant Number 20K13466 as well as by the Ministry of Education of the Republic of Korea and the National Research Foundation of Korea (NRF-2018S1A3A2075204).

Notes

1 Of course, various manuscripts were published even under Stalinism. "Frühschriften" of 1844 were published in the MEGA (1932), but they disappeared from the official Marx corpus thereafter. The *Grundrisse* was also published in the 1930s, at the height of the era of Stalin. Unfortunately, the first director of the Marx-Engels Institute and the MEGA, David Riazanov, was purged and murdered, while Paul Weller, after editing the *Grundrisse*, was sent to the war and died during the battles of the Eastern Front.
2 The sentence used to be misprinted as "The Council *is* responsible for the resolutions," which obscured Marx's intention.
3 This concept of "productive forces of capital" is essential in order to avoid Prometheanism. Today, there is increasing influence in the "stewardship" of the earth by accelerating the capitalist tendency of technological development. For instance, Aaron Bastani (2019) argues that Marx was a "fully automated luxury communist." Bastani neglects the whole critical discussion on Prometheanism over the last 20 years and falls back again to the naïve endorsement of the capitalist development of technology without paying sufficient attention to its destructive side.

References

Akashi, Hideto. 2016. "The Elasticity of Capital and Ecological Crisis." *Marx – Engels Jahrbuch* 45–58.

Bastani, Aaron. 2019. *Fully Automated Luxury Communism: A Manifesto*. London: Verso.

Benton, Ted. 1989. "Marxism and Natural Limits." *New Left Review* 178: 51–86.

Bensaïd, Daniel. 2009. *Marx for Our Times: Adventures and Misadventures of a Critique*. London: Verso.

Benton, Ted, ed. 1996. *The Greening of Marxism*. New York: Guilford Press.

Boyd, William, W. Scott Prudham, and Rachel A. Schurman. 2001. "Industrial Dynamics and the Problem of Nature." *Society and Natural Resources* 14: 555–570.

Brand, Ulrich, and Markus Wissen. 2021. *The Imperial Mode of Living: Everyday Life and the Ecological Crisis of Capitalism*. London: Verso.

Burkett, Paul. 1999. *Marx and Nature: A Red and Green Perspective*. New York: Palgrave Macmillan.

Burkett, Paul. 2006. *Marxism and Ecological Economics: Toward a Red and Green Political Economy*. Leiden: Brill.

Cole, Daniel H. 1993. "Marxism and the Failure of Environmental Protection in Eastern Europe and the U.S.S.R." *Legal Studies Forum* XVII (1): 35–72.

Commoner, Barry. 1971. *The Closing Circle*. New York: Knopf.

Crutzen, Paul, and Eugene Stoermer. 2000. "The 'Anthropocene'." *Global Change Newsletter* 41: 17–18.

Empson, Martin, ed. 2019. *System Change not Climate Change: A Revolutionary Response to Environmental Crisis*. London: Bookmarks Publications.

Engel-Di Mauro, Salvatore. 2014. *Ecology, Soils, and the Left: An Ecosocial Approach*. New York: Palgrave Macmillan.

Foster, John Bellamy. 2000. *Marx's Ecology: Materialism and Nature*. New York: Monthly Review Press.

Foster, John Bellamy, and Paul Burkett. 2016. *Marx and the Earth: An Anti-Critique*. Leiden: Brill.

Foster, John Bellamy, Richard York, and Brett Clark. 2011. *The Ecological Rift: Capitalism's War on the Earth*. New York: Monthly Review Press.

Fraser, Nancy. 2014. "Behind Marx's Hidden Abode: For an Expanded Conception of Capitalism." *New Left Review* 86: 55–72.

Fukutomi, Masami. 1989. *Keizaigaku to Shizen Tetsugaku* [*Political Economy and Philosophy of Nature*]. Tokyo: Sekaishoin.

Fukuyama, Francis. 1992. *The End of History and the Last Man*. New York: Free Press.

Giddens, Anthony. 1981. *Contemporary Critique of Historical Materialism*. Berkeley: University of California Press.

Gorz, André. 2018. *Ecologica*. Calcutta: Seagull Books.

Gunderson, Ryan. 2011. "The Metabolic Rifts of Livestock Agribusiness." *Organization & Environment* 24 (4): 404–422.

Hickel, Jason, and Giorgos Kallis. 2020. "Is Green Growth Possible?" *New Political Economy* 25 (7576): 1–18.

Honneth, Axel. 2017. *The Idea of Socialism: Towards a Renewal*. Cambridge: Polity Press.

Immler, Hans. 2011. "Vergiss Marx, entdecke Schelling." In *Marx und die Naturfrage. Ein Wissenschaftsstreit*, edited by Hans Immler and Wolfdietrich Schmied-Kowarzik, 9–12. Hamburg: Kassel University Pres.

Kapp, Karl William. 2000 [1963]. *Social Costs of Business Enterprise*. Nottingham: Russell Press.

Klein, Naomi. 2014. *This Changes Everything: Capitalism vs. the Climate*. New York: Simon & Schuster.

Klein, Naomi. 2019. *On Fire: The (Burning) Case for a Green New Deal*. New York: Simon & Schuster.

Kovel, Joel. 2007. *The Enemy of Nature: The End of Capitalism or the End of the World?* 2nd ed. London: Zed Books.

Lessenich, Stephan. 2019. *Living Well at Others' Expense: The Hidden Costs of Western Prosperity*. Cambridge: Polity.

Liedman, Sven-Eric. 2018. *A World to Win*. London: Verso.

Lipietz, Alain. 2000. "Political Ecology and the Future of Marxism." *Capitalism Nature Socialism* 11: 65–85.

Longo, Stefano, Rebecca Clausen, and Brett Clark. 2015. *The Tragedy of the Commodity: Ocean Fisheries, and the Aquaculture*. New Brunswick, NJ: Rutgers University Press.

Löwy, Michael. 1997. "For a Critical Marxism." *Against the Current* 71. https://againstthecurrent.org/atc071/p832/ (accessed 19 March 2021).

Löwy, Michael. 2015. *Ecosocialism: A Radical Alternative to Capitalist Catastrophe*. Chicago: Haymarket Books.

Löwy, Michael. 2017. "Marx, Engels, and Ecology." *Capitalism Nature Socialism* 28 (2): 10–21.

Lukács, Georg. 1971. *History and Class Consciousness: Studies in Marxist Dialectics*. Cambridge MA: The MIT Press.

Mancus, Philip. 2007. "Nitrogen Fertilizer Dependency and Its Contradiction." *Rural Sociology* 72 (2): 269–288.

Marx, Karl. 1976 [1867] *Capital*. Vol. 1. London: Penguin Books.

Marx, Karl. 1992 [1863–1865]. "Das Kapital (ökonomische Manuskript 1863–1865): Drittes Buch." In *Marx-Engels-Gesamtausgabe (MEGA²)*, vol. II/4.2. Berlin: Dietz.

Marx, Karl, and Friedrich Engels. 2009. "Werke, Artikel, Entwürfe September 1867 bis März 1871." In *Marx-Engels-Gesamtausgabe (MEGA²)*, vol. I/21. Berlin: Akademie Verlag.

Mészáros, István. 1972. *The Necessity of Social Control*. London: Merlin Press.

Mészáros, István. 1995. *Beyond Capital*. New York: Monthly Review Press.

Miyamoto, Kenichi. 1967. *Shakaiteki Shihon Ron* [*On Social Capital*]. Tokyo: Yuhikaku.

Moore, Jason W. 2016. *Anthropocene or Capitalocene? Nature, History, and the Crisis of Capitalism*. Oakland: PM Press.

Neyrad, Frédéric. 2019. *The Unconstructable Earth: Ecology of Separation*. New York: Fordham University Press.

O'Connor, James. 1998. *Natural Causes: Essays in Ecological Marxism*. New York: The Guilford Press.

Pollin, Robert. 2018. "De-Growth vs. A Green New Deal." *New Left Review* 112: 5–25.

Saito, Kohei. 2017. *Karl Marx's Ecosocialism: Capital, Nature and the Unfinished Critique of Political Economy*. New York: Monthly Review Press.

Saito, Kohei. 2019a. *Daikouzui no Maeni: Marx to Wakusei no Bussitsutaisha* [*Before the Great Flood: Marx and the Planetary Metabolism*]. Tokyo: Horinouchi Shuppan.

Saito, Kohei. 2019b. "Marx and Engels: The Intellectual Relationship Revisited from an Ecological Perspective." In *Marx's Capital After 150 Years: Critique and Alternative to Capitalism*, edited by Marcello Musto, 167–183. London: Routledge.

Shiina, Shigeaki. 1976. *Nougaku no Shiso: Marx to Liebig* [*Agricultural Theory: Marx and Liebig*]. Tokyo: Tokyo University Press.

Streeck, Wolfgang. 2017. *Buying Time: The Delayed Crisis of Democratic Capitalism*. London: Verso.

Tanuro, Daniel. 2003. *Green Capitalism: Why It Can't Work*. London: Merlin.

Tsuru, Shigeto. 1976. *Towards a New Political Economy*. Tokyo: Kodansha.

Yoshida, Fumiikazu. 1980. *Kankyo to Gijyutu no Keizaigaku: Ningen to Shizen no Busshitsutaisya no Riron* [*Political Economy of Environment and Technology: A Theory of Metabolism Between Humans and Nature*]. Tokyo: Aoki Shoten.

Weston, Del. 2014. *The Political Economy of Global Warming: The Terminal Crisis*. London: Routledge.

York, Richard and Brett Clark. 2010. "Critical Materialism: Science, Technology, and Environmental Sustainability." *Sociological Inquiry* 80 (3): 475–499.

14

ECOSOCIALIST UNDERPINNINGS OF ECOLOGICAL CIVILIZATION

Arran Gare

Introduction

In November 2007, "ecological civilization" was incorporated into the Central Commission Report to the Chinese Communist Party's 17th National Congress and embraced as a central policy objective by the government, and in 2012, the Party included the goal of achieving ecological civilization in its constitution, including this goal within its five-year plan. Then in 2017, the 19th Congress of the Party called for an acceleration of ecological civilization construction. Ecological civilization is associated with the quest for a circular economy "where one facility's waste, including energy, water, materials – as well as information – is another facility's input" (Geall and Ely 2018, 1189). Expenditure on technology to ameliorate environmental damage, reduce pollution, and reduce greenhouse gas emissions has been massively increased, although environmentalists believe far more is required. Hardly surprising is the highly contested nature of the meaning of ecological civilization, given the centrality accorded to ecological civilization in Chinese political culture (Gordon 2018).

Ecological civilization is often characterized as what comes after industrial civilization. This can also be interpreted as dealing with ecological problems generated by industrialization by utilizing market-driven technological solutions, much as in Western capitalist countries. A more radical view is that the centralization of power engendered by capitalism and industrialization needs to be challenged. Ecological civilization requires a more democratic structure, creating institutions to control markets and empower people at local levels. This is the view defended by Pan Jiahua, director of the Institute for Urban and Environmental Studies at the Chinese Academy of Social Sciences (Pan 2016; Marinelli 2018, 380ff.). More broadly, as argued by Zhang Yunfei (2019) of Renmin University, ecological civilization characterizes all civilizations to different degrees, with those societies that fail to achieve a sufficient level of ecological civilization destroying the conditions of their existence. Ecological civilization at present is relatively weak, and it is necessary to recover and advance the lost wisdom of earlier eras. Not necessarily inconsistent with this, it is argued that global ecological civilization now has to be the goal of humanity, to overcome the current global ecological crisis, along with addressing local ecological problems. Since the dynamics of capitalism are seen as the main driving force for ecological destruction on a global scale and for paralysis in efforts to avert such destruction, this more radical view is often, although not always, explicitly linked to the struggle for ecosocialism as

10.4324/9780429341427-17

socialist ecological civilization. This is the view of Pan Yue, the vice minister of China's State Environmental Protection Administration from 2008 to 2015 and now executive vice president of the Central Institute of Socialism Beijing Municipality and alternate member, 19th CPC, Central Committee, who is the leading exponent of ecological civilization at the government level. He has been strongly supported by, among others, Huan Qingzhi from the Research Institute of Marxism, Peking University (Pan 2005; Huan 2016; Gare 2010).

For these ecosocialists, the logic of capital is the prime culprit in ecological destruction. Consequently, Pan Yue argued, "we must use Marxist theoretical weapons to 'fight against any forms of production and lifestyle that deviate from ecological civilization.'" He claimed that "socialism is more likely to provide system motivation and system security for ecological civilization" (Wang 2014, 10). In accordance with this, Lu Feng from Tsinghua University argued that ecological civilization and its practice will negate and transcend modern and urban civilization, being connected to new kinds of economic, social, and cultural institutional frameworks through which people will be able to live more meaningful lives (Huan 2016, 55). In this case, ecological civilization is equated with an advanced form of ecosocialism.

So, while ecological civilization has been strongly linked to ecosocialism, the way the term is used does not involve equating ecological civilization with ecosocialism. To show that ecological civilization is underpinned by and implies ecosocialism, it is necessary to understand the historical background to the development of this concept.

The Russian source of ecological civilization

The Chinese term for ecological civilization was first used by Qianji Ye, an agricultural economist. In 1984 he published an article in the scientific socialism edition of *The Journal of Moscow University*, and in 1987, this was translated in a Chinese newspaper (Huan 2016, 52). The initial term was "ecological culture," and it was translated as "ecological civilization" (*shengtai wenming*), but in Chinese, the words for "culture" (*wenhua*) and "civilization" (*wenming*) are sometimes seen as synonymous, and the way the word "civilization" is used in China corresponds more closely to the way the word "culture" is used in Russia. The notion of "ecological culture" was initially promoted by Marxists in the Soviet Union and widely used from the 1970s onwards, for instance, by Yu. M. Manin (1983) and V.S. Lipitsky (1983). Then a leading government figure, Ivan T. Frolov, along with T.V. Vasileva, V.A. Elk, and others took up the notion of ecological culture (Frolov, Vasileva, and Los 1984), and Vasileva defended a thesis on this topic in the same year.

Frolov was a philosopher of science specializing in biology, a leading figure in the Soviet Union, and an advisor to Mikhail Gorbachev. He later became editor of the main ideological journal in the Soviet Union, *Kommunist*, and then of the main newspaper, *Pravda*. He argued at a conference at the Center for Philosophy and History of Science at Boston University in 1985, just before Gorbachev became general secretary of the CCCP, that confronting the global ecological crisis could and should unite humanity in a common goal, overcoming the Cold War. Detailing the implications of this ecological orientation, he argued that "it would be a mistake to conceive of the biosphere merely as a source of resources or a 'disposer' of wastes" (Weiner 1999, 399). It is equally important to reintegrate both aesthetics and ethical values into our way of relating to the world and into our science, Frolov argued. He called for a switch from anthropocentricism to biospherocentrism. Elsewhere, arguing against sociobiology, according to which social behaviour is determined by genes, the doctrine that revived social Darwinism and legitimated the rise of neoliberalism, Frolov invoked Marx's characterization of humans as ensembles of social relations, arguing that humans are essentially cultural beings (Frolov 1986).

Although the proponents of ecological culture might not have been aware of this, the place they accorded to culture was really the continuation of a tradition of Marxism originating in the 1920s. It was part of a radical form of Marxism promoted by the *Vpered* (Forward) wing of the Bolsheviks, and it included the commissar for enlightenment, Anatoly Lunacharsky, who had been placed in charge of environmental protection by Lenin. Rejecting the crude interpretation of the base-superstructure model of society as technological determinism, the interpretation that led Marx to proclaim that if there was one thing he knew, it was that he was not a Marxist (Engels 1962, 486), and arguing that technology and ideology were different components of culture, these radical Marxists believed that to create a socialist society, it is necessary to create a new culture, including a new form of science, to overcome the deficiencies and distortions of understanding generated by capitalism and to counter the cultural hegemony of the bourgeoisie and their managers. This was endorsed by Lenin in 1918, although he wanted a more practical orientation (White 2019, 392).

The movement for a new socialist culture, *Proletkult*, was inspired originally by Lunacharsky's brother-in-law, Aleksandr Bogdanov (Gare 1994; White 2019, chapter13). In characterizing the base-superstructure model of society in the Preface to *Contribution to the Critique of Political Economy*, Marx had argued that "It is not the consciousness of men that determines their existence, but their social existence that determines their consciousness" (Marx 1970, 19). Bogdanov pointed out that social existence is conscious existence and merged these in the category "culture." Consciousness is involved in the technological component of culture and also in coordinating people, the ideological component of culture. While Marx had written critiques of the bourgeois mode of production, showing how the categories of economics were not eternal but expressions of historically specific forms of being structuring relations between people, except in unpublished works such as the *1844 Manuscripts* and the *Grundrisse* (1973), he had only hinted at the categories required to replace these categories. Developing some observations by Marx, in *The Philosophy of Living Experience* and later *The Science of Social Consciousness* (White 2013), Bogdanov (2016) argued that science is the organized collective experience of humanity and showed how categories articulating relations between people in their organization of production are entrenched and developed by being used as metaphors or "substitutions" for nature, which are then used to interpret and explain society and legitimate its existing social relations, helping reproduce these social relations (Bogdanov 2016, 47ff.; Gare 1994; White 2019, chapter 11).

In making this argument, Bogdanov was not denying the achievements of science, even in a capitalist society where science appeared to be legitimating capitalism. However, he suggested that new advances in science would be stifled under capitalism as they challenged the interests of its ruling class, and to advance science further would involve a struggle to overcome capitalist culture (Gare 2000). The development of *Proletkult* was not merely the basis for the proletariat to unite and act effectively; it was to be a culture that would overcome the cognitive deficiencies of all past cultures while incorporating all that was best within them and would prevail by virtue of the superiority of this culture. This view of *Proletkult* was very different from the top-down imposition of correct views associated with the cultural revolution from 1928 to 1931, which gave rise to Lysenkoism, as Sheila Fitzpatrick pointed out (1978, 10). Influenced by Marx and Engels's *Eleven Theses on Feuerbach*, with its emphasis on praxis and concern to change the world, Bogdanov believed that this new culture was required not only to advance our comprehension of the world but also to provide the concepts through which people could redefine their place in nature and their relations to each other, enabling them to organize themselves to create the future.

This would be a future in which the division between manual and intellectual labour would have been overcome; workers, with the help of these concepts, would be able to manage their own work; and people would understand themselves as part of nature. In other words, humans would overcome their alienation from each other, from nature, from humanity, and from their own creative powers. This would involve overcoming Descartes's dualism and the mechanistic view of nature, expressions of class divisions of capitalist society with a sharp differentiation between the conscious ruling class and workers, who, along with nature, tend to be objectified as things, mere instruments or obstructions in the struggle by the ruling class for greater profits (surplus value). Scientists, or philosopher/scientists, who in their work had already overcome the opposition between intellectual and manual labour and were able to organize themselves, were advancing science accordingly. They were coming to recognize that they were active agents in the world they were striving to understand and beginning to overcome this dualism in their scientific theories. Building on advanced science – most importantly, thermodynamics and relativity theory, in which scientists being part of nature had come to be appreciated – Bogdanov (1984) called for and set out to develop a general theory of organization, *Tektology*, as the basis for an integral worldview. From this perspective, "[t]he entire world consisted of an organising process, an infinitely developing series of complexes of different forms and levels of organization in their mutual relations, in their struggle or their unification" (White 2019, 289). This overcame the opposition between the natural sciences and the human sciences and between science, history, and the arts, while providing people with not only the means to understand their place in nature, society, and history but also the means to organize and govern themselves rather than being organized by managers (Gare 2000). *Tektology* inspired general systems theory and was a precursor to complexity theory.

It was this notion of cultural hegemony and the need to challenge it that had a formative influence on the Italian Marxist Antonio Gramsci. Gramsci became acquainted with these ideas when he lived for two years in Moscow in 1922–1923 and 1925 (Brandist 2012). To counter the way opposition movements tend to mirror the ways of thinking, practices, and organization of those they are opposing, he called not just for a "counter-hegemony" but an alternative cultural hegemony based on a different conception of the world (Ahearn 2013). An alternative hegemonic culture would include developing a superior science: superior not only because it serves the interests of workers, but also because it overcomes the deficiencies of science generated by but then constrained and limited by capitalist social relations.

Proletkult, *tektology*, and ecology

Biology had a major role to play in creating this new culture. Initially, the biology that gained favour in the Soviet Union was anti-vitalist and anti-Idealist, and was essentially positivist and reductionist materialist. However, as Engels's *Dialectics of Nature* began to exert its influence, Soviet biology, along with psychology, became a major centre of what came to be known as the Third Way – neither vitalist nor mechanist: that is, an anti-reductionist naturalism. Ecology with its focus on the inter-relatedness of organisms and its challenge to previous disciplinary boundaries (most importantly, between physics, geology, chemistry, and biology) had an important place in advancing this new science (Gare 1994) and was strongly supported in the 1920s by Lunacharsky, who also supported the work of Vladimir Vernadsky and his concepts of the biosphere and noosphere (Rispoli 2014). Such ideas were very much in accordance with Bogdanov's *tektology* (Gare 1993). In the 1920s research in ecology in the Soviet Union, incorporating thermodynamics along with ideas from Engels, was highly original and more advanced than anywhere else in the world (Weiner 1988, chapter 6).

This did not last. With the triumph of Stalin and the implementation of what Bogdanov had warned against, "war communism" rather than "worker socialism," freedom of enquiry was severely limited. Lunacharsky resigned as commissar for education in 1929 in protest of government interference in education. While many ecologists were persecuted by Stalin and his followers, who were hostile to any claims that nature could not be completely dominated, this movement of radical science was not completely destroyed and was sustained in the Soviet Union in what Weiner characterized in the title of a later book as *A Little Corner of Freedom* (1999).

While neither Bogdanov, Lunacharsky, nor Vernadsky put forward the idea of a global ecological civilization, their work provides the background against which the introduction into China and the prominent place it has gained there can be understood and also the failure to achieve a consensus about what is meant by it. Frolov and other Russians calling for an ecological culture were carrying forward Bogdanov's conception of culture as the forms of consciousness on the basis of which people produce and organize themselves. From this perspective, socialism requires the development of a new culture, overcoming the deficiencies of previous cultures while incorporating all that is best in them. The development of post-reductionist science is central to this development. Bogdanov and Lunacharsky and many Soviet ecologists were socialists or, as with Vernadsky, were liberals sympathetic to socialist ideals (Tolz 1997). Socialist ecologists, notably Vladimir Stanchinsky, saw their work as a challenge to mainstream science. Radical scientists, inspired by *Proletkult*, saw such developments as central to creating genuine socialism and saw genuine socialism involving a new appreciation of nature (Weiner 1988; Gare 1993). Stanchinsky reformulated his already-radical ecological theories to accord with Engels's dialectics of nature and argued that ecologists should play a central role in economic planning, arguing for the importance of conservation, development, and enrichment of the environment (Gare 1996, 266ff.). Ecological civilization is underpinned by this radical socialist tradition within the sciences. As a result, it involves a very fundamental challenge to the culture of capitalism and its legitimacy as a natural form of life, and conversely, it legitimates and maintains the trajectory of movements, institutions, and governments set up to challenge capitalism, instituting socialist forms of life to create a socialist world order.

Because capitalism is dependent on science, it is here that the challenge to the hegemony of capitalist culture can be most effective, and the advance of science is showing that the worldview on which capitalism is based and which legitimates it is being invalidated. Efforts to neutralize this challenge by "managing" science to make it serve the economy is destroying it (Charlton 2012). Climate science and ecology, at least in their radical forms, are now spearheading this challenge to prevailing assumptions. This challenge has the potential to rescue science from its current fragmentation, with Robert Ulanowicz arguing that ecology should become the reference point for defining science, overcoming the conceptual logjams that currently hinder progress in understanding evolutionary phenomena, development biology, the rest of the life sciences, and, conceivably, even physics (Ulanowicz 1997, 6). Although struggling against the corrupting influence of capitalist culture and managerialism supporting traditional reductionist approaches in the sciences and supporting its reductionist proponents, such a science can be developed, it is as likely to reveal the limits to how much nature can be controlled as to show how to control nature, while at the same time facilitating an appreciation of its intrinsic significance. Through the development of the concepts of resilience, it should also provide guidelines for how to diagnose the sickness of modern civilization and how to maintain the health of ecosystems and create healthy societies, replacing economics with human ecology as the core framework for formulating public policies (Gare 2002; Ho and Ulanowicz 2005; Hornborg 2019a).

Once the seeds of radically new ways of thinking have gained a foothold, especially when they are included in narratives defining communities, they can set in motion the system innovations that can totally transform societies and civilizations. Making ecological civilization the official narrative in China might look to some like a public relations exercise. However, having this in place has resuscitated the grand narrative of socialism in its ecosocialism form. As Sam Geall and Adrian Ely argued (2018), this narrative is likely to gain strength and influence pathways to a sustainable social order, both in China and internationally, over the coming years, a view supported by Marinelli (2018, 375ff.). What is emerging is a new, reinvigorated grand narrative of socialism as ecological civilization that can challenge and replace the reductionist, materialist, social Darwinist grand narrative of capitalism, including its current neoliberal variant.

Conclusion

This history of the background to the quest for ecological civilization explains its rise in China and the diversity of its interpretations. To avoid being subjugated, China has had to embrace and assimilate huge chunks of culture from European civilization, and this was undertaken by embracing Marxism. Mobilizing the Chinese to develop the forces of production utilizing Western science and technology facilitated the industrialization of China while enabling the Chinese to maintain a critical distance from European traditions. However, the history of Marxism has been confused and often misunderstood, and in some cases, this has led to an almost uncritical adoption of core components of Western culture, notably its scientism and hidden agenda of achieving total technological control of nature and people (Toulmin 1994; Gare 1996), despite all the problems engendered by this quest. In these cases, ecological civilization can be understood as little different from forms of environmental protection characteristic of Western societies. However, there are still strong Chinese cultural traditions that have survived, traditions that, through the influence of Chinese philosophy on Leibniz and the influence of Leibniz on later philosophers, indirectly influenced the work of Marx and Engels, as Joseph Needham revealed (Gare 1995, 321). This has placed the Chinese in a position to appreciate Marx's concern for nature and the resonances of the notion of ecological culture developed in the Soviet Union with socialism without full knowledge of the ecosocialist roots of this notion. Arguing that ecological civilization has some presence in all societies is a way of recovering and defending superior aspects of past cultures, including Chinese culture, that have been suppressed with commodification, standardization, homogenization, and debasement of reality associated with the advance of capitalism. Defending cultures of the past is not inconsistent with defending socialist ecological civilization and calling for a global ecological civilization, and this is the grand narrative that is now emerging as a global cultural force (Gare 2017). Even without referring to ecosocialism, the unfolding of the core ideas of ecological civilization as it orients people for action will inevitably reveal its ecosocialist roots.

Once the full implications of ecological civilization are understood, it should be clear that there is no need to speak of "socialist" ecological civilization since in the modern world, ecological civilization could not be anything other than socialist. In fact, ecological civilization not only brings into focus the ultimate failure of capitalism and the ultimate reason it must be replaced; it also clarifies what socialism is and what humanity should be striving to create. It can provide the coherence required for an alternative hegemonic culture capable of overcoming the hegemony of capitalist culture (Gare 2017). Civilization has always been defined in opposition to barbarism and decadence, and in late capitalism we are facing a combination of hi-tech barbarism with the decadence of consumerism (Stiegler 2011). Ecology, focusing on the system of "homes" or "households" of organisms, including people – that is, the conditions of their existence – examines how the interaction between these organisms succeeds or fails to provide the conditions under which they can develop in such a way that they can augment these households and thereby the resilience of their immediate biotic communities and broader communities of these communities. Extended by radical forms of human ecology, anti-reductionist forms of ecology are providing the forms of thinking required to rethink economics and the other human sciences, ethics, and politics (Gare 2010, 2017; Hornborg 2019b, 2019c). A socialist society is one in which the self-realization of each individual, each organization, and each community will be aligned to augment the conditions for the self-realization of all individuals, organizations, and communities contributing to augmenting life, including the health of the biotic communities within which humanity has evolved and on which it is dependent, including the current regime of the global ecosystem.

References

Ahearn, Gerard. 2013. "Towards an Ecological Civilization: A Gramscian Strategy for a New Political Subject." *Cosmos and History* 9 (1): 317–326.

Bogdanov, Alexander. 1984. *Essays in Tektology: The General Science of Organization*. Translated by George Gorelik. 2nd ed. Seaside, CA: Intersystems Publications.

Bogdanov, Alexander. 2016. *The Philosophy of Living Experience*. Translated and edited by David G. Rowley. Chicago: Haymarket Books.

Brandist, Craig. 2012. "The Cultural and Linguistic Dimensions of Hegemony: Aspects of Gramsci's Debt to Early Soviet Cultural Policy." *Journal of Romance Studies* 12 (3): 24–43.

Bukharin, Nikolai, ed. 1971. *Science at the Crossroads*. London: Frank Cass.

Charlton, Bruce G. 2012. *Not Even Trying: The Corruption of Real Science*. Buckingham: The University of Buckingham Press.

Engels, Friedrich. 1962. "'Engels to C. Schmidt' London, August 5, 1890." In *Marx Engels: Selected Works*, vol. II, 486. Moscow: Foreign Languages Publishing House.

Fitzpatrick, Sheila. 1978. "Cultural Revolution as Class War." In *Cultural Revolution in Russia, 1928–1931*, edited by Sheila Fitzpatrick, 8–40. Bloomington: Indiana University Press.

Frolov, Ivan T. 1986. "Genes or Culture: A Marxist Perspective on Humankind." *Biology and Philosophy* 1: 89–107.

Frolov, Ivan T., Tatiana V. Vasileva, and Victor A. Los. 1984. *Экологическая пропаганда в СССР* [*Ecological Propaganda in the USSR*]. Moscow: Nauka.

Gare, Arran. 1993. "Soviet Environmentalism: The Path Not Taken." *Capitalism Nature Socialism* 4 (3): 69–88.

Gare, Arran. 1994. "Aleksandr Bogdanov: Proletkul't and Conservation." *Capitalism Nature Socialism* 5 (2): 65–94.

Gare, Arran. 1995. "Understanding Oriental Cultures." *Philosophy East and West* 45 (3): 309–328.

Gare, Arran. 1996. *Nihilism Inc. Environmental Destruction and the Metaphysics of Sustainability*. Como: Eco-Logical Press.

Gare, Arran. 2000. "Aleksandr Bogdanov's History, Sociology and Philosophy of Science." *Studies in the History and Philosophy of Science* 31 (2): 231–248.

Gare, Arran. 2002. "Human Ecology and Public Policy: Overcoming the Hegemony of Economics." *Democracy and Nature* 8 (1): 131–141.

Gare, Arran. 2010. "Toward an Ecological Civilization: The Science, Ethics, and Politics of Eco-Poiesis." *Process Studies* 39 (1): 5–38 [Chinese translation in *Marxism and Reality* 8 (1): 191–202].

Gare, Arran. 2017. *The Philosophical Foundations of Ecology Civilization: A Manifesto for the Future*. London: Routledge.

Geall, Sam, and Adrian Ely. 2018. "Narratives and Pathways Towards an Ecological Civilisation in Contemporary China." *China Quarterly* 236: 1175–1196.

Gordon, Caroline. 2018. "Ecological Civilization and the Political Limits of Chinese Concept of Sustainability." *China Perspectives* 4: 39–52.

Ho, Mae-Wan, and Robert Ulanowicz. 2005. "Sustainable Systems as Organisms?" *Biosystems* 82: 39–51.

Hornborg, Alf. 2019a. *Nature, Society and Justice in the Anthropocene: Unraveling the Money-Energy-Technology Complex*. Cambridge: Cambridge University Press.

Hornborg, Alf. 2019b. "The Money-Energy-Technology Complex and Ecological Marxism: Rethinking the Concept of 'Use-Value' to Extend Our Understanding of Unequal Exchange: Part I." *Capitalism Nature Socialism* 30 (3): 27–39.

Hornborg, Alf. 2019c. "The Money-Energy-Technology Complex and Ecological Marxism: Rethinking the Concept of 'Use-Value' to Extend Our Understanding of Unequal Exchange: Part 2." *Capitalism Nature Socialism* 30 (4): 71–86.

Huan, Qingzhi. 2016. "Socialist Eco-Civilization and Social-Ecological Transformation." *Capitalism Nature Socialism* 27 (2): 51–66.

Lipitsky, Vasilij S. 1983. "Экологическая культура личности и пути ее формирования [Ecological Culture of Personality and Ways of Its Formation]." In *Современная экологическая ситуация и высшая школа: Методологические аспекты* [*Modern Ecological Situation and Higher School: Methodological Aspects*], edited by Georgij V. Platonov, 42–47. Moscow: Moscow State University.

Manin, Yurij. M. 1983. "Экологическая культура и коммунизм [Ecological Culture and Communism]." In *Социальные аспекты экологии* [*Social Aspects of Ecology*], edited by Kazimir P. Buslov, 113–124. Minsk: Nauka i Tehnika.

Marx, Karl. 1970. *A Contribution to the Critique of Political Economy*. Translated by Salomea W. Ryazanskaya. Moscow: Progress Publishers.

Marx, Karl. 1973. *Grundrisse*. Translated by Martin Nicolaus. Harmondsworth: Penguin.

Marinelli, Maurizio. 2018. "How to Build a 'Beautiful China' in the Anthropocene: The Political Discourse and the Intellectual Debate on Ecological Civilization." *Journal of Chinese Political Science* 23: 365–386.

Pan Jiahua. 2016. *China's Environmental Governing and Ecological Civilization*. Heidelberg: Springer.

Pan Yue. 2005. "Interview with China's Deputy Minister of the Environment: 'The Chinese Miracle Will Soon End'." *Spiegel Online International* 10, 7 March.

Rispoli, Giulia. 2014. "Between 'Biosphere' and 'Gaia': Earth as a Living Organism in Soviet Geo-Ecology." *Cosmos and History* 10 (2): 78–91.

Stiegler, Bernard. 2011. *The Decadence of Industrial Democracies*. Translated by Daniel Ross and Suzanne Arnold. Cambridge: Polity.

Tolz, Vera. 1997. *Russian Academicians and the Revolution*. Houndmills: Palgrave Macmillan.

Toulmin, Stephen. 1994. *Cosmopolis: The Hidden Agenda of Modernity*. Chicago: Chicago University Press.

Ulanowicz, Robert E. 1997. *Ecology: The Ascendent Perspective*. New York: Columbia University Press.

Wang, Zhihe, Huila He, and Meijun Fan. 2014. "The Ecological Civilization Debate in China." *Monthly Review* 66 (7): 36–44.

Weiner, Douglas R. 1988. *Models of Nature: Ecology, Conservation, and Cultural Revolution in Russia*. Bloomington: Indiana University Press.

Weiner, Douglas R. 1999. *A Little Corner of Freedom: Russian Nature Protection from Stalin to Gorbachëv*. Berkeley: University of California Press.

White, James D. 2013. "Alexander Bogdanov's Conception of Proletarian Culture." *Revolutionary Russia* 26 (1): 52–70.

White, James D. 2019. *Red Hamlet: The Life and Ideas of Alexander Bogdanov*. Leiden: Brill.

Zhang, Yunfei. 2019. "On the Historical Position of Ecological Civilization." *Capitalism Nature Socialism* 30 (1): 11–25.

PART III

Movements, prefiguration, and frameworks

15

TRAGIC MILESTONES OF THE NIGER DELTA

Nnimmo Bassey

Introduction

The Niger Delta has acquired the dubious distinction of being one of the most polluted places in the world. This did not happen by accident. This is the outcome of over six decades of merciless exploitation, crass ecological misbehaviour, and blatant environmental racism. The Niger Delta has over the years become the playground of transnational and national oil companies that damage the environment and waste the peoples without any restraint or compunction.

The Niger Delta in Nigeria is one territory where you would always find at least one ongoing oil spill at any point in time. Between 2018 and 2019, the National Oil Spills Detection and Response Agency (NOSDRA) recorded 1,300 oil spills, or an average of 5 oil spills per day (Adegboyega 2020). Statistics show that there was an average of 1,000 oil spill incidents monthly in Bayelsa State alone (Owolabi 2021). With such horrendous statistics, it is clear that the quality of the environment has been totally abused and the health of the people utterly compromised. The media-savvy oil companies have largely succeeded in shifting blame for the ecocide they have committed to the poor victims. With a minimum of 400,000 barrels of crude oil being stolen every day, it should be obvious that this is an industrial-scale activity in which the poor victims have no means of getting engaged. And for every pipeline leakage, the companies blame sabotage as the cause even before any sort of investigation is conducted.

With so many oil spills, not surprisingly, life expectancy in the Niger Delta is a paltry 41 years, compared to an equally embarrassing national average of 55 years. The point we are making is that the living who survive oil fires remain in the grip of deadly pollution, and their lives are thus highly discounted. For the living to have a fighting chance of living in dignity, the pollution from the petroleum extractive activities must urgently be remediated across the Niger Delta.

The Niger Delta is a sacrifice zone, and the blood of the people has mingled with crude oil for so long that it is futile trying to separate them. Massacres here include that of Umuechem in 1990, that of Kaiama following the Ijaw youths' declaration and demand for ecological and socio-economic justice in 1998, and, of course, the horrendous bombardment of Odi in November 1999. Others include the attacks on Odioma in 2005 and Gbaramatu in 2009.

These are just a few of the highlights of the reign of terror visited on the oil field communities. In this chapter, we will focus on two. The first is the execution of Ken Saro-Wiwa and

 10.4324/9780429341427-19

eight other Ogoni leaders over 25 years ago and the unwillingness of the state to exonerate them for a crime they did not commit. The second is the Jesse Pipeline fire tragedy that has now been marked by more than 22 years of deadly silence.

Ken Saro-Wiwa: A man of peace

For many, Ogoni has become the training ground for environmental justice. It has remained the prime territory for learning how difficult it is to undo ecological harm once it has occurred, once it has been allowed to fester and take root. The Ogoni people have also given us a clear base to understudy the workings of a people-driven non-violent revolt, the challenges, the pitfalls, and the triumphs. Ogoni has been a metaphor for ecocide and an inspiration for resistance.

Standing at the centre of the Ogoni experience are a number of personalities, one of whom is Ken Saro-Wiwa. His leadership at various levels and platforms left indelible marks on the socio-ecological struggles of the Ogoni people and others. Some of us make regular visits to the polluted sites in Ogoni to remind ourselves that ecocide in any location is a crime against Mother Earth and all our relatives. Ogoni reminds us all that corporate greed can convert a verdant land into a land where humans and other living beings are literally either sick or dead.

The literary output of Ken Saro-Wiwa helped preserve his thoughts for us and for generations yet unborn. Needless to say, his bluntness also made him controversial. That can be understood because when you are a minority fighting to breathe, those whose knees are pressed into your neck would claim that as long as you can complain, it means you can breathe. In other words, their knees will only be lifted from your neck when you fall silent. Dead. The noose snuffed the physical life from him 25 years ago, on 10 November 1995, but he still speaks. His satirical story *Africa Kills Her Sun* (1989) shows how fiction can chisel a message in stone. Writing about how a priest would approach to pray for a person about to be executed, he said: "The priest will pray for our souls. But it's not us he should be praying for. He should be praying for the living, for those whose lives are a daily torment."

His fiction was never altogether fictive. According to one Onookome Okome (2000), "These fictive characters are modelled on social types and local events. This explains why some of these characters provoked great and enthusiastic, albeit sometimes acerbic debate in Nigeria's literary history." Okome goes on to say:

> [H]is political ideas about the Nigerian Federation were even more controversial. His book on the Nigerian civil war (*On a Darkling Plain: An Account of the Nigerian Civil War*), carefully conceived around the minority/majority problems of Nigeria's ethnic groups, aroused heated hate-debate, especially among members of the three largest Nigeria ethnic groups.
>
> *(Okome 2000)*

His focus on bringing the plight of the Ogoni people to the world in the context of the unequal majority-minority relations within the Nigerian state, combined with the brutal state capture by notorious transnational oil companies, obviously earned him many adversaries, including those who eventually orchestrated his judicial murder, along with Barinem Kiobel, Saturday Dobee, Paul Levura, Nordu Eawo, Felix Nuate, Daniel Gbokoo, John Kpuinen, and Baribor Bera. Their deaths were both the epitome of the viciousness of an unholy matrimony between a rapacious transnational entity and an autocratic state and a glaring failure of international diplomacy (Naagbanton 2016).

Saro-Wiwa was conscious of the fact that the consequences of the struggle could be dire, even when prosecuted non-violently. In *Silence Would Be Treason: Last Writings of Ken Saro-Wiwa*, he stated that he signed a death warrant when he "undertook to confront Shell and the Nigerian establishment." He wrote that if his life was not cut short, he would look forward to "A few more books, maybe, and the opportunity to assist others." In a letter he wrote on 19 June 1995, he stated:

> I know they will do everything to resist us and that they may still want me out of the way. I am not careless of my safety, but I do recognise and have always recognised that my cause could lead to death. But as the saying goes, how can man die better/than facing fearful odds/for the ashes of his fathers/and the temple of his Gods? No, one cannot allow the fear of death to dent one's beliefs and actions. I only wish there were more Ogoni people on the ground. However, the cause cannot die."
>
> *Corley, Fallon, and Cox 2013)*

The matter of having more Ogoni people on the ground to keep the struggle alive remains an active concern, a task that must be done. Yes, the cause has not died, and 27 years after the expulsion of Shell from Ogoni, the oil wells are still not gushing crude. However, the spate of oil pollution remains, and the cleanup of the territory, although commenced, has its speed and mode of delivery highly contested. Having layers of leadership on the ground is essential for any movement. The Ogoni struggle has been kept alive by the deep mobilizations that have gone on over the years and by the clear understanding of the value of their environment and cultural autonomy by the majority of the people. Organizational efforts have floundered and become quite fractious at times, probably due to an alternative notion of sacrifice: superficial commitment to the ideals of the collective. It may well also be driven by impulses of indiscipline and possible conspiracy to subvert the pursuit of the common good.

The Ogoni Bill of Rights (1990) is a major milestone document, serving to coalesce the pains, dreams, and demands of the Ogoni people. It stands as a major decolonial document and was the precursor of similar pursuits by other ethnic nationalities in the Niger Delta, including the Kaiama Declaration of the Ijaws, the Ikwerre Rescue Charter, the Aklaka Declaration for the Egi, the Urhobo Economic Summit Resolution, and the Oron Bill of Rights, among others (Bassey 2013).

Article 16 of the Ogoni Bill of Rights states that "neglectful environmental pollution laws and sub-standard inspection techniques of the Federal authorities have led to the complete degradation of the Ogoni environment, turning our homeland into an ecological disaster" (1990). Three decades later, this summation remains accurate, even more poignant.

The Ogoni Bill of Rights spoke of the land turning into an ecological disaster. This position was validated by the United Nations Environment Programme (UNEP) in their report of the Environmental Assessment of Ogoniland (2011). The report submitted to the Nigerian government in August 2011 revealed extensive pollution of the soil by petroleum hydrocarbons in land areas, swamps, and sediments. The effort to remediate the Ogoni environment is handled by the Hydrocarbons Pollution Remediation Project (HYPREP), set up by the government in 2012 but coming into operation in 2016.

A September 2020 visit by this author to one of the remediation sites at Eleme, Ogoni, was quite revealing. Whereas the depth of hydrocarbon pollution was at an alarming five metres at the time UNEP conducted its study, the state of affairs has deteriorated over the years. Hydrocarbon pollution was found to have now gone as deep as an alarming ten metres at Lot 2. One other finding was that 30,000 litres of petrol was recovered from this lot. We saw a layer of hydrocarbons on the excavated pit at Lot 16 at Korokoro community, beside the tanks of recovered crude that were stored nearby.

The recovery of crude oil from the remediation sites shows that without the remediation, the pollution would obviously sink deeper, leaving the disaster more intractable. It also offers a stark warning to oil field communities that even where the land looks normal, tests need to be done at intervals of time to ensure the integrity of what lies beneath the surface.

The months of November 1990, when the Ogoni Bill of Rights was issued, and November 1995, when Ken Saro-Wiwa and the other leaders were executed, are cardinal milestones in the march for ecological and socio-political justice for the Ogoni people, with echoes among all marginalized peoples who are victims of destructive extractivism.

Twenty-five years after the judicial murders, the wounds inflicted on the Ogoni people are yet to heal. Twenty-five years after the act, the Nigerian state has still not found the place to formally exonerate the Ogoni leaders and foster healing in the land. Twenty-five years after the macabre act, even the sculpture in honour of the Ogoni 9 lies captive at the Apapa quays in Lagos, Nigeria, held by a system that is afraid to come to terms with an artistic artefact (Rustin 2015). Who will tell the Nigerian government that arresting and detaining a piece of sculpture in an effort to block the memory of crimes committed by the state is an exercise in futility?

Ken Saro-Wiwa saw it all. He felt it. He told it. He challenged all. His last public speech or *allocutus*, stands like a banner at the head of a marching column, and we do well to pay attention:

> We all stand before history. I am a man of peace, of ideas. Appalled by the denigrating poverty of my people who live on a richly endowed land, distressed by their political marginalization and economic strangulation, angered by the devastation of their land, their ultimate heritage, anxious to preserve their right to life and to a decent living, and determined to usher to this country as a whole a fair and just democratic system which protects everyone and every ethnic group and gives us all a valid claim to human civilization, I have devoted my intellectual and material resources, my very life, to a cause in which I have total belief and from which I cannot be blackmailed or intimidated.
>
> *(1995)*

Jesse pipeline of horror

Despite continued environmentalist efforts to bring ecological sanity to the oil-producing regions, spills, leaks, and disasters have continued as the state and multinational oil companies press on with their extractivist operations. One truly disastrous aspect of these operations is the rickety oil infrastructure crisscrossing the Niger Deltas lands and waterways.

Here we remember the tragic pipeline fire that occurred at Atiegwo, near Jesse, Delta State, on 17 October 1998, killing over 1,000 community members. This was a 16-inch petrol pipeline owned by the Nigerian National Petroleum Corporation (NNPC) that linked the Warri refinery to Kaduna. The fire raged for about five days and was eventually put out by American firefighters (Akande 2018).

Without any investigation, the military government and the Petroleum Products Marketing Company (PPMC), a subsidiary of the state-owned NNPC, alleged that the cause of the inferno was sabotage. However, this charge was not substantiated. The oil companies and the military government were quick to blame the victims. The pails and basins that littered the death scene were interpreted to have been taken there by pipeline vandals to scoop spilled petrol. The military government of General Abdusalam Abubakar declared that no compensation would be paid. Soon, surviving villagers became fearful that they would be prosecuted. The fear led some families to prematurely discharge their relatives from hospitals, a situation that may have contributed to an increase in fatalities. We repeat, the root cause of the conflagration is yet to be established, 22 years after the tragic incident. Twenty-two years is long enough to bring closure to this unfortunate incident.

A mother who lost her daughter, Eunice, in the inferno had this to say to environmental monitors who visited the scene:

> She said she was going to the farm. She left us happy. We were expecting some red cassava for dinner. She never came back. We saw the basin of the cassava. We saw the *karta* (head pad). We recognised our basin and her cloth. Her body we did not see. Her voice we did not hear. The fire took her from us. They say we are vandals. How? Can Eunice be a vandal? It is the oil people who have been vandalising our means of livelihood. It is the government that has stolen from us and continues to do so even to this minute.
>
> *(Environmental Rights Actions 21 October 1998a)*

What caused the fire?

Former Chief of Army Staff Major General David Ejoor (retired) was particularly piqued by the massacre and addressed the press in very strong terms (Environmental Rights Action 7 December 1998b). According to him, the evidence suggested that oil companies and the government caused the fire. He said that "when the spillage became a general knowledge, the oil companies moved in to cover the cartel that was siphoning petrol from a joint valve near Idjerhe in tankers. Towards daybreak, the saboteurs failed to put the pipes back properly and hence the spillage of petrol." According to the general, the spilled products got into farmlands as well as into the Ethiope River. This attracted the attention of the community people. "People going to their farms discovered that they were wading in petrol instead of water. There was a rush to fetch the petrol from the farm and the floating petrol in the river."

Eyewitnesses recounted that five minutes before the fire, there was a Shell Petroleum Development Company helicopter hovering overhead and urging the people to evacuate the scene. Analysts believe that since the victims were mostly Urhobo, if the officer in the helicopter had shouted the information in their language, they would have escaped the tragedy. The interpretation of this is that the employment pattern in the companies is skewed against the oil field communities.

Moreover, General Ejoor stated that after warning the people from the helicopter,

> the officials followed up their threat with firing nerve gas at the crowd, which made it impossible for them to run. Those who attempted to run could not move their limbs with agility. The horror came; the place was set on fire with the intention of killing everybody and to prevent anybody from giving evidence.
>
> *(ERA 2018b)*

Many pipeline fires have been recorded in the Niger Delta. Some can be traced to poor facility management, including the non-replacement of corroded pipelines or those that had reached their optimal lifespans. Most pipelines in Nigeria are designed for a limited lifespan of 20 years (Nnadi et al. 2014). Other incidents have been traced to vandalism or oil theft.

Recently, the general manager of the NNPC stated that oil pipelines in Nigeria are all compromised (Bademosi 2020). That is a very troubling situation. It shows that pipelines can leak volatile petroleum products at any time. Another worrying statistic came through when the NNPC stated that there were 45,347 pipeline breakages and/or explosions in Nigeria over the past 18 years. While speaking on this, the group managing director of the NNPC, Mele Kyari, fingered pipeline vandalism and crude oil theft as major challenges for the oil industry for years and attributed this to "poverty in surrounding communities, community-industry expectation mismatch, and corruption" (Olalekan 2020).

The analysis by the NNPC largely misses the point and heaps the blame on the victims, on the hapless communities. Crude oil theft is big business that requires technical knowledge and equipment, layers of security, and other protections within the system to thrive. The theft has been said to be at industrial scale. And, because the country does not really metre or measure the actual amount of crude oil extracted, the measure of the volume of crude being stolen on a daily basis remains in the realm of speculation.

The Nigerian Extractive Industries Initiative (NEITI) reckoned that Nigeria lost about $42 billion to crude oil theft between 2009 and 2018. According to NEITI, about $38.5 billion was lost to crude theft alone, $1.6 billion on domestic crude, and a further $1.8 billion was lost on refined petroleum products (Udo 2019).

Figures that have been bandied about range from 200,000 to 400,000 (Udo 2019) to as high as 1,000,000 barrels a day. A top government figure once speculated that as much oil as is being officially exported is also being stolen. One thing is clear: this humungous amount of crude oil could not be stolen by poor villagers or even by those engaged in bush refining. Indeed, it has been said that oil companies are involved in the business and that the international community is complicit (Bassey 2012, 149).

Pipelines in Nigeria have largely been carriers of pipe dreams. Water pipelines are largely dry, and those installed to convey crude oil to the refineries run largely empty as the refineries are comatose.

The loss of lives in the inferno of 1998 was, and remains, painful. However, we must not fail to mention that one regular blind spot associated with accidents of this nature is the lack of focus on what happens to the environment as a result of the incident. The environmental assessment of Ogoni by the United Nations Environment Programme (UNEP) clearly illustrated the harms of irresponsible extractive activities in the Niger Delta long after the initial incidents. The report submitted to the government in 2011 and leading to the establishment of the Hydrocarbons Pollution Remediation Project (HYPREP) showed that ground and surface waters in Ogoni were contaminated beyond acceptable levels. Ground water was found to have benzene, a known carcinogen, at 900 times above World Health Organization standards (UNEP 2011). In some places, the hydrocarbon pollution had seeped into the ground to a depth of five metres. By the time remediation was carried out in 2020, the pollution had sunk down to a depth of ten metres. These findings fully vindicated Ken Saro-Wiwa, who dared to ask for environmental sanity in Ogoniland. His legacy includes the persistence of diverse and creative non-violent ecological activism in the Niger Delta and the emergence of global networks of solidarity for social, ecological, and climate justice.

Conclusion

The sorry state of the Niger Delta is a product of colonial extractivism – the same trajectory pushed by neoliberalism. Colonialism is not yet history in Africa or in the world. The global trade architecture has been in place for centuries and has been engineered by transnational corporations and international financial institutions as the chief guardians of neocolonialism and institutionalised thievery. Their interests are assured through the preservation of these mechanisms.

Transnational oil companies, like other transnational corporations (TNCs), are products of imperial geopolitics whose levers they hold, manipulate, and tilt to suit their profit-making propensities. They have succeeded thus far because of careful modes of manipulation, erasure, and replacement of imaginations as well as histories. The strength of neocolonialism lies in the perpetuation of coloniality, wherein colonial-era relationships were maintained in post-colonial political formations.

In many instances, transnational corporations were the original colonialists, invading territories with their bands of mercenaries and harvesting profits for imperial powers. As their direct rule became expensive and untenable, they handed over political and administrative control to their home governments, which then provided the security needed for continued plunder by the corporations. That system continues today and persists under the reign of neocolonialism. And there are many subtle and not-so-subtle tools that keep the rot going. The oil companies are behaving true to type, like the leopard that does not change its spots. But they can be tamed.

They must be tamed. On 29 January 2021, the Appeal Court at The Hague gave a significant ruling 13 years after a suit filed by Nigerian plaintiffs against Shell began (Wifa and Adebola 2021). The court ruled that Shell must compensate the Nigerians whose environment they have polluted. On 12 February 2021, the Supreme Court in the UK further ruled that Royal Dutch Shell can be sued in the United Kingdom for offences committed by their subsidiary in Nigeria (HOMEF 2021). We note the world is moving towards making ecocide a crime alongside genocide, war crimes, and crimes against humanity. The tightening of the noose around malignant polluters should be a good incentive for oil companies to change their *modus operandi* and think of the people and planet before profits. In the meantime, the Niger Delta's environmental activists continue the work begun by Ken Saro-Wiwa, whose cause did not die.

We thought it was oil, but it was blood

The other day
We danced in the street
Joy in our hearts
We thought we were free
Three young folks fell to our right
Countless more fell to our left

Looking up,
Far from the crowd
We beheld
Red-hot guns

We thought it was oil
But it was blood
We thought it was oil
But this was blood

Heart jumping
Into our mouths

Floating on
Emotion's dry wells
We leapt in fury
Knowing it wasn't funny
Then we beheld
Bright red pools

We thought it was oil
But it was blood
We thought it was oil
But this was blood

Tears don't flow
When you are scarred
First it was the Ogoni
Today it is Ijaws
Who will be slain the next day?
We see open mouths
But hear no screams
Standing in a pool
Up to our knees

We thought it was oil
But it was blood
We thought it was oil
But this was blood

Dried tear bags
Polluted streams
Things are real
When found in dreams
We see their Shells
Behind military shields:
Evil, horrible, gallows called oil rigs
Drilling our souls

We thought it was oil
But it was blood
We thought it was oil
But this was blood

The heavens are open
Above our heads
Toasted dreams in a flared
And scrambled sky
A million black holes
In a burnt up sky

Their pipes may burst
But our dreams won't burst

We thought it was oil
But it was blood
We thought it was oil
But this was blood

This we tell you
They may kill all
But the blood will speak
They may gain all
But the soil will RISE
We may die but stay alive
Placed on the slab
Slaughtered by the day
We are the living
Long sacrificed

We thought it was oil
But it was blood
We thought it was oil
But this was blood.

References

Adegboyega, Ayodeji. 2020. "NOSDRA: Nigeria Records 1,300 Oil Spills in Two Years." *Premium Times*, 8 June. www.premiumtimesng.com/news/top-news/396658-nosdra-nigeria-records-1300-oil-spills-in-two-years.html+-

Akande, Segun. 2018. "In 1998, Nigeria's Worst Fire Outbreak Killed 1098 People in Delta State." *Pulse*, 16 February. www.pulse.ng/gist/jesse-pipeline-explosion-in-1998-nigerias-worst-fire-outbreak-killed-1098-people- in/cxsd6e9

Bademosi, Adetola. 2020. "Nigeria's Pipeline Networks Completely Compromised, Says Kyari." *Nigerian Tribune*, 20 September. https://tribuneonlineng.com/nigerias-pipeline-networks-completely-compromised-says-kyari/

Bassey, Nnimmo. 2012. *To Cook a Continent: Destructive Extraction and the Climate Crisis in Africa*. Oxon: Pambazuka Press.

Bassey, Nnimmo. 2013. "Two Years After the UNEP Report Ogoni Still Groans." 1 August. http://nnimmo.blogspot.com/2013/08/two-years-after-unep-report-ogoni-still.html

Corley, Íde, Helen Fallon, and Laurence Cox, eds. 2013. *Silence Would Be Treason: Last Writings of Ken Saro-Wiwa*. Ottawa: Daraja Press.

Environmental Rights Action (ERA). 1998a. "Idjerhe (Jesse) Oil Fire Disaster: Petroleum Pipeline Explosion: An Avoidable Tragedy." *Urhobo Historical Society*, 21 October. http://waado.org/Environment/IdjerheFire/idjerhe_Era.htm

Environmental Rights Action (ERA). 1998b. "Causes of the Idjerhe Fire Disaster. ERA's Environmental Testimonies #1." *Urhobo Historical Society*, 7 December. http://waado.org/Environment/IdjerheFire/CausesOfFireDisaster.html

HOMEF. 2021. "Outcome from the Polluters' Judgement Roundtable." *Health of Mother Earth Foundation*, 26 February. https://homef.org/2021/02/26/outcomes-from-the-polluters-judgment-roundtable/

Naagbanton, Patrick. 2016. *Footprints of Nkpoo Sibara, Dele Giwa and Ken Saro-Wiwa*. Vol. 1. Makurdi: DNA Traeces Empire Limited.

Nnadi, Uzoma, Zaki El-Hassan, David Smyth, and James Mooney. 2014. "Lack of Proper Safety Management Systems in Nigeria Oil and Gas Pipelines." Paper presented at the Institution of Chemical Engineers conference, Hazards 24, Edinburgh, 7–9 May. www.icheme.org/media/8910/xxiv-paper-14.pdf

Ogoni Bill of Rights. 1990. www.waado.org/nigerdelta/RightsDeclaration/Ogoni.html

Okome, Onookome. 2000. *Before I am Hanged: Ken Saro-Wiwa – Literature, Politics and Dissent*. Trenton, NJ: Africa World Press, Inc.

Olalekan, Fakojeyo. 2020. "Pipeline Explosion: Over 45,000 Incidents Recorded in 18 Years – NNPC." *Nairametrics*, 21 January. https://nairametrics.com/2020/01/21/pipeline-explosion-over-45000-incidents-recorded-in-18-years-nnpc/

Owolabi, Razak. 2021. "Sabotage by Oil Thieves, Cause of Environmental Pollution in Nigeria's South-South." *Eco News Nigeria*, 28 February. www.environewsnigeria.com/sabotage-by-oil-thieves-cause-of-enviromental-pollution-in-nigerias-south-south/

Rustin, Susanna. 2015. "Ken Saro-Wiwa Memorial Art Bus Denied Entry to Nigeria." *The Guardian*, 5 November. www.theguardian.com/environment/2015/nov/05/ken-saro-wiwa-memorial-art-bus-denied-entry-to-nigeria

Saro-Wiwa, Ken. 1989. *Adaku and Other Stories*. London: Saros International.

Saro-Wiwa, Ken. 1995. "Trial Speech of Ken Saro-Wiwa." https://en.wikisource.org/wiki/Trial_Speech_of_Ken_Saro-Wiwa

Udo, Bassey. 2019. "Nigeria Lost $42 Billion to Crude Oil Theft in Nine Years – NEITI." *Premium Times*, 19 November. www.premiumtimesng.com/news/headlines/361353-nigeria-lost-42-billion-to-crude-oil-theft-in-nine-years-neiti.html

UNEP. 2011. "Environmental Assessment of Ogoniland: Site Factsheets, Executive Summary and Full Report." *United Nations Environment Program*, 1 August. www.unenvironment.org/resources/assessment/environmental-assessment-ogoniland-site-factsheets-executive-summary-and-full?_ga=2.152372081.992977037.1602959307-1394307030.1602959307

Wifa, Eddy, and Titilayo Adebola. 2021. "Triumph for Farmers and Fisherfolks: The Hague Court of Appeal Finds Shell Liable for Oil Spills in Nigeria." *Home of Mother Earth Foundation*, 4 February. https://homef.org/2021/02/04/triumph-for-farmers-and-fisherfolks-the-hague-court-of-appeal-finds-shell-liable-for-oil-spills-in-nigeria/

16
NAKEDNESS AND POWER

Seth Tobocman, Leigh Brownhill, and Terisa E. Turner

The Nigerian peasant women's resistance to Big Oil has yielded insights which contributed to our constructing an ecofeminist, ecosocialist, theoretical-methodological framework (see chapters by Bassey, Brownhill, Isla, Giacomini, Turner, this volume). This framing is especially suited for an orthodox but extended Marxist analysis of class struggle in its racialized and gendered fullness. We call this framework "ethnicized, gendered class analysis."

In the following introduction to the comic, we address only one dimension of this theoretical-methodological framework: the revolutionary tactic that can, in a non-violent manner, disempower the large transnational corporations that are at the heart of capitalist exploitation. This tactic is the simultaneous, cross-border, coordinated, joint, direct action that strikes at capitalist production and consumption.

In the midst of a global wave of popular mobilization in the late 1980s and early 1990s, women small holder farmers in the Niger Delta in West Africa rose up against Nigeria's massive state-owned and multinational oil industry. We chronicled and assessed what was a peasant-worker uprising in an article entitled "Why Nigerian women are at war with Chevron" (Turner and Brownhill 2004). The graphic narrative, "Nakedness and Power," tells that story in the form of a comic so as to make Nigeria's feminist peasant struggles more accessible to a wider audience. The story of these mainly land and water struggles is of considerable significance for a number of reasons, not the least of which are its rich theoretical implications. For example, in the Nigerian case, peasant women moved first and then were followed by male waged workers from the general, not professional and managerial, ranks of the oil industry. This escalated into a semi-successful general strike driven in particular by a refusal of market women to supply food to the cities.

We are Marxists schooled in the tradition and oeuvre of CLR James (see Grimshaw 1992 and www.clrjames.uk). This tradition has focused since the 1930s on the global character of both capitalists and the exploited. James's prolific theory and analysis centred on the anatomy of resistance, fight back, capacity acquisition, alliance building, and the exercise of power by ordinary people globally, towards instituting socialism and pushing capitalism into the dustbin of history.

CLR James always insisted that the concept of socialism requires no adjectival modifiers because it is already and by definition ecological, democratic, feminist, and humanist. Marx's most original contribution to knowledge and social transformation, averred James, is the

 10.4324/9780429341427-20

dialectical realization that resistance from the dispossessed, exploited, and threatened is organized, disciplined, and united by the very mechanism of the process of capitalist production itself (Marx 1867 [1967]). As corporations globalize from above, so do workers globalize from below (Turner and Brownhill 2001). In this process of globalization from below, workers gain capacities for functioning within socialist or "commoning" relations, which are constantly being constructed and refined. We developed the theoretical-methodological framework that we call ethnicized, gendered class analysis while living in Africa and working closely with the African women depicted in this graphic narrative. And we shared the perspective of CLR James, who saw the globally transformational power of unity with the peaceful direct actions of Black women. This perspective helped us formulate a theorization and methodology for class struggle analysis that is integrally racialized and gendered.

Concepts that this framing embraces include the "hierarchy of labour power" with its racialized and gendered divisions within the exploited class, the "male deal," the "ethnicized gendered class alliance," and the "fight for fertility," among others. (See Terisa E. Turner's chapter in this volume.) All these concepts can be discussed with reference to the Nigerian women's actions. However, we here restrict our discussion to one key conceptual dynamic, "production-consumption strikes," which have to do directly with system change out of capitalism and into a successor system of "commoning" powered by solar, not fossil, energy. The Nigerian women's anti-oil uprising of the early 1990s showed us that there is tremendous power in coordinated, cross-border, direct action to simultaneously stop production and stop consumption.

The first to take action in the Nigerian insurgency were unwaged peasant women who protested against the foreign and state oil industry. Women stopped oil exports by occupying Africa's largest oil export terminal. At the insistence of these occupying/activist/peasant/unwaged/mobilized women, waged workers throughout the society, notably men in the oil industry, went on strike. They shut down much of West Africa's economy. In so doing, they constrained the production and export of petroleum from Nigeria for a period of time.

Co-terminus with this production strike was a widespread consumption strike. Globally networked environmental activists organized boycotts of petroleum products and specific oil companies in several countries. The oil industry was stymied because, notably for Shell and Chevron but also for other oil companies operating in Nigeria, the industry's source of value and profit – that is, crude oil – was not available. Also not available was part of its petroleum product consumer market.

The crisis of capital was intensified by a global women's nakedness demonstration spurred by the example of Nigerian women. They had declared that they would use the "curse of nakedness" to impose social ostracization (and even impotence, as was commonly believed possible) on oil men who were targeted as perpetrators of violence against people and the ecological system. Local collaborators with the oil industry were also targeted as "male dealers." The African nakedness action went viral. Women protested the war for oil, which was the US's first invasion of Iraq in 1991, using nakedness as a power. They were joined by some men.

In a number of versions of Aristophanes's 411 BC play, *Lysistrata*, women worldwide doff their clothes and use their naked bodies to spell out anti-war slogans. Such viral nakedness threatened not only a sex strike but also a labour power production strike. Women were using their power of refusal to deliver domestic services to stop capitalist men and their largely male working class allies from expanding control over fossil fuels and the global war machine.

Over the three decades since the early 1990s Nigerian uprising, there has been a tremendous expansion of the use of the powerful cross-border tactic that is the production-consumption strike. Coordinated actions of many kinds keep erupting from below. Initiatives against injustice and for self-determination regularly rise up in one place only to go global within hours or days. This was evident in the "Black Spring" of 2020 when the Black Lives Matter movement brought multi-racial millions onto the world's streets, sparked by the police murder of George Floyd in Minneapolis, USA. This globalized tactic was evident in April 2021 when a majority Black and female workforce tried to form a union in trillionaire Jeff Bezos's Amazon warehouse in Bessemer, Alabama. Their efforts launched a global movement of Amazon workers and their civil society allies who are also consumers who produced labour power (Leon 2021). Increasingly in the contemporary period, workers, waged and unwaged, producers and consumers, North and South, take action against capitalists and their captured regimes when these are weakened by one uprising. They do so by launching another uprising and then by multiplying anti-corporate actions on both national and global scales.

Four observations about production-consumption strikes are relevant here. First, there is a potential in them for the non-violent disempowerment of specific corporate entities whose value chains are disrupted and challenged. Non-violence is essential for a number of reasons, including the building of participatory democracy, the benefits of family-friendly mobilizations, the potential to disarm forces of repression, and the maintenance of and respect for peace. For these and other reasons, non-violence is the path chosen by those who initiate these strikes.

The second observation is that women – almost always Black, Brown, and Indigenous women – are the social actors who, in almost all the cases, take the initiative in launching campaigns to refuse capitalist exploitation. The pathbreakers of an ecofeminist, ecosocialist transition are those at the very bottom, at the base and foundation, of the hierarchy of labour power worldwide. When women in struggle are on the move, the whole hierarchy above them is threatened by collapse. This is because the capitalist epoch cannot persist without the work of billions of women, most of whom are unwaged and who produce the labour power and almost all the inexpensive "wage goods" that facilitate capitalists' refusal to pay a living wage. The COVID-19 pandemic and attendant lockdowns brought home to all the indispensability to capital of women's unpaid work.

Our third observation about production-consumption strikes is that the social connections between actors who engage in them are the very social connections required for direct deals and all kinds of cooperation among people within a nascent ecosocialism. This new society and relationship with nature is what will help us survive the fascism that capitalists and their political agents resort to as their profits and capacity to control labour and oil (both necessary for profits under fossil capitalism) become tenuous. Direct deals among commoners are both very old and very new. With the acute rupture and dissolution of capitalist value chains caused by the COVID-19 pandemic, these direct deals are increasing in number. Direct cooperation and social solidarity are, for many, the preferred or only options given the 2020–2021 erosion of for-profit value and supply chains.

Capitalist sanctions and boycotts seek to criminalize direct exchange but with questionable and intermittent success. Governments and private actors have long entered into direct exchanges (called counter trades in some sectors, such as the oil industry) in the face of capitalist debt burdens. Commodified buying and selling that depends on the seller receiving money are not attractive if that seller expects any income to be confiscated and used to pay debt.

Direct deals can take place only between parties who have control over the goods or services that are being exchanged. As corporate control is interrupted by production-consumption strikes, worker control is expanded. This in turn potentially expands opportunities for free association and democratic practice. These are practices that make it possible to heal the earth and prioritize needs of the most afflicted, who in their activism shake the foundations of the hierarchy of labour power from below.

Our fourth and final observation about production-consumption strikes is that their transformational and system-transition potential depends on a deep revolution of consciousness. Just such a revolution of consciousness has been given an immense boost by the life-altering consequences of the COVID-19 pandemic. The illness or loss of a loved one or acquaintance is now a universal reality because of the virus. COVID-19 itself emerged from habitat destruction due to continuing commodification via capitalist expansion into every nook and cranny left to wild creatures. And as horrific as the pandemic is, exacerbated by the neoliberal handling of the health crisis, the disaster of climate chaos looms even larger.

Global society has been shocked into a new consciousness about racialized and gendered class relations and class struggle. In the United States in 2021, two individuals controlled wealth equal to that of the bottom 90 percent. This profanity of inequality has fueled a new social awareness characterized by the recognition of the unsustainability of this dramatic polarization of wealth. The capitalists who control COVID-19 vaccines' intellectual property rights are refusing to socialize them so that the world's people can be freely, universally, and rapidly vaccinated and protected. Profits are prioritized instead. Meanwhile new and more dangerous strains of the COVID-19 virus emerge from the infected and unvaccinated. Unless everyone is vaccinated, no one is safe from continued infections and deaths. Capital's value chains will continue to rupture as danger precludes school, work, travel, and all aspects of the global capitalist process of exploitation and domination. In recognition of this death spiral, a leap in social consciousness is emerging that includes awareness of the imperative need for transformative system change.

This fast-evolving social consciousness was expressed by the Western Massachusetts chapter of the US-based organization CodePink (Women for Peace) in their 8 March 2021 salute to International Women's Day:

> We live or die together: that is the new consciousness that has gripped the world in 2021.
>
> Women; especially Black, Brown, and Indigenous women, are at the forefront of the worldwide fight to stay alive in the face of the Covid pandemic. The examples are numerous: teachers and parents resist forced, in-person work in dangerous schools, Amazon workers try to unionize and provoke global solidarity, detainees mount hunger strikes, and India's women farmers feed encampments hundreds of millions strong. The 'most vulnerable' face-to-face workers, predominantly women of colour, (and overwhelmingly so on a global scale), literally produce and sustain life, practices not off-shoreable. The potential the pandemic has revealed is the power of women to stand up and refuse to sustain life under deadly, dire work conditions, for the boss and for the state. Instead, women are protecting our own and others' lives.
>
> With the collapse of capitalists' supply and value chains, there emerge life-supporting practices and alliances, collectivity, and mutual support based on a new, intense global consciousness of ourselves as one people. These practices are anti-capitalist because they are social, not individual. They are alternatives to the collapsing global supply chains. The social awareness of 2011's 'Occupy Movement' that "We are the 99%," has expanded globally into the planetary realization that we are each other's keepers and that acting in accord with this new consciousness is a matter of life or death.
>
> This International Women's Day we salute women, prominent among whom are women of color, who are empowering people of all genders to survive through practical solidarity. We celebrate those women who, under urgent circumstances, are showing us all how to realize the potential of the refusal to be exploited by amplifying collective life-support and protection.

NAKEDNESS AND POWER
1992, FREEDOM CORNER, NAIROBI, KENYA, WOMEN WERE ON HUNGER STRIKE BECAUSE THEIR SONS HAD BEEN PUT IN PRISON BY THE DICTATOR, DANIEL ARAP MOI.
MOI DOES NOT LET US GROW FOOD. HE MAKES US GROW COFFEE FOR EXPORT INSTEAD.
BETTER TO STARVE IN PUBLIC.
POLICE
ATTACKED
RUTH WANGARI COVERED HER HEAD TO PROTECT AGAINST TEAR GAS.
BUT WHEN SHE LOOKED UP, SHE SAW
POLICE ABOUT TO
SHOOT A YOUNG MAN.
RUTH WANGARI TORE OFF
HER CLOTHES.

NAKEDNESS AND POWER
TO SOME AFRICANS IT IS A POWERFUL CURSE
FOR A WOMAN TO EXPOSE HER VAGINA TO A MAN WHO HAS MADE HER ANGRY. THE WEAPON OF NAKEDNESS IS SAID TO CAUSE IMPOTENCE, MADNESS, OR DEATH. IT IS A SOURCE OF LIFELONG SHAME FOR A MAN TO BE CONFRONTED IN THIS WAY. THIS CURSE IS BASED ON THE IDEA THAT WOMEN ARE THE CREATORS AND SUSTAINERS OF LIFE AND SO HAVE THE POWER TO REVOKE THE LIVES OF OFFENDING MALES.
THE POLICE FLED.
THE VIGIL AT FREEDOM CORNER CONTINUED FOR A WHOLE YEAR.
THE WOMEN FREED 51 PRISONERS.
INSPIRED BY THE ACTION AT FREEDOM CORNER, FARMERS RIPPED OUT COFFEE AND PLANTED FOOD TO FEED THEIR FAMILIES. THE LANDLESS POOR ALSO SEIZED LAND ON WHICH TO GROW FOOD.

THEN THEY TORE OUT THE DICTATORSHIP OF DANIEL ARAP MOI,
AND PLANTED A NEW GOVERNMENT, MORE REPRESENTATIVE OF THE PEOPLE, WHICH INCLUDED SOME OF THOSE WHO HAD PARTICIPATED IN THE PROTEST AT FREEDOM CORNER.
THE STRUGGLE OF RURAL AFRICAN WOMEN HAS A MESSAGE
FOR THE WORLD.
MANY PEOPLE IN NIGERIA FEED THEIR FAMILIES THROUGH HUNTING, FISHING, AND
FARMING.
BUT IN THE 1970s THE OIL BOOM DISRUPTED THIS WAY OF LIFE. THE GOVERNMENT FORCED FOLKS TO ALLOW COMPANIES TO LAY PIPELINE RIGHT THROUGH FARMS AND VILLAGES. PIPES LEAKED, CAUSING OIL FIRES TO BURN DAY & NIGHT, FOR YEARS, SCARING AWAY ANIMALS, POLLUTING LAND, AIR, WATER.
FOLKS MUST HARVEST CROPS AMID SMOKE & FLAMES. FISH DIE FROM OIL SPILLS. IN SOME VILLAGES THERE IS NO CLEAN DRINKING WATER. SINCE THE DISCOVERY OF OIL THE NUMBER OF PEOPLE LIVING IN POVERTY HAS TRIPLED.

OGONI
FOR YEARS NIGERIANS HAVE RESISTED THESE CONDITIONS.
ON JULY 8th, 2002, 600 WOMEN TOOK OVER THE
CHEVRON TEXACO EXPORT TERMINAL AT ESCRAVOS, NIGERIA.
THE TAKEOVER LASTED 10 DAYS. WOMEN NEGOTIATED 26 DEMANDS WITH CHEVRON.
BUT ONE DEMAND
Chevron
CHEVRON WOULD NOT DISCUSS;
WHEN WE WERE HERE WITHOUT CHEVRON, LIFE WAS NATURAL & SWEET, WE WOULD GO TO THE RIVERS FOR FISHING, THE FORESTS FOR HUNTING. BUT TODAY THE EXPERIENCE IS SAD. I AM SUG-GESTING THEY SHOULD LEAVE OUR COMMUNITY COMPLET-ELY AND NEVER COME BACK.
WOMEN ROSE UP ALL OVER THE NIGER DELTA, TAKING 12 OIL FACILITIES, THREATEN-ING TO USE THE CURSE OF NAKED-NESS.
THEY SHUT DOWN 40% OF NIGERIA'S OIL PRODUC-TION.
IT COST THE NIGERIAN STATE $11 MILLION AND THE COMPANIES $2.5 MILLION PER DAY.
IN RETALIATION, OIL COMPANY SECURITY GUARDS RAPED DOZENS OF WOMEN DURING A PROTEST AT AN OIL COMPANY HEADQUARTERS.
THE WOMEN HELD A PRESS CONFERENCE TO DEMAND REPARATIONS FOR THE RAPES, THREATENING TO USE THE CURSE OF NAKEDNESS.
WE HAVE DECIDED TO DIE AT CHEVRON'S GATE INSTEAD OF SLOWLY FROM OIL SPILLS.

SOME WOMEN IN THE U.S. & EUROPE HEARD ABOUT THE NIGERIAN WOMEN.
THEY SAW:
THAT THE SAME OIL COMPANIES THAT WERE POLLUTING NIGERIA WERE POLLUTING THE WHOLE WORLD.
THEY SAW THAT AN OIL MAN WAS TAKING AMERICA TO WAR IN IRAQ.
THEY ORGANIZED A
CHEVRON TEXACO
BOYCOTT
BUT THEY DID MORE:
FROM SAN FRANCISO TO SOUTH AMERICA, WOMEN TOOK OFF THEIR CLOTHES TO PROTEST AGAINST THE COMING WAR IN IRAQ, PAINTING SYMBOLS OF PEACE ON THE HILLSIDES WITH THEIR BODIES.

PROTESTERS? THAT'S LIKE LISTENING TO A FOCUS GROUP.
BUSH INVADED IRAQ
AND DECLARED VICTORY.
BUT IN NIGERIA, THE ACTIONS OF RURAL WOMEN HAD INSPIRED UNIONS TO STRIKE. THERE WERE STRIKES BY UNIVERSITY EMPLOYEES AND RAILWAY MEN. NIGERIAN OIL WORKERS SEIZED TRANSOCEAN AND HALLIBURTON OFFSHORE OIL RIGS, TAKING BRITISH WORKERS HOSTAGE.
SENDING E-MAILS TO THEIR FAMILIES, HOSTAGES EXPRESSED FEAR OF, BUT ALSO SOME SYMPATHY FOR, THEIR CAPTORS.
BECAUSE BLACK WORKERS HAD TO COME TO WORK EVERY DAY IN DANGEROUS MOTOR BOATS
WHILE WHITES WERE FLOWN TO WORK.
THAT'S WHY LABOR UNIONS FROM MANY COUNTRIES EXPRESSED SUPPORT FOR THE
NIGERIAN OIL WORKERS.

THE NIGERIAN NAVY WAS READY TO STORM THE OIL RIGS. IT WAS RUMORED THAT THEY WERE BRINGING BRITISH MERCENARIES. BUT THE OIL WORKERS UNION THREATENED TO SHUT DOWN ALL OIL PRODUCTION IN NIGERIA IF FORCE WAS USED.
NIGERIAN NAVY
SO THE OIL COMPANIES AGREED TO SOME OF THE WORKERS' DEMANDS. THE STAND OFF
ENDED PEACE-FULLY.
THE WORLD BANK
PRESSURED PRESIDENT OBASANJO of NIGERIA TO RAISE THE PRICE OF GAS & KEROSENE FOR DOMESTIC CONSUMPTION BY 55%.
THIS WAS THE KEROSENE FOLKS USED TO COOK
FOOD
SO UNIONS CALLED A
GENERAL STRIKE
MARKET WOMEN CLOSED THE MARKET TO KEEP THE STRIKE GOING.
THE STRIKE LED TO RIOTS.
THE STRIKE ENDED AS THE UNIONS ACCEPTED A 35% INCREASE AS A COMPROMISE.

MEANWHILE, THE WAR IN IRAQ WAS LOOKING LESS LIKE A VICTORY.
WITH THE MIDDLE EAST IN FLAMES, THERE WAS INTEREST IN DEVELOPING OTHER SOURCES OF OIL.
ON JULY 11th, 2003, BUSH WENT TO NIGERIA TO OPEN A NEW OIL FIELD.
HE FOUND THAT WOMEN HAD TAKEN THE AMPUKE OIL FLOW STATION, OWNED BY SHELL.
WOMEN TIED UP PRODUCTION ACROSS THE DELTA.
SO BUSH MOVED
U.S. TROOPS FROM BASES IN GERMANY TO AFRICA.
THAT'S RIGHT, WE MAY SOON BE AT WAR IN NIGERIA TOO.
BUSH HAS HIS ARMIES.
HE ALSO HAS THE W.T.O. (WORLD TRADE ORGANIZ--ATION), THROUGH WHICH HE TRIES TO INFLUENCE THE ECONOMIC POLICIES OF OTHER COUNTRIES.

IN SEPTEMBER OF 2003, THE W.T.O. MET IN CANCUN. PEOPLE CAME FROM ALL OVER THE WORLD TO PROTEST AGAINST THE MEETING.
NOWTO
THE TRADE MINISTER OF KENYA LED THE MAJORITY OF THE WORLD'S NATIONS IN WALKING OUT ON THE W.T.O. MEETING. THE TALKS COLLAPSED.
IF FOLKS IN AFRICA CAN STOP THE FLOW OF OIL,
IF FOLKS IN THE U.S. AND E.U. CAN
BOYCOTT
GAS,
THEN THESE 2 GROUPS CAN DEAL DIRECTLY WITH EACH OTHER
AND BYPASS THE OIL COMPANIES.
IF WE THE PEOPLE TAKE CHARGE OF THE TRADE IN ENERGY, WE CAN SWITCH FROM OIL TO CLEANER FORMS OF ENERGY, LIKE SOLAR POWER.
THERE WOULD BE NO MORE WAR FOR OIL. IMAGINE A WORLD WHERE BUSH HAS TO ASK ORDINARY FOLKS FOR THE GAS TO PUT IN HIS WAR MACHINES.
THIS COMIC CREATED BY SETH TOBOCMAN, LEIGH BROWNHILL, TERISA TURNER, LAIRD OGDEN

References

Grimshaw, Anna, ed. 1992. *The CLR James Reader.* Oxford: Blackwell.

Leon, Luis Feliz. 2021. "A Worldwide Workers' Revolt Against Amazon Has Begun." *In These Times*, 15 April. https://inthesetimes.com/article/workers-world-unite-amazon-union-busting-organizing-labor-rights

Marx, Karl. 1867 [1967]. *Capital: A Critical Analysis of Capitalist Production, Volume One: The Process of Capitalist Production*, edited by Frederick Engels. New York: International Publishers. See in particular Chapter 32.

Turner, Terisa E., and Leigh Brownhill. 2001. "Introduction: Gender, Feminism and the Civil Commons: Women and the Anti-Corporate, Anti-War Movement for Globalization from Below." *Canadian Journal of Development Studies* 21: 805–818.

Turner, Terisa E., and Leigh Brownhill. 2004. " 'Why Women are at War with Chevron': Nigerian Subsistence Struggles Against the International Oil Industry." *Journal of Asian and African Studies* 39 (2–3): 63–93.

17

ECOSOCIALIST ACTIVISM AND MOVEMENTS IN SOUTH AFRICA

Vishwas Satgar and Jacklyn Cock

Introduction

South Africa has a rich history of radical politics. Environmental justice has been a crucial current in this ferment, confronting racialized and patriarchal ecologies reproducing a carbon-centric capitalism. The undergrowth of community organizations, trade unions, and networks championing environmental justice has been variegated over time and spatially dispersed. Such struggles have included confronting health issues facing mine workers, toxic chemicals affecting workers, pollution affecting communities, women confronting extractivism, hunger, and land struggles. Mining and industrial development have engendered numerous environmental flashpoints and counter-movements. At this moment of worsening planetary ecological crisis, environmental justice politics are challenged in South Africa to advance a just transition capable of dismantling carbon capitalism and rebuilding a society according to ecosocialist principles. Survival depends on ending our dependence on coal as a source of jobs, electricity, and foreign exchange and understanding it as a driver of the climate crisis. In this complex, unpredictable process, the notion of "democratic ecosocialism" is potentially a conceptual tool to crack open the "frozen imaginaries" of people who have not envisaged radical alternatives and those who accept the inequalities and ecological damage of carbon capitalism as natural and inevitable. It is also a crucial ideological resource to advance post-national liberation politics and resistance to shallow and technocratic approaches to the just transition, keen to appropriate, weaken, and stall mass power demanding system change.

In this chapter we bring to the fore various democratic ecosocialist activist engagements and movement-building initiatives. First, we locate activist and radical scholarly engagements with the most dangerous contradiction we face: the climate crisis. We set out the contestations underway to ideologically frame the climate crisis as a crisis of ongoing carbon-capitalist accumulation within the Anthropocene. Second, we situate the political economy of South Africa's carbon capitalism, drawing on Marxist analyses of the minerals-energy complex. Third, we locate the debate and critique underway in South Africa of historical socialist alternatives and their limits to confront an ecocidal logic of carbon capitalism. Fourth, we provide a synoptic mapping of the making of a democratic ecosocialist imagination in contemporary frontline struggles. Finally, we conclude with the challenges facing the advance of a democratic ecosocialist project in South Africa.

10.4324/9780429341427-21

Climate ecocide and the global carbon capitalist Anthropocene

The possible extinction of the human race is now commonsense in global climate emergency discourses. However, what is driving and causing human and non-human extinction because of worsening climate crises is not commonly agreed. Climate ecocide is not understood as imbricated in the reproduction of carbon capitalism. In South Africa, the carbon-capitalist state has consistently treated nature as capital to be exploited, polluted, extracted, and valorized. In this regard there is a continuity between the colonial, apartheid, and post-apartheid socio-ecological order. The post-apartheid state, in its modernizing disposition to "catch up," has also gone further to become diffusionist in its approach to global discourses, which really does not amount to much in terms of the lived experience of the people but serves a performative function. In 2002, the South African government hosted the World Summit on Sustainable Development and used this platform to affirm an identity of concern for the environment. Throughout the 2000s, institutions in the global power structure, including UN agencies, the ILO, the FAO, and even the OECD, have been affirming "green growth," "green jobs," and another round of "green revolutions" in agriculture. The South African state has been a disciplined interlocutor in these discourses and has translated this into South African policy speak. The hypocrisy of declaring "green development and growth" has not been lost on the radical academy and grassroots forces. The critique has been consistent to highlight how "green growth" merely reproduces the carbon-based minerals-energy complex, further advancing mining-related pollution and dispossession and narrow approaches to the "just transition" (Satgar 2014; Cock 2014, 2018).

In terms of the climate crisis, the South African carbon state locked into Obama's Copenhagen Accord in 2009, hosted the UN Conference of the Parties in 2011, and is now a signatory to the Paris Climate Agreement. Given the marketized and financialized solutions embedded in the UN climate negotiations, this means that climate justice is not central to its discourses, policy understandings, and practices (Bond 2012). Moreover, official UN discourses on the Anthropocene, premised on generic human causality, are now part of the South African carbon states imaginary. For radical scholarship in South Africa, this has meant challenging the ontology of the notion of the Anthropocene. In its official rendering by climate scientists and physicists, the narrative is straightforward. There is a history to why the earth is heating. For the past 150 years, industrial societies have been at the forefront of extracting, burning, and emitting carbon through coal, oil, and gas. Over the past 50 years, there has been a "golden spike" and what climate scientists call the hockey curve feature of carbon emissions, meaning there has been a consistent and intensive increase in carbon emissions. The scientific consensus is simple: human beings are a geological force shaping the planetary conditions that sustain life. We are causing climate change. This is now known as the age of the Anthropocene.

For Marxist and critical scholarship in South Africa, transplanting such a discourse is tantamount to negating the racialized and gendered violence, dispossession, and oppression implicated in the making of a deeply unequal carbon-centric South African capitalism. It also completely occludes an understanding of industrial capitalism, mass consumerism, and fossil fuel corporations in the global North in the making of the climate crisis. In the contemporary context, Anthropocene discourse is inadequate to explain the fracking boom in the US (now making it one of the leading carbon-extracting nations in the world) and the aversion of every US president to locking in to a serious approach to the climate crisis. So rather than rejecting the Anthropocene, radical scholarship has treated it as the capitalist Anthropocene with a longer history, of about 500 years, tied to imperial ecocide as part of militarized mercantilism (Satgar 2018a). Today, the logic of ecocidal accumulation is central to the reproduction of carbon capitalism. It is the harbinger of deleterious impacts on women of colour and subaltern communities barely surviving in the peripheries of capitalism. The impacts of cyclones Idai and Kenneth (both in 2019) on Mozambique rendered this visible with disproportionate impacts on women and children. The Anthropocene is about a global carbon capitalism. It is reproduced by the industrialized countries, carbon capital (including in the global South, as in South Africa), the US imperial state, and the lack of a climate justice agenda within the UN multilateral system. Climate ecocide, the destruction of all of us – human and non-human life – through climate change, is being led by these forces. South Africa's contribution to planetary ecocide is explored further next.

South Africa's carbon capitalism

South Africa is one of the most unequal countries in the world according to any measure and since apartheid. Ironically, this is a conclusion of the World Bank in its recent 2018 report on South Africa. The Southern Africa Labour and Development Research Unit (SALDRU) has made these observations since 2014. Their research has shown that the top 10 percent of households gets two-thirds of South Africa's income while half of all South Africans are chronically poor, living in households with a per-capita income of R1 149 ($77.04) or less per month. South Africa's drought (2014–2021), our first major climate shock, made these inequalities worse through high food prices. New climate inequalities have also been created through the privatization of water. The working class, unemployed, and poor have borne the brunt of the drought and enclosures.

Alongside racialized and gendered super exploitation, high unemployment, and increasing poverty, South Africa is a carbon-intensive economy, based largely on coal. The current post-apartheid state is heavily involved in promoting extractivism (Burton 2018). The coal-dominated electricity sector is an important component of the minerals-energy complex (MEC), the system of accumulation which continues to dominate the economy and which has historically relied on cheap coal and cheap labour (Fine and Rustomjee 1996). It "encompasses critical links and networks of power between the financial sector, government, the private sector and parastatals such as the Industrial Development Corporation and Eskom" (Baker et al. 2015, 8). With coal providing 95 percent of its electricity, South Africa is the highest carbon emitter in Africa and the 14th highest in the world. It has a per-capita carbon footprint higher than China, India, or Brazil. The latest government policy document, the Integrated Development Plan (IDP), indicates that coal will be a significant part (59 percent) of South Africa's energy mix until at least 2030. There is no sense of the urgency of the climate crisis in this document. It is silent on the two key social categories: those who have contributed the most and benefitted the least from building the economy, especially coal workers, and people living in mining-affected communities in which the coal mines and coal-fired power stations are situated.

A powerful body, the Mining Council (earlier called the Chamber of Mines) promotes the notion of coal mining and burning as the essential road to economic growth and stability and is seldom challenged. A recent report is deliberately aimed at countering "negative public opinion about coal" and emphasizes the creation of jobs, procurement spending, and the "adverse" effects of strict environmental laws (COM 2018, 13). Eskom, the state-owned energy provider, is at the centre of the tensions between closing the coal mines to reduce carbon emissions and protecting existing jobs. This is a dilemma for a government of "fragile stability" which is under pressure to deliver on a commitment made in 2009 to reduce carbon emissions drastically. Eskom, which has always been committed to coal, is in crisis. Mismanagement, large-scale corruption, and cost over-runs in building two of the largest coal power stations in the world,

Kusile and Medupi, have pushed Eskom to the point of collapse with heavy debt (R450 billion at present) and frequent blackouts (euphemistically termed "load-shedding") throughout the country. With the worsening climate crisis, South Africa is a carbon criminal state, contributing to the greenhouse effect and the extinction of the human species and other life forms. It is reproducing an 'ecocidal' capitalism destroying the conditions that sustain life.

The limits of historical socialist alternatives

The scale of transformation to advance socio-ecological restructuring of society, to deal with the climate crisis, and to break with ecocidal capitalism requires a new Left project. Lobbying, elite solutions, and incrementalism are not going to work, given the urgency. A new Left imaginary that makes red green and green red or environmental justice activists socialists and socialists environmental justice activists is crucial. Activist practice and scholarship are alive to this challenge in South Africa, which also entails confronting the limits of historical socialist alternatives (Satgar 2018b). South Africa has a Marxist-inspired tradition of socialist thought that is over a century old. The Marxist-inspired liberation movement has had a diverse socialist imagination which has included Sovietized socialism (even Trotsky's minimum programme), revolutionary nationalism, and social democracy. The ANC-Alliance, which has ruled South Africa since 1994, is shaped by all three versions of twentieth-century socialism. While these socialisms have not come to the fore in South Africa in the post-apartheid period, but lurk in the national liberation imagination, they have been theorized in a manner that grounds them in historical experience of these socialisms and, in particular, assumptions about nature.

From a Marxist ecology perspective, these socialisms have the following problems: (a) a blindness to the fact that Marx was an original systems thinker who connected human social relations with nature. Marx understood that the labour process mediated the relationship with nature (Burkett 2014). Further, the human-nature relationship underpinned a "metabolic relationship" with nature as a whole. This means the more capitalism undermined natural cycles and ecosystems, the more the antagonism with nature deepened. (b) An absence of thinking about value creation as grounded in both nature and labour. While labour was "priced in," all these socialisms externalized the costs of nature in the production process. Pollution, climate change, species extinction, and ecosystem destruction, for example, are not taken into account in how production is organized. Nature must be conquered, rather than respected as in agroecology. Nature as having inherent value was denied. (c) These socialisms are all productivist. They copied capitalism's obsession with growth, which meant that accumulation and wealth creation were based on the assumption of endless resources. There were no ecological constraints. (d) All these socialisms are obsessed with technology as progress. Technology is not neutral and is embedded in class relations. For corporations, science and research are about profit making. Unleashing the "forces of production" will not necessarily meet the needs of society and, worse, will have destructive consequences for nature. For example, the introduction of genetically engineered seeds in industrial agriculture is massively increasing the use of agrotoxins and producing new superpests and superweeds (African Centre for Biodiversity 2019).

In the South African context, the renewal of socialism as democratic ecosocialism is also central to the Democratic Marxism intellectual project. This is a book series that rejects a vanguardist Marxism, recognizes the unfinished character of Marxist thought and the need to engender perspectives in dialogue with other currents of radical thought, and affirms deeply democratic practices as the basis for transformative change. In the third volume in the series, *The Climate Crisis: South African and Global Democratic Eco-Socialism*, there is an emphasis on positioning Marxist ecology in dialogue with Indigenous thought, ecofeminism, and radical democratic theory.[1] The democratic ecosocialist imagination emerging seeks a dialectic of praxis, which means that the building of grassroots convergence is crucial for the theorizing of democratic ecosocialism. The Democratic Marxism series is shaped by perspectives from below. We now turn to a mapping of such activist movements and forces developing a democratic ecosocialist horizon of struggle in South Africa.

Democratic ecosocialist struggles to define the deep just transition from below

With the unravelling of the national liberation project in South Africa, there are a host of social forces contesting the political field. One of the animating concepts in these struggles from below is "environmental justice." This is complicated by a fault line which runs through the South African environmental movement. During the apartheid regime, environmentalism operated effectively as a conservation strategy which was mainly concerned with the protection of threatened plants, animals and, wilderness areas and neglected social issues (Beinart and Coates 1995; Kahn 1990; Cock and Koch 1991; Agyeman 2005). Furthermore, for many Black South Africans, environmentalism was linked to dispossession as thousands were forcibly removed to create national parks as well as "protected areas" and, in the process, lost their land and livelihoods (Walker 2008).

The notion of environmental justice represents a dramatic shift away from this traditional authoritarian concept of environmentalism. It is linked to social justice as "an all encompassing notion that affirms the value of all forms of life against the interests of wealth, power and technology" (Castells 1997, 132). It provides a radical alternative to the discourse of ecological modernization, which has been criticized for its reformism and overemphasis on consensual politics (Warner 2020). Environmental justice puts the needs of the poor and excluded at the centre of its concerns, particularly in relation to the impacts of the climate crisis. It rejects the market's ability to bring about social justice or environmental sustainability and thus represents a powerful challenge to the economic growth model and the increasing commodification and financialization of nature packaged as the mythical "green economy" (Martinez-Alier et al. 2014). In the context of intensifying climate crisis and ecosocialist renewal, environmental justice discourse is furnishing a crucial ideological bridge for democratic ecosocialist convergence. Currently the battle lines are being drawn between two broad approaches:

1 A minimalist position that is primarily defensive, emphasizing social protection of vulnerable workers and shallow, reformist change with "green jobs," "green growth," and retraining and consultation
2 An alternative notion of a just transition involving transformative change with totally different forms of producing and consuming to create a more just and equal society

The following crucial struggles are deepening and clarifying a democratic ecosocialist frontier in contemporary South Africa, from below, to enable a transformative just transition:

- Mining impact and pollution struggles: Many grassroots organizations are converging around these struggles. One example is the Highveld Environmental Justice Network and Mining Communities United in Action (MACUA). Central to this network is an NGO called Earthlife Africa, which was launched 30 years ago and which is mobilizing diverse grassroots communities on a variety of rights and claims, some of which have a constitutional grounding as expressed in section 24 of the Bill of Rights. The Bill states that "everyone has the right to an environment which is not harmful to their health and well-being." Many of these forces are converging in a "Life after Coal" alliance, which includes GroundWork and the Centre for Environmental Rights.[2] As a movement it is fueled by the growing contradiction between the discourse of rights and the experience of unmet needs among the millions of South Africans living in poverty (GroundWork 2018).

- Labour and the just transition: Beginning in 2010 the labour movement in South Africa, particularly the labour federation COSATU (made up of 22 different unions), played a key role in introducing and promoting a transformative understanding of a just transition. The concept was grounded in peoples' lived experiences and was aspirational; it was at the heart of a powerful narrative of hope for a more just and sustainable world, a compass for alternative forms of producing, consuming, and relating to human and non-human nature. Central to this shift in organized labour was the research and policy work done by the National Union of Metalworkers (NUMSA) on socially owned renewable energy (Satgar 2015). However, in the past few years, the labour movement has split and retreated into a defensive position, focused on protecting existing jobs. This protective stance is wholly comprehensible in light of South Africa's unemployment rate of 40 percent (one of the highest in the world), massive job losses in the mining industry (particularly gold and platinum), and high levels of poverty. One of the new labour federations in South Africa, the South African Federation of Trade Unions, has a strong commitment to the just transition, and the Federation of South African Trade Unions has also done some policy work on the just transition.
- Anti–fossil fuel extractivism: Several struggles are being waged across South Africa to prevent the expansion of the minerals energy complex through fracking, off-shore gas drilling, and more coal mining. However, these struggles have recently been given a boost by a legal judgement affirming the right of a community to say no to titanium mining. This struggle is popularly known as the "Xolobeni struggle." The villagers of Xolobeni reject being described as "poor" because it diminishes their sense of independence and self-reliance. Their present livelihoods involve strong social networks and access to the natural resources for subsistence farming for some 200 households (Bennie 2017). The community's environmental imaginary involves a deep attachment to the land and a recognition of their right to live close to their ancestors and to protect the land as the source of their identities and livelihoods.
- Climate jobs: In 2009 the one million climate jobs campaign emerged out of the organizing efforts of eight trade unions in the United Kingdom. The campaign focused on the need to connect the climate crisis to the challenge of high levels of unemployment, particularly among youth. The proposal envisaged the decarbonization of work through renewable energy; efforts to increase the energy efficiency of homes and public buildings by insulating them; the implementation of cheap, clean public transport; and the development of "green skills" through training programmes. Moreover, the campaign for climate jobs was then embraced in South Africa in 2011 by an environmental, labour, and social movements coalition. A similar process of research, trade union education, and public engagement has emerged in South Africa (Ashley 2018). In short, the climate jobs alternative is about decent work while seeking to bring down carbon emissions and address the climate crisis as part of the deep just transition.
- Food sovereignty and the solidarity economy: The South African Food Sovereignty Campaign (SAFSC) emerged in early 2015, after a national consultation, as a national campaigning platform to facilitate a loose convergence of organizations (movements, community organizations, NGOs, trade unions, and networks). In total, just over 50 organizations have

converged around the SAFSC platform since its formation. The alliance base continues to grow around different interventions made by the SAFSC. Central to its practice has been a critique of the carbon-based and globalized food regime, symbolic actions to spotlight the drought and its impact (including hunger tribunals, bread marches, and drought speak outs), developing tools for food sovereignty pathway building in communities, villages, towns, and cities through commoning and building convergences with anti-capitalist logics such as solidarity economy networks and similar initiatives to create a counter logic (Satgar and Cherry 2020).

- The Climate Justice Charter process: In the midst of the worst drought in the history of South Africa, the Cooperative and Policy Alternative Centre (COPAC) advanced the idea of a Climate Justice Charter for South Africa. The Centre is a vibrant alliance partner of the South African Food Sovereignty Campaign. In 2019 organizers hosted intensive constituency-based dialogues with drought-affected communities; the media; labour; faith-based communities; and youth, social justice, and environmental justice organizations.[3] Various solidarity protest actions were also organized through this process. A Climate Justice Charter, advancing a vision of an emancipatory ecology that tackles the legacies of deprivation under apartheid and a just transition benefitting the subaltern, has come together. The charter includes a focus on engendering a climate justice future through grassroots-led systemic alternatives. On 16 October 2020, the charter, a climate science document (prepared by some of South Africa's leading climate scientists) and demands from communities to end hunger, water stress, pollution, and climate harm were handed over to South Africa's parliament. The demand made to the parliament was that the charter be adopted as per section 234 of the South African constitution, which provides for charters to be adopted. Currently, partner organizations are engaging with the charter in different parts of the country as the basis to engage with local and provincial governments. At the same time, the charter is guiding and inspiring deep just transition processes in communities and workplaces The Climate Justice Charter Movement is emerging and being constituted in this process to anchor these deep just transition struggles to ensure decarbonization, system change, and a people-driven climate justice state.

Challenges for a democratic ecosocialist future

There are four major challenges confronting ecosocialist activists and movements in South Africa. First is the question of convergence. The climate justice movement globally had a high point at the Cochabamba Peoples Summit in 2010. Subsequently, a resurgent climate activism led by #NoDAPL in 2016 and then in 2018 by Extinction Rebellion and Greta Thunberg's #FridaysForFuture has unleashed a social media–driven crowd politics on the ground. The globalization of the latter forms of resistance has inspired resistance but also created a mimetic dynamic which does not allow for convergence with rooted forces engaged in environmental justice struggles. The ecosocialist forces mapped in this chapter, who have been on the ground in South Africa and engaged in struggles for many years, are also challenged by how to build convergences with these more recent forces. Ensuring these forces converge in a manner that respects independence and self-initiative is going to be crucial while building a solidaristic and strategic politics of mass grassroots power. Institutionalizing a mass-based climate justice movement is a necessary condition to advance a democratic ecosocialist project.

Second, South Africa is steeped in neo-corporatist forms of social dialogue. Elite processes have affirmed green modernization and specifically green neoliberal discourses. If the neo-corporatist approach – involving government, labour, and business – prevails to define the modalities of the just transition, then this will further divide mass forces in South Africa. This could also be described as the difference between a "social dialogue" and a "social power" approach. Writing of the global scene, Sean Sweeney (2018) maintains that

> the political success of this approach is due largely to the just transition being defined in a benign and non-confrontational way which poses little or no challenge to the mainstream, pro-growth, business dominated narrative, a narrative that was largely created by the liberal wing of the global corporate elite.
>
> *(Sweeney and Treat 2018, 27)*

In their view the social dialogue approach is simply not up to the task of bringing about the kind of revolutionary transformative change that the climate crisis requires. In South Africa it is the growing dysfunctionality of the state, racial divisions within business, and a fragmenting labour movement that make social dialogue unworkable.

A third concern is how to anchor just transition struggles in coal-based mining communities and around just transition plans for Eskom and Sasol, two of South Africa's biggest carbon emitters. Coal-based mining communities are extremely precarious. At least half of the 80,000 coal workers in South Africa are migrants employed on short-term contracts negotiated by labour brokers living a precarious and transient existence. Most of the poor, Black communities living close to the operational coal-fired power stations and open-pit working or abandoned mines are experiencing the direct loss of their health due to air pollution. They are dealing with forced removals; social dislocation and dispossession; loss of their land-based livelihoods, including subsistence food production threats; limited access to clean water; the violation of ancestral graves; and inadequate consultation in the awarding of mining licenses. The needs of these communities and workers have to be front and centre in just transition interventions to push for the restructuring of Eskom and Sasol. If this is not the starting point, then such transitions will be shallow and will be against the needs of such communities and workers. Moreover, the historical ecological debt owed by these corporations will be occluded. A democratic ecosocialist orientation will be supplanted by a top-down cost-cutting exercise to achieve "net zero," and the outcome will be an unjust transition.

Fourth is the challenge of preventing the normalization of the climate crisis through government technocratic speak and the politics of recognition. More than 25 years of ANC rule have demonstrated how the state uses various tactics to neutralize mass power, divide, coopt, and even repress mass forces. Currently, the ANC state does not deem the climate crisis to be an emergency, despite climate science identifying Southern Africa as a "climate hot spot," its people experiencing directly the worst drought in the history of the country, heatwaves, flash flooding, and tornadoes. The ruling party is tied into a nexus of corrupt interests that reaches deep into the minerals-energy complex. It is the biggest obstacle to a deep just transition. In this context, difficult strategic dilemmas have to be addressed by climate justice and, more generally, democratic ecosocialist forces in the South African context. Mass power to achieve social justice tipping points to slow down the climate crisis is crucial. A corrupt and failing state cannot deliver on the challenges of the deep just transition. Hence, the logic of democratic ecosocialist counterhegemonic struggles will have to engage with the question of state power as a terrain of struggle.

Conclusion

South Africa's extractivist carbon capitalism has assigned nature an instrumental meaning. Nature is mainly viewed as external, as a store of natural resources, what Oelschlaeger (1991) has termed "resourcism" and often used by capital instrumentally to externalize production costs. Naomi Klein has decried the

> expansionist extractive mindset which has so long governed our relationship to nature . . . we need a new civilizational paradigm, one grounded not in dominance over nature but in respect for natural cycles of renewal and acutely sensitive to natural limits.
>
> (Klein 2015, 80)

The environmental imaginary of the democratic ecosocialist forces in South Africa takes this even further to stress our shared connections in an ecological community. This contrasts sharply with the values of neoliberalism, including its green variant, such as possessive individualism and acquisitiveness. Instead, the ecosocialist forces emerging value participatory grassroots democracy and accountability. As for many ecosocialists, the goal is "living well" rather than striving to live better at the expense of others.

There is no blueprint for a democratic, deep just transition from fossil-fuel capitalism to ecosocialism. Such an alternative is being driven from below by those most affected and not responsible for the crisis. At present, for most South Africans, the idea of a just transition is elusive and has no meaning for the two categories of people who built the coal economy: coal workers and people living adjacent to coal mines and coal-fired power stations. As one of the most unequal societies in the world, South Africa demonstrates the need for a massive redistribution of wealth and power. It is in danger of illustrating an *unjust* transition. However, this is not inevitable, given the contingencies of resistance from below. The ecosocialist forces rising in the country portend a different future and a transformative possibility. As Jason Moore writes, "Capitalism will give way to another model – or models – over the next century" (Moore 2015, 294). Our challenge is to ensure that those models are just and sustainable, which are the two central pillars of ecosocialism.

Notes

1 See http://roape.net/2020/02/13/marxism-and-the-climate-crisis-african-eco-socialist-alternatives/.
2 See https://lifeaftercoal.org.za/.
3 See www.safsc.org.za/wp-content/uploads/2019/11/Climate-Justice-Charter-Draft1-2019.pdf.

References

African Centre for Biodiversity. 2019. *Objection Against General Release of Three 2,4-D GM Maize Varieties*. Johannesburg: African Centre for Biodiversity. www.acbio.org.za/sites/default/files/documents/Objection%20against%20general%20release%20of%20three%202%204%20D%20GM%20maize%20varieties.pdf

Agyeman, Julien. 2005. *Sustainable Communities and the Challenge of Environmental Justice*. New York: New York University Press.

Ashley, Brian. 2018. "Climate Jobs and Two Minutes to Midnight." In *The Climate Crisis: South Africa and Global Democrative Eco-Socialist Alternatives, edited by* Vishwas Satgar, 272–292. Johannesburg: Wits University Press.

Baker, Lucy, Jesse Burton, Catrina Godinho, and Hilton Trollip. 2015. *The Political Economy of Decarbonisation: Exploring the Dynamics of South Africa's Energy Sector*. Energy Research Centre, University of Cape Town, South Africa. www.erc.uct.ac.za/sites/default/files/image_tool/images/119/Papers-2015/15-Baker-etal-Political_economy_decarbonisation.pdf

Beinart, William, and Peter Coates. 1995. *Environment and History: The Taming of Nature in the USA and South Africa*. London: Routledge.

Bennie, Andrew. 2017. "Resistance Is Fertile: Amadiba Agriculture Challenges Elite Mining Agenda." *Daily Maverick*, 20 July. www.dailymaverick.co.za/article/2017-07-20-op-ed-resistance-is-0fertile-amadiba-agriculture-challenges-elite-mining-agenda/ (accessed 28 March 2021).

Bond, Patrick. 2012. *Politics of Climate Change: Paralysis Above, Movement Below*. Pietermaritzburg: University of KwaZulu-Natal Press.

Burkett, Paul. 2014. *Marx and Nature: A Red and Green Perspective*. Chicago: Haymarket Books.

Burton, Jesse. 2018. *Coal Transitions in South Africa*. Cape Town: Energy Research Centre, University of Cape Town.

Castells, Manuel. 1997. *The Power of Identity*. London: Blackwell.

Chamber of Mines (COM). 2018. *Annual Report of the Chamber of Mines*. Johannesburg: Chamber of Mines.

Cock, Jacklyn. 2014 "The 'Green Economy': A Just and Sustainable Development Path or a 'Wolf in Sheep's Clothing'." *Global Labour Journal* 5 (1): 23–24.

Cock, Jacklyn. 2018. "The Climate Crisis and a 'Just Transition' in South Africa: An Eco-Feminist-Socialist Perspective." In *The Climate Crisis: South Africa and Global Democrative Eco-Socialist Alternatives*, edited by Vishwas Satgar, 210–230. Johannesburg: Wits University Press.

Cock, Jacklyn, and Eddie Koch. 1991. *Going Green: People, Politics and the Environment in South Africa*. Cape Town: Oxford University Press.

Fine, Ben, and Zavareh Rustomjee. 1996. *The Political Economy of South Africa: From Minerals-Energy Complex to Industrialisation*. Cape Town: Hurst Publications.

GroundWork. 2018. *Coal Kills: Research and Dialogue for a Just Transition*. Pietermaritzburg: Groundwork. https://lifeaftercoal.org.za/wp-content/uploads/2018/10/Coal_Kills.pdf

Kahn, Farieda. 1990. "Involvement of the Masses in Environmental Politics." *Veld and Flora* 76 (2): 36–38. https://journals.co.za/doi/10.10520/AJA00423203_1942

Klein, Naomi. 2015. *This Changes Everything: Capitalism vs. the Climate*. New York: Simon and Shuster.

Martinez-Alier, Juan, Isabelle Angeulovskia, Patrick Bond, Daniela Del Bene, Frederico Demaria, Julien-Francois Gerber Lucie Greyl et al. 2014. "Between Activism and Science: Grassroots Concepts for Sustainability Coined by Environmental Justice Organizations." *Journal of Political Ecology* 21 (1): 20–60.

Moore, Jason. 2015. *Capitalism in the Web of Life: Ecology and the Accumulation of Capital*. New York: Verso.

Oelschlaeger, Max. 1991. *The Idea of Wilderness*. New Haven, CT: Yale University Press.

Satgar, Vishwas. 2014. "South Africa's Emergent 'Green Developmental State'?" In *The End of the Developmental State*, edited by Michelle Williams, 126–153. Pietermaritzburg: University of KwaZulu-Natal Press.

Satgar, Vishwas. 2015. "A Trade Union Approach to Climate Justice: The Campaign Strategy of the National Union of Metal Workers." *Global Labour Journal* 6 (3).

Satgar, Vishwas. 2018a. "The Anthropocene and Imperial Ecocide: Prospects for Just Transitions." In *The Climate Crisis: South Africa and Global Democratic Eco-Socialist Alternatives*, edited by Vishwas Satgar, 47–68. Johannesburg: Wits University Press.

Satgar, Vishwas. 2018b. *The Climate Crisis: South African and Global Democratic Eco-Socialist Alternatives*. Johannesburg: Wits University Press.

Satgar, Vishwas, and Jane Cherry. 2020. "Climate and Food Inequality: The South African Food Sovereignty Campaign Response." *Globalizations* 17 (2): 317–337.

Sweeney, Sean, and John Treat. 2018. *Trade Unions and Just Transitions: The Search for a Transformative Politics*. New York: Trade Unions for Energy Democracy.

Walker, Cherryl. 2008. *Landmarked: Land Claims and Land Restitution in South Africa*. Johannesburg: Jacana.

Warner, Rosalind. 2020. "Ecological Modernization Theory: Towards a Critical Ecopolitics of Change." *Environmental Politics* 19 (4): 538–556.

18
AGROECOLOGY AS ECOFEMINIST ACTIVISM

Terran Giacomini

Ecofeminists, Indigenists, allied scholars, and activists have developed the theoretical foundations for understanding the importance of women's resistance to capitalist relations and affirmation of life-enhancing alternatives (Mies 1986 [1998], 1996; Allen 1986; Mies and Shiva 1993; Bennholdt-Thomsen and Mies 1999; Kovel 2002 [2007]; Federici 2004, 2019; LaDuke 2005; Brownhill and Turner 2020; Lake, Greensfelder, and Colligan 2014; Isla 2015; also see chapters in this volume from Leigh Brownhill, Ana Isla, Mary Mellor, Ariel Salleh, and Terisa E. Turner). We learn from these scholars and activists that patriarchy, colonialism, white supremacy, and androcentrism are core relations that underpin accumulation. Maria Mies and her co-authors have demonstrated that capitalist accumulation requires a waged-unwaged hierarchy. Women, peasants, colonized and Indigenous peoples, and more-than-human beings are defined as a part of "nature," outside the formal economy. Their labour and creativity is naturalized so that it can be more easily controlled and exploited. The exploitation and control over the unwaged are "the prerequisite for the emergence of the central relation of production between wage labour and capital, which in turn allows women and the colonies to be appropriated as 'natural resources'" (Mies, Bennholdt-Thomsen, and von Werlhof 1988, 4). Especially, important to capital is the work done by women who produce and maintain labour power – capital's most strategically important commodity. Women's resistance to their own exploitation, and to that of others and the land, is crucial because of the power women have, linked to their social location, to deny capital what it needs to continually expand and survive. Any contraction, any loss or threat of loss of control over its indispensable profit base represents a serious threat to the system as a whole.

Women's resistance and alternatives are important for another reason. Women's work, as well as that of other subsistence producers, is very different from the work involved in producing other commodities like cars and software. It is primarily focused on the production and maintenance of life, not the "never-ending accumulation of dead money" (Bennholdt-Thomsen and Mies 1999, 17). Subsistence production – caring for nature, including people, by centrally securing food, water, fuel, and medical and other needs – is central to creating a world beyond capitalism. As Silvia Federici points out, "reproductive work, insofar as it is the material basis of our life and the first terrain on which we can practice our capacity for self-government, is the 'ground zero of revolution'" (2012, 196). This reproductive work and the knowledge and social relations that sustain life require and enable direct access to the means of life and networks of social cooperation,

10.4324/9780429341427-22

that is, commoning. This commoning is oriented toward not only physical survival but also the realization of intrinsic value – love, joy, beauty, community, autonomy, spirituality: in short, that which is "the deepest level of resistance to capital" (Kovel 2008, 9) and which allows us to "evolve into more *human* human beings" (Boggs 2016, 255).

Neoliberal capitalism has undermined the most basic functions of society. With the rise of predatory finance capitalism, a new alliance between banks, extractive industries, and the state is waging a war on subsistence, plundering communities across the globe along similar lines as the conquest and colonization of the Americas, Africa, and Asia (Acosta 2015; Hildyard and Sol 2017). Global firms and their states are promoting carbon markets and corporate technologies that expand commodification and delay meaningful action on the climate crisis (Amorelli, Gibson, and Gilbertson 2021). Capitalist agriculture is a primary cause of social and ecological crises and cannot be ignored. A small handful of firms dominate nearly every link of the food chain from farm to table (Hendrickson et al. 2020).[1] The wealth generated by people and the land is channeled upward to global firms' bottom lines at an unprecedented scale. The extreme extractivism of industrial agriculture is transgressing ecological bounds like never before, fundamentally altering relationships among peoples and between people and the land (Patel and Moore 2017). The enclosure and commodification of land and other means of food production weaken and undermine the self-reliant economies of small- and medium-scale farmers, peasants, and Indigenous peoples, especially the women among them. In many places around the world, women provide the majority of household food needs. Women are also the main protectors of the crop biodiversity necessary to face of climate crisis (Shiva 2016).

As neoliberal capitalism reaches more deadly extremes and tipping points, actions to defend and scale out life-centred food systems and alternatives are more urgent and more obviously necessary than ever before. Around the world organizers from Black, Brown, Indigenous, and people of colour communities and their allies, especially women, are converging around a shared horizon of struggle that offers radical hope for survival and deep transformation towards ecosocialism. The global Via Campesina movement for food sovereignty and agroecology is broadening such a horizon.

The Via Campesina is the world's largest globally networked social movement with more than 200 million members in 81 countries on five continents. They are fighting for grassroots democratic control over the means of life. The Via Campesina's politics of food sovereignty and agroecology are grounded in the life priorities, political economies, and cosmovisions of small- and medium-scale farmers, peasants, Indigenous peoples, workers, migrants, women, youth, and other food producers who are on the frontlines of exploitation and dispossession in the capitalist food chain. According to the Via Campesina (2015) agroecology, grounded in food sovereignty, is not just a method of food production: "[a]groecology is political; it requires us to challenge and transform structures of power in society. We need to put the control of seeds, biodiversity, land and territories, waters, knowledge, culture and the commons in the hands of the peoples who feed the world."[2] Gendered class struggle is central to La Via Campesina's resistance and transformation:

> We women among us are not objects of policies that want to empower us, but rather we are active subjects of Agroecology and guardians of Biodiversity. We want our central role in food production and in the reproduction of life, as well as in the economy of our families and communities, to be visible and recognized. Agroecology means that our rights as women are protected and realized, not just as mothers and caregivers of our homes. Agroecology implies our full participation in the social and political life of our communities, ensuring our access to land, water, seeds

> and means of production with autonomy and freedom. Our equal participation in decision-making spaces is essential.
>
> *(Via Campesina 2018a)*

This chapter explores the ecofeminist character of the political priorities and activism of specific women and non-binary people who are fighting for food sovereignty and agroecology in La Via Campesina. I understand ecofeminist activism as that which stands against relations of exploitation and dispossession, especially of women and non-human nature, and for life-affirming, commoning alternatives (Giacomini et al. 2018). Drawing on participant observation and interviews with women and non-binary peasants in La Via Campesina, I show that their food provisioning and resistance practices are crucial to social transformation in at least two ways: first, by saving seeds, refusing fossil fuel consumption, and building solidarity across key power hierarchies, their agroecology challenges links of the capitalist food chain and builds and strengthens nodes of an alternative, commoner food web. The web, millennia older than the chain, is the interconnected set of relationships that organize the production and consumption of food and caretaking of territory (Giacomini 2020; Brownhill and Turner 2020). For generations, the web has been defended by peasant and Indigenous communities, especially by women, whose work and struggles have been crucial to sustaining life and the commons. Second, the Via Campesina women and non-binary farmers in this study are engaging in campaigns and direct actions to resist extractive projects and greenwashing. They seek deep systemic change that goes beyond limited reformism and the "greening" of capitalism (Isla 2015). They are fighting not only for physical survival and to end exploitation but also to realize intrinsic value that is core to humanity and our relationships.

Next, I highlight four ways in which specific Via Campesina women and non-binary farmers are resisting accumulation and affirming alternatives across multiple links of the capitalist food chain – seeds, markets, fossil inputs, and the climate – and economic sectors: food and energy.

Seed commons versus hybrids and GMOs

The first link of the food chain is the seed. In the capitalist food chain, transnational corporations have attempted to establish increasing levels of control over seeds by forcing costly agrotoxins and bioengineering technologies on the world's farmers. Across the globe, Via Campesina food producers and their allies are expanding local networks of seed saving to defend peasant and Indigenous seed systems. These networks are a core requirement for food systems based on food sovereignty and agroecology. They allow farmers to refuse commodification and be autonomous from the chain. In 2013 La Via Campesina launched a global campaign for seed sovereignty to protect peasants' and Indigenous peoples' seed-saving, use, and exchange practices and their customs and cultures threatened by patents and other forms of enclosure. Women are at the forefront of this campaign (Via Campesina 2013, 2015, 2018b; Shiva 2016). With the commodification of seeds, women's use of and control over seeds is restricted (Pionetti 2005), and their contribution to agriculture is either destroyed or devalued (Shiva 1989). With seed sovereignty, women have power and autonomy. They can, for example, feed their families and communities and access land unhindered by corporate market relations.

On 18 May 2019, I spoke to Vicky Akello, a small-scale farmer and member of the Eastern and Southern Africa Small-scale Farmers' Forum, Uganda (ESAFF-Uganda), a member organization of La Via Campesina. Since the mid-2010s, Akello and her allies have been in a countrywide battle to stop the introduction of genetically modified organisms (GMOs) into

Uganda's food system (Jaramogi 2017). Among the main proponents of GMOs in Uganda are multinational companies and research institutions developing a disease-resistant variety of GMO banana (Mwesinga 2017; Slow Food 2017). Uganda is home to more than 50 varieties of bananas. Bananas are an important source of income and nutrition for subsistence producers. With the introduction of GMOs into Uganda's food system, small-scale farmers and their communities are at risk of losing control not only over bananas but also over other important food sovereignty crops, including maize, cassava, potatoes, rice, soybeans, and sweet potatoes. The introduction of GMOs threatens Uganda's biodiversity-based subsistence-positive food system.

In 2017, in response to the passage of the National Biotechnology and Biosafety Bill (known as the "GMO Bill") that would have legalized GMOs, ESAFF-Uganda and their allies coordinated a national campaign against the bill. Akello (2019) said the primary goal of the campaign was to prevent the introduction of GMOs: "[O]ur resolution, we made it: any GMO in Uganda must be labeled. Any GMOs must be approved to enter Uganda, and then we do not want any GMO to enter Uganda without any label on it." In the world of GMOs, labeling is a *de facto* ban on GMOs. By January 2018, under tremendous pressure from farmers, president Yoweri Museveni refused to sign the bill into law. This success, hard won, is continually being defended. Uganda is still under pressure to legalize GMOs (AFSA 2019).

Akello reported that the mobilizations against the GMO Bill impelled an increase in seed commoning:

> Farmers have started leaving the hybrid seeds and coming back to Indigenous seeds because they know the effect. Now they have come to the organization of ESAFF-Uganda to find out, "how have you managed to bring these seeds? Where were the seeds from?"
>
> We started using Indigenous seeds, agroecology, especially in the North. I have started multiplying Indigenous seeds to fight climate change. I think farmers brought over a hundred seeds that were disappeared. Now we have started multiplying in favour of doing agroecology.
>
> *(Akello 2019, personal communication)*

Women are key defenders of the seed commons in Uganda. According to Akello,

> The women are now starting to put their own seed at the fireplace for preservation, in pots. If you lack seed you have to go to your friend to get the seed from them. They have seeds, and if you ask for seeds from woman to woman they do not reject. Go and plant that one and you can continue giving others those indigenous seeds.
>
> *(Akello 2019, personal communication)*

Akello and other farmers in ESAFF-Uganda are resisting GMOs and defending and expanding their Indigenous seed commons. For Akello and her community, seeds are not commodities to be owned, bought, and sold. Having and sharing seeds are key to food sovereignty. The seed commons under women's protection is a crucial element of the food web and a basis for resistance to commodification in the chain.

Local food sovereignty versus the corporations' global chains

Plantation agribusiness and multinational food companies depend for their profits on the super exploitation of racialized and migrant workers, especially women. Many of the Via Campesina women I interviewed are positioning their agroecology as a refusal of the racialized and gendered class exploitation underpinning agribusiness profits. Roz Corbett from the Landworkers' Alliance (LWA) in the United Kingdom said that her community's local agroecology is helping eaters "delink" from the global food chain in solidarity with racialized and migrant food producers. According to Corbett:

> The international perspective is really important to agroecology and food sovereignty. For example, the migrant workers in Spain who work fourteen hours in a plastic desert picking tomatoes for Tesco, and the migrant workers in Romania who come here because they can earn more but still they have to live four of them in a caravan – we are doing this work because it is not right to make someone else, somewhere else do the work, and they have the right to do the work they want to do that is properly recognized.
>
> *(Corbett 2020, personal communication)*

The agroecology Corbett and her allies are fighting for is about much more than simply expanding local food production. It seeks to transform relationships of power and domination. Corbett views agroecology as a means to oppose the racism and ethno-nationalism that she sees existing in her community. By seeing "everything in food sovereignty from a global perspective," she and other activists are

> [t]ak[ing] what could potentially be seated in a nationalist viewpoint, as well as "local food for local people," and "our standards are better" kind of thing, and we are kind of pushing against it. This is not what food sovereignty is for. It is about everyone globally, and our environment globally and all the associated campaigns around that. This feels really important when you come from a country where nationalism is strong and deep-rooted.
>
> *(Corbett 2020, personal communication)*

Corbett recognizes that agroecology is not only a way to produce food more ecologically. For her, agroecology, grounded in food sovereignty, aims to disrupt the global chain's plantation logics and build commoner food webs within which workers around the world "freely express themselves in self-determined, social labor" (Kovel 2005, 3). By growing food locally, Corbett and her allies in the global North are building food sovereignty in cooperation with producers in the South who are fighting for the same. They also make it possible for producers and consumers to engage in simultaneous production-consumption strikes and boycotts that have the potential to strengthen the web vis-à-vis the chain (see Tobocman, Brownhill, and Turner in this volume).

Agroecology and direct action against "false solutions" to the climate crisis

Many of the Via Campesina members I interviewed are involved in climate justice activism to stand up against "false solutions" to the climate crisis. Using their market and political power, transnational corporations and their states are positioning corporate markets and technologies as the best and only "solution" to the ecological catastrophe that their enclosures produced (Amorelli, Gibson, and Gilbertson 2021). These false solutions include so-called climate smart agriculture (CSA). Many big fertilizer companies, including Syngenta and Yara International, and food giants like Danone are working through the UN-endorsed Global Alliance on Climate Smart Agriculture (FAO 2021) to push a failed paradigm of industrialization under the guise of climate smart approaches and technological innovation.

The World Bank, the UN Food and Agriculture Organization (FAO), and the International Fund for Agricultural Development (IFAD) propose climate smart agriculture as a solution to the climate crisis that also promotes "transformative change" for women (World Bank, FAO, and IFAD 2015, 5). These agencies claim that transformative policies encourage the adoption of "high-yielding" seeds as well as land titling and privatization. "Climate smart" policies seek to tie smallholder farmers more tightly to capitalist value chains including carbon markets (Taylor 2018). There is a high risk that such policies will intensify the conditions under which women are exploited as invisibilized, unwaged producers of labour power on the one hand and, on the other, as cheap waged goods producers labouring under conditions of microcredit debt bondage. The Via Campesina rejects the World Bank's liberal gender inclusion policies and instead insists that true transformation comes from community control over the means of life and decision-making processes.

On 1 February 2019, I interviewed Elizabeth Mopfu, a small-scale farmer from Zimbabwe and the general coordinator of La Via Campesina. She denounced climate smart agriculture as part of a global greenwashing of chemical industrial agriculture and a new green revolution in Africa. She said,

> Look at how CSA [climate smart agriculture] is being promoted and why we are really fighting against CSA. People should really know what exactly it is about. Some people are really talking about it in a way [in] which we peasants are not happy about because we know it is promoting the Green Revolution. They really want to promote the massive use of fertilizers, which we peasants cannot afford. If I get the inputs in the form of fertilizer, herbicides, I use it, now look, there is no rain. But I purchased those inputs and in return they want me to pay something. That is why we are really witnessing the suicides committed by some of the peasants like in China and India.
>
> *(Mpofu 2019, personal communication)*

Mpofu and her allies recognize that confronting the climate crisis and defending small-scale farmers requires a major shift in power from the food chain to the food web. She and her allies are fighting for a fundamentally different food system in which life and health are at the centre, not corporate profits. Peasants and Indigenous peoples, especially women, already feed more than 70 percent of the world's population using only a quarter of the agricultural resources and 10 percent of agriculture's fossil energy (ETC Group 2017). Scaled out and with government support, food sovereignty and agroecology can reduce carbon emissions by half within a few decades (GRAIN 2016). Agroecology is an important part of the "true solution" to the crisis (Amorelli, Gibson, and Gilbertson 2021; Rosset and Altieri 2017). Peasants and Indigenous peoples' practices to caretake the soil, avoid and reduce agrotoxins, lower food miles, avoid expensive and risky technologies and other false solutions, and fight to give land back to peasants and Indigenous peoples are part of the "just transition" from capitalism to commons (Via Campesina 2018c; Giacomini 2020). For many Via Campesina members, transitioning away from fossil fuel–based inputs is an important part of their climate justice activism.

Agroecology versus fossil agriculture

The capitalist food chain is heavily dependent on hydrocarbon energy (hereafter referred to as fossil fuels). Export-based industrial monocultures require large amounts of fossil fuel–based agrotoxins and heavy machinery. Industrial agriculture is a "linerized system" that, at one end, is "a receptacle for petro-industrial inputs. Therefore, the other end has become an ejector for those inputs' byproducts" (Qualman 2019, 27). In contrast, one of the core characteristics of agroecology is its reliance on animal, solar, microbial, and other sources of energy rather than fossil fuels (Rosset and Altieri 2017).

On 27 March 2017, I spoke to Paula Gioia, a peasant from La Via Campesina's Arbeitsgemeinschaft bäuerliche Landwirtschaft (AbL) in Germany. They told me about the on-farm practices they and their community are using to avoid fossil fuel consumption and contribute to "cooling down the earth" (Via Campesina 2007). These practices include diversifying production, using draft animal power, and localizing value chains. Gioia explained the benefits of these practices:

> With agroecology and food sovereignty there is much less reliance on fossil fuels. Production is not specialized. It is more diverse, and diversified production reduces the need for heavy machinery and fossil fuels. Concretely, here on our farm, we try to use horses as much as we can instead of the tractor. We use horse-drawn carts for transporting the harvest. Our production is oriented toward local markets. With agroecology, the long-distance transportation system that brings products from one corner of the world to the other is also reduced or minimized. The food chains are shorter and consumption is more adapted to local tastes and seasons. Refrigeration, packaging, plastics, and other harmful materials used in long-distance transport are also reduced. With agroecology we are not only supporting the environment but also local economies – the street markets rather than the supermarkets. We avoid intermediaries by selling directly to the small shops that are not part of the big supermarket chains.
>
> *(Gioia 2018, personal communication)*

By practicing agroecology, Gioia and other Via Campesina members are challenging the logics of extractivism that degrade the land, fuel the climate crisis, and destroy local communities. They are also taking responsibility for the "negative commons," not treating nature as a dumping ground but working in harmony with and caretaking nature's cycles for present and future generations (Bennholdt-Thomsen and Mies 1999, 143).

While many Via Campesina members view their agroecology as a form of direct action, many also recognize the importance of other kinds of non-violent actions to stop extractivism and affirm life-centred alternatives. Next, I discuss a non-violent direct action in which the priorities of women were prominent.

Non-violent civil disobedience against the fossil fuels and for life

Since its founding in 1993, La Via Campesina has built convergences and solidarities with other movements "from below" (Martinez-Torres and Rosset 2010; Giacomini 2016; Calvário, Desmarais, and Azkarraga 2019). In the fight against false solutions, La Via Campesina is joining with their allies in other sectors of the value chain. They are engaging in campaigns and direct actions to challenge capitalist processes and neoliberal policies promoted at the annual

UN climate meetings, the Conference of the Parties (COP) to the United Nations Framework Convention on Climate Change (UNFCCC). On 3 December 2017, parallel and in opposition to COP23 in Bonn, Germany, La Via Campesina members and social movement activists from across Europe and the world participated in a large-scale civil disobedience action to shut down the Hambach open pit coal mine, the largest of its kind in Europe and one of the continent's major sources of greenhouse gas emissions (Watts 2017).

Activists travelled 50 kilometres from Bonn to the Rhinelands, West Germany, where the Hambach mine is located. The action was organized by Ende Gelände (in English, "Here and No Further"), a joint initiative by German and other European organizations fighting for "an immediate stop to coal mining and a just transition to a social and ecological society" (Ende Gelände 2016). Four thousand activists from across Europe and the world participated. The civil disobedience action forced the mine's owners, RWE, to temporarily halt the operation of three massive coal excavators and one conveyor belt. Organizers celebrated their action, proclaiming that, "[f]or one day, the destruction ceased" (Ende Gelände 2017). Months later, in June 2018, under pressure from civil society and social movements, the German government announced plans to phase out coal-powered energy in the country (Amelang 2018).

The action built solidarity across racialized and gendered class hierarchies in at least two ways. First, it was significantly influenced by queer feminism. Queer feminism centres the priorities of two-spirit, LGBTQIA+ and queer Indigenist feminists fighting for gender self-determination, decolonization and inclusive, life-centred mode of relationship with one another, and the land (Hall 2017). The participants in the coal mine shutdown were organized into five "fingers," each of which blockaded a different location of the mine to strategically avoid massive disruption of the action by law enforcement. The five fingers contained multiple "blocks" of smaller numbers of people. One of them was called the queer feminist block. Feminists and their allies within Ende Gelände created the queer feminist block as a response and an alternative to a direct-action culture that is aggressive, macho, and individualistic (An 2018). Rather than simply blockading the mine at all costs, the primary objective of the queer feminists and allies was to complete the action together through prefigurative collective decision-making, caretaking, and solidarity. Queer feminists practice emphasized sharing power and caretaking one another's physical and emotional well-being. Their action philosophy was grounded in an understanding that strength is rooted in cooperation and mutualism, not competition and individualism (Giacomini 2018).

Second, the participants in the coal mine shutdown affirmed South-North solidarity. In the action statement, Ende Gelände organizers explained that "[c]ountries such as Germany must face their historical responsibility in global warming." This requires that "[p]hasing out coal and all fossil energies cannot be negotiated. This measure is necessary to give those affected by climate change – now and in the future – a chance to live a good life" (Ende Gelände 2017). Before heading into the mine, there was a big rally at which speakers from communities on the frontlines of the climate crisis took the main platform and developed the action's political message and intention. Among them were Indigenous peoples fighting to protect their territories from carbon trading in the Amazon, Indonesia, and the USA; activists fighting coal in South Africa; and members of La Vía Campesina, who affirmed food sovereignty and agroecology as alternatives to false corporate solutions.

The action showed that people are not waiting for governments to take the lead on the climate crisis. They are taking matters into their own hands, drawing on grassroots democratic organizing practices and solidarities across racialized and gendered class hierarchies. These solidarities were ecofeminist and ecosocialist in important ways. They sought not only to strengthen opposition to fossil capital and false solutions but also to transform our relationships with one another to protect the "mutual flourishing of life on earth" (Clark 2019, 132). Such solidarities are contributing to the development of a fundamentally different set of global power relations from below, one that exists in opposition to the corporate "captured" UNFCCC (Corporate Accountability 2017).

Conclusion

Ecofeminist analysis draws attention to the ways in which women's life priorities and activism are crucial to social transformation. Ecofeminist activism stands against the exploitation of all creatures, human and non-human. It seeks a major shift in power and control over the means of life so that "ultimately, the earth is no longer privately owned" and so that people can "self-determine their transforming of nature" (Kovel 2007, 160). The Via Campesina women and non-binary farmers profiled in this study are seeking to transform sites of enclosure and exploitation in the capitalist food chain into sites of resistance and the construction of commoner food webs. Their seed saving, local food systems, refusals of fossil inputs, campaigns, and direct actions are creating the social and material infrastructure necessary for food producers to remain on the land and break free from global value chains. Women are among the food web's primary architects. Their knowledge, social networks, and material commons are crucial to building locally rooted "soil not oil" food webs on a global scale (Shiva 2008).

A global food web is emerging from initiatives by women and their allies to take action in solidarity with others across key power hierarchies. These Via Campesina women and their allies are coordinating their activism across multiple links – seeds, markets, the climate, and fossil fuels – and sectors – food and energy – of the global food chain. As capitalist processes of enclosure and commodification wage a war on subsistence and undermine the integrity of ecosystems, women and their allies are fighting back and practicing alternatives to defend life.

Notes

1 Briefly, the food chain refers to the infrastructure and relationships that organize the production, distribution, and consumption of food. There are different kinds of food chains under the control of different interests. The capitalist food chain is based on the commodification of food and private, for-profit control over the means of life. In contrast, food chains exist in peasant and Indigenous communities. These food chains (webs) are much older, are based on life needs and on shared control over the means of life. The capitalist food chain is parasitic on pre-existing food chains. For more on food chains, see ETC Group 2017.

2 This chapter addresses the political and social movement dimensions of agroecology. For more on the scientific dimension, see Altieri 1987; Gliessman 2007.

References

Acosta, Alberto. 2015. "Post-Growth and Post-Extractivism: Two Faces of the Same Cultural Transformation." *Rebelión*, 27 March. https://rebelion.org/dos-caras-de-la-misma-transformacion-cultural/

AFSA. 2019. "Open Letter of Appreciation to the President of the Republic of Uganda." *Food Sovereignty Alliance in Africa*, 2 September. https://afsafrica.org/open-letter-of-appreciation-to-the-president-of-the-republic-of-uganda/

Akello, Vicky. 2019. Personal Communication. Harare, Zimbabwe, 18 May.

Allen, Paula Gunn. 1986. *The Sacred Hoop: Recovering the Feminine in American Indian Traditions*. Boston: Beacon Press.

Altieri, Miguel. 1987. *Agroecology: The Scientific Basis of Alternative Agriculture*. Boulder, CO: Westview Press.

Amelang, Sören. 2018. "Germany Starts Coal Exit Talks in Bid to Improve Patchy Climate Record." *Clean Energy Wire*, 6 June. www.cleanenergywire.org/news/germany-starts-coal-exit-talks-bid-improve-patchy-climate-record

Amorelli, Lucia, Dylan Gibson, and Tamra Gilbertson. 2021. "Hoodwinked in the Hothouse: Resist False Solutions to Climate Change, Third Edition." https://climatefalsesolutions.org/wp-content/uploads/2021/04/HOODWINKED_ThirdEdition_On-Screen_version.pdf (accessed 22 April 2021).

An. 2018. Personal Communication. WhatsApp Audio, 13 February.

Bennholdt-Thomsen, Veronika, and Maria Mies. 1999. *The Subsistence Perspective: Beyond the Globalised Economy*. New York: Zed Books.

Boggs, Grace Lee. 2016. *Living for Change: An Autobiography*. Minneapolis: University of Minnesota Press.

Brownhill, Leigh, and Terisa E. Turner. 2020. "Ecofeminist Ways, Ecosocialist Means: Life in the Post Capitalist Future." *Capitalism Nature Socialism* 31 (1): 1–14. https://doi.org/10.1080/10455752.2019.1710362

Calvário, Rita, Annette Aurélie Desmarais, and Joseba Azkarraga. 2019. "Solidarities from Below in the Making of Emancipatory Rural Politics: Insights from Food Sovereignty Struggles in the Basque Country." *Sociologia Rualis* 60 (4): 857–879. https://doi.org/10.1111/soru.12264

Clark, John P. 2019. *Between Earth and Empire: From the Necrocene to the Beloved Community*. Oakland, CA: PM Press.

Corbett. 2020. Personal Communication. WhatsApp Audio, 7 May.

Corporate Accountability. 2017. "Polluting Paris: How Big Polluters Are Undermining Global Climate Policy." https://corporateeurope.org/sites/default/files/attachments/pollutingparis_cop23report_2017.pdf (accessed 22 April 2021).

Ende Gelände. 2016. "Facts about Ende Gelände: Coal and the Rhineland." www.ende-gelaende.org/wp-content/uploads/2016/02/Facts-about-Ende-Gel%C3%A4nde-and-the-Rhineland.pdf (accessed 22 April 2021).

Ende Gelände. 2017. "Ende Gelände in the Hambach Opencast Mine." *Ende Gelände*, 24 November. www.ende-gelaende.org/en/news/ende-gelande-in-the-hambach-opencast-mine/

ETC Group. 2017. "Who Will Feed Us?: The Peasant Food Web vs. the Industrial Food Chain, Third Edition." www.etcgroup.org/sites/www.etcgroup.org/files/files/etc-whowillfeedus-english-webshare.pdf (accessed 20 April 2021).

FAO. 2021. "Global Alliance for Climate Smart Agriculture: Member List." www.fao.org/gacsa/members/members-list/en/ (accessed 20 April 2021).

Federici, Silvia. 2004. *Caliban and the Witch: Women the Body and Primitive Accumulation*. Brooklyn, NY: Autonomedia.

Federici, Silvia. 2012. *Revolution at Point Zero: Housework, Reproduction and Feminist Struggle*. Oakland, CA: PM Press.

Federici, Silvia. 2019. *Re-Enchanting the World: Feminism and the Politics of the Commons*. Oakland, CA: PM Press.

Giacomini, Terran. 2016. "System Change Grounded in Food Sovereignty at the 2015 Climate Talks in Paris." *La Via Campesina*, 27 January. https://viacampesina.org/en/system-change-grounded-in-food-sovereignty-at-the-2015-climate-talks-in-paris/

Giacomini, Terran. 2018. "The 2017 United Nations Climate Summit: Women Fighting for System Change and Building the Commons at COP23 in Bonn, Germany." *Capitalism Nature Socialism* 29 (1): 89–105. https://doi.org/10.1080/10455752.2018.1434217

Giacomini, Terran. 2020. "An Ecofeminist Analysis of Grassroots Activism for a Just Transition." In *Climate Justice and Community Renewal: Resistance and Grassroots Solutions*, edited by Brian Tokar and Tamra Gilbertson, 194–205. London: Routledge.

Giacomini, Terran, Terisa E. Turner, Ana Isla, and Leigh Brownhill. 2018. "Ecofeminism Against Capitalism and for the Commons." *Capitalism Nature Socialism* 29 (1): 1–6.

Gioia, Paula. 2018. Personal Communication. WhatsApp Audio, 27 March.

Gliessman, Stephen R. 2007. *Agroecology: The Ecology of Sustainable Food Systems*. Boca Raton, FL: CRC Press.

GRAIN. 2016. *The Great Climate Robbery: How the Food System Drives Climate Change and What We Can Do About It*. Toronto: Between the Lines.

Hall, Laura. 2017. "Indigenist Intersectionality: Decolonizing an Indigenous Eco-Queer Feminism and Anarchism." *The Institute for Anarchist Studies* 15 January. https://anarchiststudies.org/indigenist-intersectionality-decolonizing-an-indigenous-eco-queer-feminism-and-anarchism-by-laura-hall/

Hendrickson, Mary K., Philip H. Howard, Emily M. Miller, and Douglas H. Constance. 2020. "The Food System: Concentration and Its Impacts." *Family Farm Action Alliance*. https://farmactionalliance.org/wp-content/uploads/2020/11/Hendrickson-et-al.-2020.-Concentration-and-Its-Impacts-FINAL.pdf (accessed 17 April 2021).

Hildyard, Nicholas, and Sol. Xavier. 2017. "A Critical Introduction to Infrastructure Mega-Corridors." *Counter Balance*. www.thecornerhouse.org.uk/sites/thecornerhouse.org.uk/files/Mega%20Corridors.pdf (accessed 17 April 2021).

Isla, Ana. 2015. *The "Greening" of Costa Rica: Women, Peasants, Indigenous Peoples, and the Remaking of Nature*. Toronto: University of Toronto Press.

Jaramogi, Patrick. 2017. "Will GMOs Solve Hunger Problems in Uganda?" *ESAFF News*, 10 March. https://esaff.org/index-php/will-gmos-solve-hunger-problems-in-uganda/

Kovel, Joel. 2002 [2007]. *The Enemy of Nature: The End of Capitalism or the End of the World?*. 2nd ed. London: Zed Books; Halifax: Fernwood.

Kovel, Joel. 2005. "The Ecofeminist Ground of Ecosocialism." *Capitalism Nature Socialism* 16 (2): 1–8. https://doi.org/10.1080/10455750500108146

Kovel, Joel. 2008. "Ecosocialism, Global Justice, and Climate Change." *Capitalism Nature Socialism* 19 (2): 3–14. https://doi.org/10.1080/10455750802091123

LaDuke, Winona. 2005. *Recovering the Sacred: The Power of Naming and Claiming*. Chicago: Haymarket Books.

Lake, Osprey Orielle, Claire Greensfelder, and Elaine Colligan. 2014. "Women's Climate Action Agenda." *Women's Earth and Climate Action Network International*. https://e01c23b4-9f2e-4830-9320-a86de06b013e.filesusr.com/ugd/bf92a9_5bf690f628354586a58189ee17704c6a.pdf (accessed 12 April 2021).

Martinez-Torres, Maria Elena, and Peter M. Rosset. 2010. "La Vía Campesina: The Birth and Evolution of a Transnational Social Movement." *The Journal of Peasant Studies* 37 (1): 149–175. https://doi.org/10.1080/03066150903498804

Mies, Maria. 1986 [1998]. *Patriarchy and Accumulation on a World Scale: Women in the International Division of Labour*. London and New York: Zed Books.

Mies, Maria, Veronika Bennholdt-Thomsen, and Claudia von Werlhof. 1988. *Women the Last Colony*. London: Zed books.

Mies, Maria, and Vandana Shiva. 1993. *Ecofeminism*. London: Zed books.

Miles, Angela. 1996. *Integrative Feminisms: Building Global Visions, 1960s–1990s*. New York: Routledge.

Mpofu. 2019. Personal Communication. Harare, Zimbabwe, 1 February.

Mwesinga, Alon. 2017. "Can a GM Banana Solve Uganda's Hunger Crisis?" *The Guardian*, 12 December. www.theguardian.com/global-development/2017/dec/12/gm-genetically-modified-banana-uganda-hunger-crisis

Patel, Raj, and Jason W. Moore. 2017. *A History of the World in Seven Cheap Things: Guide to Capitalism, Nature and the Future of the Planet*. Oakland: University of California Press.

Pionetti, C. 2005. *Sowing Autonomy: Gender and Seed Politics in Semi-Arid India*. London: Institute for Environemnt and Development.

Qualman, Darrin. 2019. *Civilization Critical: Energy, Food, Nature and the Future*. Halifax and Winnipeg: Fernwood Publishing.

Rosset, Peter, and Miguel Altieri. 2017. *Agroecology: Science and Politics*. Winnipeg: Fernwood Publishing.

Shiva, Vandana. 1989. *Staying Alive: Women, Ecology and Development*. London: Zed Press.

Shiva, Vandana. 2008. *Soil Not Oil: Environmental Justice in a Time of Climate Crisis*. Amsterdam and New York: Elsevier.

Shiva, Vandana. 2016. *Seed Sovereignty, Food Security: Women in the Vanguard of the Fight Against GMOs and Corporate Agriculture*. Berkeley, CA: North Atlantic Books.

Slow Food. 2017. "The GMO Bill Puts the Future of Uganda Under Threat: Slow Food Offers a Different Way Forward." *Slow Food*, 19 December. www.slowfood.com/gmo-bill-puts-future-uganda-threat-slow-food-offers-different-way-forward/

Taylor, Marcus. 2018. "Climate-Smart Agriculture: What Is It Good for?" *Journal of Peasant Studies* 45 (1): 89–107. https://doi.org/10.1080/03066150.2017.1312355

Turner, Terisa E., and Leigh Brownhill. 2006. "Ecofeminism as Gendered, Ethnicized Class Struggle: A Rejoinder to Stuart Rosewarne." *Capitalism Nature Socialism* 17 (4): 87–95. https://doi.org/10.1080/10455750601004574

Via Campesina. 2007. "Small Scale Sustainable Farmers Are Cooling Down the Earth." *La Via Campesina*, 9 November. https://viacampesina.org/en/small-scale-sustainable-farmers-are-cooling-down-the-earth/

Via Campesina. 2013. "La Via Campesina: Our Seeds, Our Future." *La Via Campesina*, 18 June. https://viacampesina.org/en/our-seeds-our-future/

Via Campesina. 2015. "Seed Laws that Criminalise Farmers: Resistance and Fightback." https://viacampesina.org/en/wp-content/uploads/sites/2/2015/04/2015-Seed%20laws%20booklet%20EN.pdf (accessed 20 April 2021).

Via Campesina. 2018a. "Declaration at the II International Symposium on Agroecology April, 2018." *La Via Campesina*, 8 April. https://viacampesina.org/en/declaration-at-the-ii-international-symposium-on-agroecology/

Via Campesina. 2018b. "In rural Paraguay, Women Are on the Frontlines of a 'Race Against Time' to Save Native Seeds." *La Via Campesina*, 13 September. https://viacampesina.org/en/in-rural-paraguay-women-are-the-frontlines-of-a-race-against-time-to-save-native-seeds/

Via Campesina. 2018c. "La Via Campesina in Action for Climate Justice." In *Radical Realism for Climate Justice*. Berlin: Heinrich Boll Foundation.

Watts, Jonathan. 2017. "Germany's Dirty Coalmines Become the Focus for a New Wave of Direct Action." *The Guardian*, 8 November. www.theguardian.com/environment/2017/nov/08/germanys-dirty-coalmines-become-the-focus-for-a-new-wave-of-direct-action

World Bank, FAO, and IFAD. 2015. *Gender in Climate-Smart Agriculture: Module 18 for the Gender in Agriculture Sourcebook. Agriculture Global Practice*. Washington, DC: World Bank Group. www.fao.org/3/a-i5546e.pdf (accessed 20 April 2021).

19

"YOU CAN BLOW YOUR BRAINS OUT AND YOU AIN'T GETTING NOWHERE"

Jazz, collectivism, and the struggle for ecological commons in Louisiana's sugar parishes

Benjamin Barson

Building off saxophonist and self-described matriarchal ecosocialist Fred Ho, who posited that jazz aesthetics both offer resistance to commodification and embody an "Eco-Logic[al] Aesthetic," I consider jazz as the product of an ecosocial set of relations that could be considered a precursor to ecosocialism (Ho 2011). I posit that New Orleans jazz cannot be understood apart from the agrarian commons fought for by the workforce of Louisiana's sugar plantation economy (Scott 2005; Gould 1984). Sugar informed the lives of early jazz musicians, from the folkways of communities working the crop to the migrations and markets that pro-sugar imperial policy engendered to the anti-plantation resistance this economy engendered. A generation of jazz, including Ben and Chris Kelly, bassist Pops Foster, and trombonist Kid Ory, grew up on Louisiana sugar plantations.

In the following pages, I examine how Black Louisianan music cultures reflected the same drive for autonomy observed in the act of growing one's own food. By tracing the politics of land, labour, and music making in the sugar belt in the decades following the Civil War and Reconstruction, I suggest that we can hear in the New Orleans brass band the sounds and struggles of the commons. I foreground the variety of ways that Black musicians and activists responded to the infrastructure and social relations that the production of sugar required. I take a regional approach so as to locate Black Louisianan sugar workers' resistance within a broader Afro-Caribbean struggle against the sugar system. In addition to marronage and, in the post-emancipation years, strikes, it is through the proliferation of gardens that a coherent Black agrarian praxis rooted in communal land tenure systems – a predecessor of contemporary US Black southern variants of ecosocialism – can be seen (Akuno 2019).

Tracing social relations in sonic production can be a difficult and perhaps overly speculative task, but it is striking that musicians themselves often emphasized communal phenomenology as essential to creating good music. As trombonist Jim Robinson, born on the Deer Range

10.4324/9780429341427-23

plantation, explained, collectivity, communication, and interdependence are values in early jazz that are more important than individualistic virtuosity:

> Now if them people don't work together, you can blow your brains out and you can't – you ain't getting nowhere. Now, if that banjo and that drum are going just as fast as a cyclone, [and] you can't execute your horn like you want to. . . . It's too fast, understand? But as long as that drum and that piano and that banjo and bass, that must work together.
>
> *(Robinson 1958)*

In Jim Robinson's words, the sublime in this music is not the quest for virtuosity or seamless individual self-expression. In contrast to the vision of singular genius composer or virtuoso in Western art music, here the art lies in the collective management of a shared sense of time and space, one which propels communication, synergy, and communally reinforcing expression. This is all the more striking in that such euphoric and even utopic sounds were elaborated within the dystopian working conditions of sugar plantations, which were marked by relatively high levels of industrialization with deadly impacts on the workforce. The communalistic and empathetic music experience that Jim Robinson described was an apt phenomenological practice necessary for the kinds of autonomous institution building that gardening portended, both as the sonic embodiment of the replacement for the world in which they toiled and as an incubator for its new social relations. I explore this theme in more detail next.

Non-linearity and communal economics in gardening

> I only marveled at the way the garden is for me an exercise in memory, a way of remembering my own immediate past, a way of getting to a past that is my own (the Caribbean Sea) and the past as it is indirectly related to me (the conquest of Mexico and its surroundings).
>
> *– Jamaica Kincaid (2001)*

In the long durée of plantation slavery, the gardens of Afrodescendants in the US South were some of the most concrete manifestations of a world where many worlds fit. These systems of autonomous production were vital to social reproduction and the cultural identity of the Caribbean, to which Louisiana was more organically tied than to the US South. Writing in 1774, the Jamaican plantocrat Edward Long (1774, 537) calculated that 20 percent of Jamaica's hard currency was circulating in the slaves' internal economy. Architect Benjamin Latrobe opined in 1820 that Louisiana would starve without the foodstuffs bought and sold at the slave markets in *Place des Nègres*, also known as Congo Square (Latrobe 1951, 47). On Sundays, several hundred enslaved Afrodescendants gathered here for collective dancing and singing, and the space is now memorialized as the Afrocentric precursor to jazz. But its initial *raison d'être* was as a critical circuit in the independent agro-botanical economy practiced by enslaved Black sugar hands (Johnson 1991). In addition to being vital for social reproduction, these spaces sustained ecologies that may have gone extinct otherwise. "Nurtured in even the harshest circumstances," writes Donna Haraway, "slave gardens not only provided crucial human food, but also refuges for biodiverse plants. . . . Slave gardens are an underexplored world" (2015, 165).

Slave gardens are an underexplored world, and so is the music created therein. Black music studies initially focused on songs created on plantation grounds and not those from the self-administered provision grounds of enslaved Afrodescendants (Baraka 1963; Southern 1971). Important as Black music was to resisting plantation labour regimes, including work "slow down" songs, it is also important not to lose sight of the vital cultural connections and social relations of these ecological refuges. A study of the landscape architecture and the radical polycultures of these gardens reveals some striking corollaries with the distinct phenomenologies found in Black Atlantic music, which has historically promulgated what musicologist Olly Wilson (1992, 327) calls a "heterogenous sound ideal." Heterogeneity was similarly commented on by the planter Matthew Lewis in 1834, in Westmoreland Estate, Jamaica. While walking through what he called the "negro village," Lewis reported that the divisions between home, garden, and personal property were blurred by a rich botanical abundance: "[T]he whole village is intersected by lands, bordered with all kinds of sweet-smelling and flowering plants" (Lewis 1834, 85; Higman 2001, 262). Women were at the centre of these autonomous systems of production and the independent markets that they generated, a product of a radical egalitarianism along gendered and racial lines that could be termed ecofeminism (Mintz 1955; Brownhill and Turner 2019).

These gardens were striking because they disrupted the plantation's rectangular geometry and its commodity logic. The plantation was marked by straight, neatly ordered lines demarcating monocultural grids. In contrast, noted the white Jamaican attorney James Simpson in 1832, it was "impossible" to "make any survey of the land by the negroes, and they generally cultivate it in a straggling way" (Higman 2001, 36). These "straggling" landscapes were both fiscally and geometrically illegible to Euro-American planters, and they did not conform to their gardening ideal. In 1844, the influential landscape architect Andrew Jackson Downing opined that "the mind can only attend with pleasure and satisfaction, to one object, or one composite sensation, at the same time" and therefore concluded that "there is something unpleasing in the introduction of fruit trees among elegant ornamental trees" (Downing 1841, 80–81). Yet mixtures of fruit trees and ornamental trees, not to mention dozens of other plant species, characterized the "negro gardens" that Matthew Lewis encountered. This intercropped sublime was an important manifestation of social and ecosocial harmony as envisioned by enslaved Afrodescendants. Not only crucial to the food sovereignty of the enslaved communities (and the subsistence of white Caribbeans), it reflected an autonomous path of economic and cultural development.

These gardening practices are broadly resonant with other nineteenth century Afro-Atlantic philosophies. Several scholars of Black culture have noted how "randomizing the flow of paths" is a Senegambian practice that reappears in the Americas as a means to slow the movement of evil (Thompson 1983, 222; Georgia Writer's Project 1940; Tobin and Dobard 2011, 49). Evil travelled in straight lines, and the oppressive geometry of plantations proved this Black Atlantic axiom. Endless divisions of rectangular plantation landscapes reflected the planter ideal. The straight lines of plantation boundaries and their endless fields of vertical cane penetrating both the sky and soil stood as geoengineered monuments to Afrodiasporic commodification and the planter pursuit of profit and power – expressions of a Euclidean ego antithetical to the cooperative ideal.

The gardens of enslaved Afrodescendants manifested a different flow, and its epistemological grammar reinforced communitarian trends in Black music. The "impossible" incongruity of these gardens and their sophisticated intercropping find corollaries in the seemingly constant overlapping rhythms, known as hemiolas, in Black Atlantic music. From Santiago de Cuba to Haiti to New Orleans, this music thrives on the experience of multiple time signatures weaving in and out of perception (Floyd 1999). As with Black gardens, these sonic and rhythmic practices generated a range of responses, from wonderment to hostility, from white observers. "So strangely they vary their time, as 'tis a pleasure to the most curious ears" remembered one otherwise racist Barbados slaveowner, while New Orleans architect Benjamin Latrobe declared about Congo Square dances and their hypnotic array of interlocking rhythms: "I have never seen anything more brutally savage" (Dubois 2016, 114). The contemporary Haitian American writer Eugene Danticat describes the experience of these interlocking rhythms during Haitian carnival with considerably more nuance:

> At last, my body is a tiny fragment of a much larger being. I am part of a group possession, a massive stream of joy. . . . In that brief space and time, the carnival offers all the paradoxical elements I am craving: anonymity, jubilant community, and belonging.
> *(Danticat 2015, 147)*

The inclusivity and multi-sensorial aspect of Black Atlantic music and the diverse polycultures of Black Atlantic gardens shared a communitarian grammar and a commitment to radical inclusivity: of life forms, of forms of being, and of the euphoria simulated by multiple experiences of time and consciousness.

The connections between Black music and land practices do not only operate at the level of phenomenology and consciousness. One can also analyze them historically, as these cultural forms literally grew together. In Haiti, the *kombit* labour system of collective land management was announced and coordinated by large percussion and vocal ensembles (Métraux 2016). Across the plantation Caribbean, communal sound led sugar workers to their semi-autonomous commons from plantation enclosure. In Jamaica, the conch shell signaled to slaves when their shift was done so that they could return to these provision grounds, playing melodies and rhythms associated with another world. These motives became etched not only into the sonic repertoire of enslaved peoples, but also in the very name of these gardens: they were referred to as the "shellblow grounds" (Struge and Harvey 1838, 224–225; Higman 2001, 263). These connections between sound and independent social relations were made even more apparent during the winter of 1831–1832 in Jamaica. In Montego Bay, one of the largest slave uprisings in Jamaican history broke out, and planters were terrified at the sudden effectiveness of these conch shells. When plantations were burnt *en masse* by dissident labourers, the shells no longer announced scheduled shifts to the provision grounds but coordinated a collective arson of biblical proportions – a different kind of deliverance. "The conch shell was heard to blow in every quarter, accompanied by huzzas and shouts from the infatuated slaves," reported an exasperated solider (Mathieson 1926, 212), while other witnesses reported that "whole fields, each perhaps contained twenty, thirty, forty acres or upwards were thus ignited," and "the sky became a sheet of flame, as if the whole country had become a vast furnace" (Senior 1835, 179; Zoellner 2020, 111–112). The aurality of the conch shell thus contained not only the utopian vision of the provision grounds but also the will to eliminate the demonic grounds of the plantation complex and its global enclosure economy. This resonant body martialed collective resistance in this revolt, containing the seed of a new world cultivated in its dialectic with the Black Atlantic gardening complex.

Music, like any form of culture, was not mere entertainment among Afro-Caribbean plantation workers. It embodied the social relations and visions of liberation produced by an enslaved workforce. This held true as well for what is often referred to as the northern outpost of the Caribbean, New Orleans (Vidal 2019), and its surrounding sugar plantations. Several accounts by former Louisianan slaves describes how illicit Black music at night reflected the complex dialectic between scenes of subjection and spaces of collective pleasure (Hartman 1997). For instance, Elizabeth Ross Hite described how sugar houses – industrial processing centres, regulated by the rhythm of profit, which often claimed the lives and limbs of overtired workers – were late-night dance halls: "De slaves had balls in de sugar house. Dey would start late an' was way out in de field whar de master could not heah dem. Not a bit of noise could be heahed" (Hite n.d.). To avoid alerting the master, slaves danced by a low candlelight. Music transformed these spaces, re-socializing the anti-social and dehumanizing world of the planation. The nighttime sugar factory was now a place where "de slave had some fine times" doing "de buck dance and de shimme" and "sake[ing] dere skirts" (Hite n.d.) to the left and the right. Black music and its communitarian grammar not only contributed to the formation of a new commons but also created a culture of pleasure and solidarity to defend them.

Musical transfiguration: Brass bands as escape agriculture

With this background in mind, we would do well to refocus our attention on the time and place of post-Emancipation Louisiana and the musicians who grew up among the generations of Black rural labourers following the abolition of slavery. I argue that it is no coincidence that many early New Orleans jazz musicians were agricultural workers who both worked for and resisted Louisiana's sugar barons. With the convergence of multiple social movements and revolutionary traditions in post-emancipation Louisiana, both communal music making and alternative economic systems proposed by freedpeople imagined an ecosocial and ecosocialist transformation. To understand how this came to be so, it is necessary to briefly overview the tumultuous years following the abolition of slavery, referred to as Reconstruction (Foner 1988; Du Bois 1935).

In the aftermath of slavery, plantations were seized throughout Louisiana's sugar country by freedpeople who created new commons, expanding their gardens to larger proportions and elevating self-sufficiency to a higher priority than the growth of commodity crops. The records of military officials of the Freedman's Bureau, set up to supervise the transition to wage labour, are revealing in this regard. In January 1865, a community of freedmen requested of the Freedman's Bureau "the opportunity of working the Place among ourselves," and their request was granted. But freedmen did not want to grow what they called "slave crops," such as cotton and sugar. According to one military observer, they hated these crops because they "had enriched the masters, but had not fed them" (Pearson 1906, 181; Foner 1988, 51). During this period, reports abound of Louisianan freedpeople living off the land on their own terms, despite planter obstinacy. One planter in Bayou Black complained, "Much trouble has resulted from negroes being allowed to keep horses and hogs," making them less in need of plantation wages (Berlin et al. 1990, 556). More conservative Freedmen's Bureau agents were committed to "breaking up" these "colonies of squatters"; more radical agents marveled at their success and argued that only the "small farms" of freedmen could "destroy the serfdom of capital" (Hahn et al. 2008, 578).

Records kept by these agents reveal these were collective endeavours, spearheaded by different families who had toiled on the plantations as slaves. One Freedman Bureau's agent noted: "A crop had been planted by the people on the place, in which all were to share" (Hahn et al. 2008, 627). The opinion of Freedman's Bureau agents varied widely. Perhaps the most famous Freedman's Bureau agent was African American activist Martin R. Delany, best known as a proponent of the idea that Afro-descended people should endeavour to create a Black state outside the United States. In these decades, he became an advocate for a Black land commons in the South: "Get up a community and get all the lands you can – if you cannot get any singly" (Hahn et al. 2008, 632). In Austin, the Black-run *Free Man's Press* wrote in 1868: "A good way is to club together and buy a piece of land and divide it up into lots" (*The Free Man's Press* 1868). Land reform had swept the collective psyche of this emancipated generation.

From Western North Carolina to Virginia to Louisiana, Black communes proliferated. An innumerable thousand emerged where freedpeople refused to grow the slave crops of cotton and sugar and instead built regenerative ecosystems (Dulken 2019). In Louisiana, these enunciations were explicitly tied to overthrowing the tyranny of capital. In 1865, the New Orleans *Tribune*, the nation's first daily Black newspaper, called for a "Bank of Laborers" to collectively manage and finance farming implements for occupied plantations.

> As capitalists needed capital to work the plantations, let the people themselves make up this capital. Our basis for labor must now be put on a democratic footing. There is no more room, in the organization of our society, for an oligarchy of slaveholders, or property holders.
>
> *(Tribune 1865)*

They framed their struggle for postwar autonomy as an "emergency," one which could not be solved with "the money of the few" but rather only through a "democratic association [that] looks to small shares, with a large concourse of adherents." They proposed mechanisms for profit sharing with a sophistication that anticipated modern-day cooperatives. But they had powerful enemies, including from the North. Louisiana Governor Warmoth was a pro-business, Illinois-born Union Army veteran who was terrified of these proposals to end plantation exploitation. He claimed later in his autobiography that land reform activists "thought to establish an African State Government . . . [and] urged the negroes of Louisiana to assert themselves and follow Hayti" (Warmoth 2006, 51–52). Warmoth's reactionary denunciation was on the mark in one respect: many of the key organizers of this imitative were of Haitian descent and maintained familial connections in the Black republic. Several among this cadre, such as Rodolphe Desdunes (whose son, Daniel, was a popular jazz musician and civil rights activist) self-consciously invoked the French and Haitian Revolutions in their Reconstruction activism (Brickhouse 2007; Scott 2007; Bell 2011; Otto 2011). They embodied Louisiana's legacy in an Afro-Caribbean, pan-African rejection of plantation labour in favor of a cooperative and ecological alternative.

As a reign of white supremacist terror called "Redemption" overthrew Reconstruction and disrupted these commoning initiatives across the South, organized rural labour endured significantly longer in Louisiana's sugar districts than in cotton country – by some estimates, by more than decade. The unique conditions of sugar plantations partly explain the success of collective organizing strategies among Louisiana freedpeople relative to Black activists elsewhere. Unlike sharecropping arrangements in the cotton-picking districts, on sugar plantations, hundreds of freedpeople lived in direct proximity. Institutions, mutual aid societies, communal agriculture, and mass meetings were organized with relative ease (Scott 2005, 83).

These same conditions made communal music making particularly powerful. One of the most concrete ways this can be evinced is how bands helped fundraise for Black-administered plantation schools (Koenig 1997, 87). In 1885, for the "Teachers Institute program," "an audience of more than 500 people assembled in the hall" to hear "music by the colored band that volunteered its services Mr. Shieb spoke upon the subject of education after the band had played several excellent pieces Prof. Puckette addressed the audience on the subject of spelling" (Koenig 1997, 91). Black bands were also connected to Reconstruction activism, as one freedwomen remarked in 1864: "Some will look upon these times as if nothing but politics, mass meetings, drums and fifes and gilt muskets were all the go" (*Christian Recorder* 1864). This fusion of public sound and mass assembly can be best characterized as *brassroots* democracy.

Plantation music education, political activism, and the proximity of sugar workers – who lived in miniature industrialized cities called plantations – help explains why a very high proportion of early New Orleans jazz musicians were born in the sugar districts. The map in Figure 19.1 measures levels of sugar production across the locations of some seventeen influential jazz musicians active at the turn of the century New Orleans.

On sugar plantations, brass bands were deeply tied to mutual aid associations. Sonny Henry explained that membership in a band was impossible without joining "[t]he fraternity lodges,

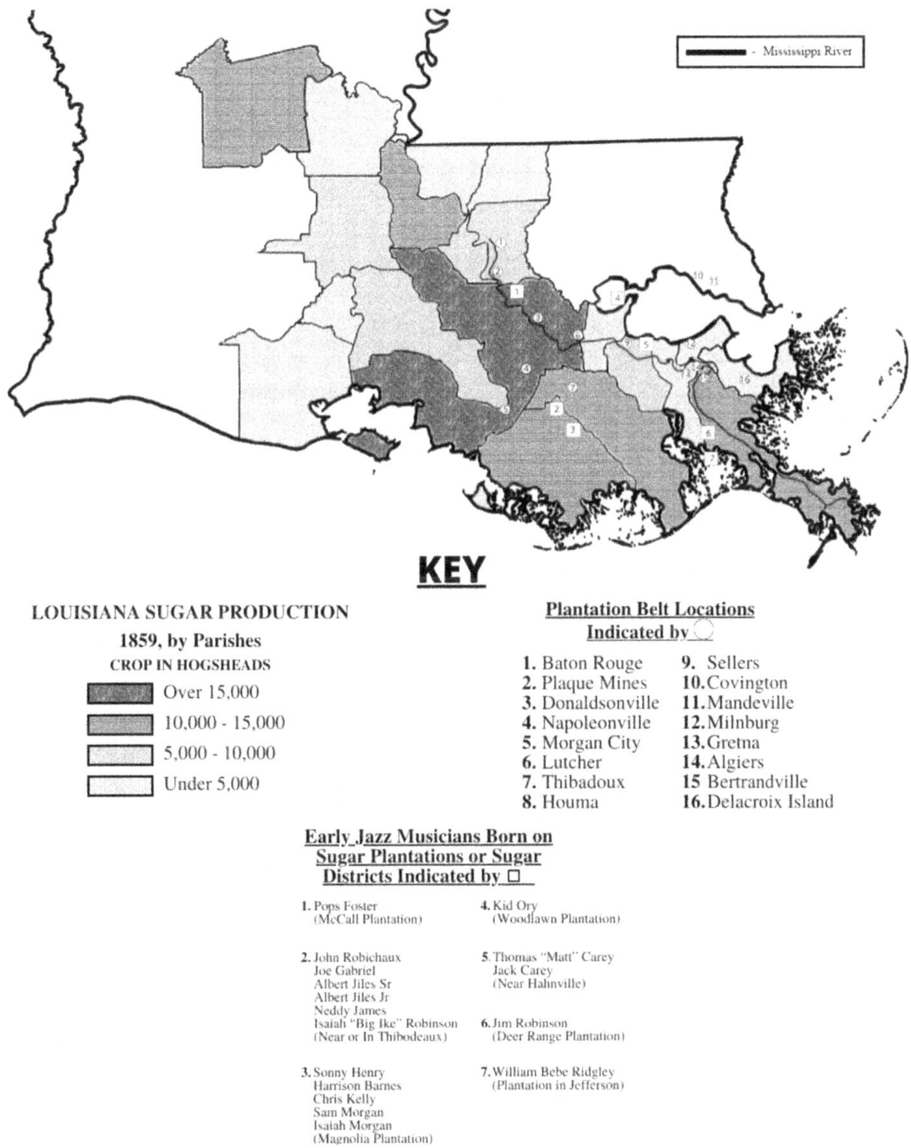

Figure 19.1 Louisiana sugar production 1859, by parishes

Source: Stitterson (1953, 48)

that's the onliest way they'd be in it." These institutions provided social services when the state or the plantation bosses failed to do so. "When you're sick they give you a little benefit," he explained, and "they gave meetings sometimes twice a month" that addressed topics ranging from politics to community affairs (Henry 1959a, 1959b).

Land reform was dealt a death blow by opportunist Republican capitalists like Governor Warmoth, who took land redistribution off the table during Louisiana's otherwise radical Constitutional Convention in 1868 (Connor 1980; Shugg 1937). He became a planter not long after (Rodrigue 2006). Nine years later, the Compromise of 1877 meant that white supremacist terror would rule the South. Planters would enforce a brutal social discipline reminiscent of slavery throughout the nineteenth century and into the twentieth (Henry 1959a).

Denied land of their own and massacred when they attempted to strike, Black workers still wielded a potent weapon: they could refuse to come to work. As early as February 1866, the West Baton Rouge *Sugar Planter* called attention to the plantations lying idle from lack of hands. "The fact, long ere this, must have become patent to every reflecting mind," the editor wrote, "that a substitute for negro labor must soon be procured" (*Sugar Planter* 1866). In 1871, Robert Somers wrote, "The great law of demand and supply in the matter of labour operates here under curious circumstances, the supply neither knowing what it is worth nor what it wants" (Somers 1871, 222). The majority of workers refused to fulfil contracts for the year. According to one observer in April 1869, instead of working, they congregated "in and around the cities and towns, where they scarcely ever pretend to engage in regular labor; they talk politics and lead profligate and corrupt lives and subsist in the most wretched manner" (Bouchereau and Bouchereau 1869, vii; Sitterson 1953, 237). They tried to organize other workers, too: in 1874 the New Orleans *Picayune* reported that "Large sections of the State are overrun by lawless bands of negroes, who visit plantations, stop all work, threatened the lives of the peaceful and contended laborers, and fill the county with terror" (Halpern 2004, 20).

Instead of raising wages, planters began hiring musicians to play for and, in some cases, offer musical training to rural labourers. Music was especially important for attracting migrant workers during cutting season. The influential trombonist Kid Ory, who grew up on the Woodland plantation, remembered that the great band of Henry Peyton came from New Orleans during the harvest season. Festivities of drink and music, funded by planters, became the scene where new contracts were signed. In other cases, they enticed migrant workers to settle down. Ory himself travelled to a plantation across the state because he "thought he might get a chance to hear Claiborne Williams's band" (McCusker 2012, 31). He was duped, the band never arrived, and Ory was stuck in a predatory contract.

Other planters took an addition step, such as ex-Governor Warmouth, who provided musical training for his workers. According to one musician, "The plantation owners wanted to keep the people interested and wanted them to learn something other than farming" (Koenig 1997, 4). These programmes were not the outgrowth of planter altruism, but rather a means to stabilize an exploitative labour regime. Nonetheless, Black and Creole of colour working-class communities were not "duped" by these sonic institutions. Their very existence embodied a contested social contract and reflected the contradictory desires of planters and their reluctant workers. For Black Louisianans, they found in these bands a means to continue a dream denied. With their historic struggle for the commons interrupted by violence and political machination, bands became all the more important to fashion self-identity, create collective action, and institutionalize cultural resistance. They created egalitarian cultures, demonstrated in their inclusion of women. Women musicians in parades or second lines were unheard of in late nineteenth- and early twentieth-century New Orleans, but in the sugar parishes, Black women were frequently observed in brass bands (Tucker 2016, 265).

The Afro-Creole musician James Humphrey was key to this process. A Civil War veteran and "one of the best solo cornetists in the city in his day" (Koenig 1982, 24), Humphrey was hired by Warmoth to teach sugar workers at Magnolia plantation. Several of Humphrey's pupils became innovative New Orleans jazz musicians, including Chris Kelly and Sam Morgan (Julian 1974, 8). Gene Miller notes that Humphrey's "students at Magnolia and elsewhere became some of the best jazz musicians in New Orleans; many of them eventually led their own bands and made recordings" (Miller 1994, 9).

The "Professor," as he was affectionately called, gave his students some of their first gigs in New Orleans; Humphrey's grandson Willie Humphrey remembers that "Jim Humphrey used to bring his country boy bands down [to New Orleans] to play big Mardi Gras parades," which were so cold that "sometimes the valves would freeze on the horns" (Koenig 1981, 25). The scenes described by his pupils are strikingly picturesque against the grim backdrop of the plantation work camps. Humphrey would be met by students eager to accompany him to practice or, sometimes, his quarters – he might have to spend the night if things went late. Humphrey taught bands all across the sugar plantation belt, and by the time Sonny Henry started learning how to play at 15 years old (the same age when his schooling ended), Humphrey had developed a system. "I gone tell you," explained Henry, "the way he taught the boys, I think it was the right way" (Henry 1959a).

But while the Professor shared musical knowledge and professional opportunities, he also experienced personal growth in his collaborations. Most often, would-be musicians did not have cash to pay and instead exchanged crops grown in their gardens. Willie Humphrey recalled that "[m]y grandpa used to bring us pecans, sweet potatoes, sugar cane and all from the country. He also made a garden and from the figs in it he made enough to pay the taxes" (Koenig 1981, 24). The story reflects that Black workers in the sugar parishes were still cultivating gardens and provision grounds, perhaps the same garden plots their families worked on during slavery (Gould 1984, 44). Humphrey's garden shows that he was a student himself, learning crop preparation techniques with the plantation workers whose sound he helped develop. His grandson remembered that "[Humphrey] had a garden for food and I think teaching music was a labor of love" (Koenig 1982, 16). These forms of Black rural agency point to an underlying synergy between two spaces: the agro-botanical and the musical. Both were spaces of autonomous expression and self-development.

Regimented musical training and brass musical instruments made a profound impact on generations of Black youth, and Black workers sought them out whenever possible. Later in life, the skills that these mobile schools afforded became stepping stones for ex-canecutters to enter New Orleans's prestigious second-line bands. The experience of Sonny Henry's first gig in New Orleans, with the Excelsior Band, is instructive in this regard. "I didn't have a uniform," he remembered, "[so] I say [to bandleader George Moret] 'I haven't got no uniform.' . . . [H]e says, 'Can you read?' I told him, 'Yeah.' He says, 'Well, I don't want the uniform, I want the man.'" Henry recalled the difficulty of his first march – "I couldn't hardly walk [march in time]," but veteran Vic Gaspard reassured him: "Now, listen, you take it easy, 'cause this is your first time. . . . It's difficult for you to, you know, catch the step" (Henry 1959a, 1959b). Sonny Henry reclaimed and repurposed the music he was exposed to in the plantation schools to join a community of urban Black musicians, who entrusted him with a kinetic literacy that was exclusive to New Orleans musicians. A system of plantation control had been turned into a means of exodus, allowing working-class Black structures of affiliation, work, and community to take root in and against a society that had attempted to extinguish these spaces.

Brass bands were transformed by Black sugar workers as a kind of compensatory commons to the land they had been denied by racial capitalism. Building off a centuries-long inheritance, such bands reproduced a commons that created new geographies of Black social life and created opportunities for planation workers to contribute to an emerging Creolized culture that would come to be called jazz. As Paul Friedrich (1977, 191) has written about Mexican *cuerpos filarmónicos* during this same period, "musical specialization compensated for restricted access to the means of subsistence, in this instance because of the ruling faction's preferential access to *ejidal* [communal] land." The plantation brass bands created a different kind of subsistence by capturing the phenomenology of the provision grounds and the Black Atlantic commons and forcefully articulating these currents into early jazz. This was why, as the former sugar worker and New Orleans trombonist Jim Robinson explained, "you can blow your brains out and you ain't getting nowhere." Collective improvisation with brass, string, and percussion instruments was a novel and powerful innovation that collapsed the boundary between individual and community. To be a self-interested virtuoso negated the point of the music. Its power comes from that great dialectic of self and other, of history and spirit. In the long durée of plantation slavery and its afterlives, that history of struggle was the confrontation of an embryonic ecosocialism, rooted in the concrete practices of self-sufficiency in the provision grounds and gardens of Afrodescendant plantation workers, in opposition to racial capitalism and slavery.

Conclusion

Over centuries, Black music has reflected the activism, agency, and revolutionary proposals of enslaved peoples and their descendants. When the Great Migration engendered a massive rural-to-urban shift, the lessons from the struggles in the agricultural districts lingered. As jazz became circulated globally, it held a paradoxical power as a projection of US soft power and Black revolutionary dissidence, leading to widespread adoption by colonized peoples struggling against regimes of capital and accumulation not dissimilar from those found in the Black belt South (Denning 2015; Eschen 2009). This paradox is not dissimilar to the paradox found in the plantation brass band: funded by planters to fend off revolt, yet a meaningful circuit of resistance that provided a way to escape the plantation for many of the musicians responsible for developing early jazz. They brought with them knowledge of distinct agro-botanical systems, dreams of the post-emancipation commons, and a process of communal re-creation in brassroots democracy. These lessons have borne important consequences for the movement for anti-racist democracy and ecosocialism, which have assumed both a mass and a musical character. This understanding squares with that of activist and ecosocialist Colia Clark (2014), a Student Nonviolent Coordinating Committee (SNCC) activist. For Clark, music and the forms of commoning practiced by Black peoples in the plantation belt can be understood as a fusion of a spiritual-religious ecological consciousness and a belief in communal self-determination – what she calls ecosocialism. I close with her words in order to connect readers to a form of knowledge that cannot be reduced to historiography:

> Music is the spirit of a people. Music is the thing that gets you up in the morning. Of course, Africans have a song for everything. [sung] *I woke up this morning with my mind. Stay on freedom, Hallalelu.* It's the thing that gets you up in the morning but it's the thing that puts you to bed at night. It's the thing that, in troubled hours, just comes and tells you: if you stay on the battlefield, stay on the battlefield, it's gonna be alright, that there will be a victory. That it's coming, and its music for your dark hours. As when my friend Medgar [Evers] was killed 51 years ago. For me, it was just the darkest hour. But the music tells you that [sung] *the darkest hour is just before day. Sun's gonna*

come and wash the darkness away. You know, it's the music that keeps us going. And now we got the eco-music: *Eco, tell it on the mountain. Over the hills and everywhere. Go, tell it on the mountain. Ecosocialism is born.*

References

Akuno, Kali. 2019. "It's Eco-Socialism or Death." *Cooperation Jackson*, 17 February. https://cooperation-jackson.org/blog/ecosocialismordeath

Baraka, Amiri. 1963. *Blues People: Negro Music in White America*. New York: W. Morrow.

Bell, Caryn Cossé. 2011. "'Rappelez-Vous Concitoyens': The Poetry of Pierre-Aristide Desdunes, Civil War Soldier, Romantic Literary Artist, and Civil Rights Activist." In *Rappelez-Vous Concitoyens: La Poésie de Pierre Aristide Desdunes*, edited by Pierre-Aristide Desdunes, 1–19. Shreveport, LA: Les Éditions Tintamarre, Centenary College of Louisiana.

Berlin, Ira, Thavolia Glymph, Steven F. Miller, Joseph P. Reidy, and Julie Saville, eds. 1990. *Freedom: A Documentary History of Emancipation, 1861–1867: The Wartime Genesis of Free Labor: The Lower South*. Vol. 2.1. New York: Cambridge University Press.

Bouchereau, Louis, and Alcée Bouchereau. 1869. *Statement of the Sugar and Rice Crops Made in Louisiana in 1868–69*. New Orleans, LA: Pelican Book and Job Printing Office.

Brickhouse, Anna. 2007. "'L'Ouragan de Flammes' ('The Hurricane of Flames'): New Orleans and Transamerican Catastrophe, 1866/2005." *American Quarterly* 59 (4): 1097–1127.

Brownhill, Leigh, and Terisa E. Turner. 2019. "Ecofeminism at the Heart of Ecosocialism." *Capitalism Nature Socialism* 30 (1): 1–10.

Christian Recorder. 1864. (Philadelphia, PA).

Clark, Colia. 2014. "Black Red and Green Revolutionary Eco-Music Tour." *Old Dog Documentaries*. https://vimeo.com/91707067 (accessed 20 April 2021).

Connor, William P. 1980. "Reconstruction Rebels: The New Orleans Tribune in Post-War Louisiana." *Louisiana History: The Journal of the Louisiana Historical Association* 21 (2): 159–181.

Danticat, Edwidge. 2015. *After the Dance: A Walk Through Carnival in Jacmel, Haiti*. New York: Knopf Doubleday Publishing Group.

Denning, Michael. 2015. *Noise Uprising: The Audiopolitics of a World Musical Revolution*. New York: Verso Books.

Downing, Andrew Jackson. 1841. *A Treatise on the Theory and Practice of Landscape Gardening: Adapted to North America; with a View to the Improvement of Country Residences. Comprising Historical Notices and General Principles of the Art, Directions for Laying Out Grounds and Arranging Plantations, the Description and Cultivation of Hardy Trees, Decorative Accompaniments to the House and Grounds, the Formation of Pieces of Artificial Water, Flower Gardens, Etc. With Remarks on Rural Architecture*. New York and London: Wiley and Putnam.

Dubois, Laurent. 2016. *The Banjo: America's Africa Instrument*. Cambridge, MA: Harvard University Press.

Du Bois, W.E.B. 1935. *Black Reconstruction: An Essay Toward a History of the Part Which Black Folk Played in the Attempt to Reconstruct Democracy in America, 1860–1880*. 1st ed. New York: Harcourt, Brace and Co.

Dulken, Danielle. 2019. "A Black Kingdom in Postbellum Appalachia." *Scalawag*, 9 September. https://scalawagmagazine.org/2019/09/black-appalachia-kingdom/

Eschen, Penny Von. 2009. *Satchmo Blows Up the World: Jazz Ambassadors Play the Cold War*. Cambridge, MA: Harvard University Press.

Floyd, Samuel A. 1999. "Black Music in the Circum-Caribbean." *American Music* 17 (1): 1–38.

Foner, Eric. 1988. *Reconstruction: America's Unfinished Revolution, 1863–1877*. New York: Harper & Row.

The Free Man's Press. 1868. (Austin, TX).

Friedrich, Paul. 1977. *Agrarian Revolt in a Mexican Village*. Chicago: University of Chicago Press.

Georgia Writers' Project. 1940. *Drums and Shadows: Survival Studies Among the Georgia Coastal Negroes*. Athens: The University of Georgia Press.

Gould, Jeffrey. 1984. "The Strike of 1887: Louisiana Sugar War." *Southern Exposure* 12: 45–55.

Hahn, Steven, Steven F. Miller, Susan E. O'Donovan, John C. Rodrigue, and Leslie S. Rowland, eds. 2008. *Freedom: A Documentary History of Emancipation, 1861–1867: Land & Labor, 1865*. Vol. 1.3. Chapel Hill: University of North Carolina Press.

Halpern, Rick. 2004. "Solving the 'Labour Problem': Race, Work and the State in the Sugar Industries of Louisiana and Natal, 1870–1910." *Journal of Southern African Studies* 30 (1): 19–40.

Haraway, Donna. 2015. "Anthropocene, Capitalocene, Plantationocene, Chthulucene: Making Kin." *Environmental Humanities* 6 (1): 159–165.
Hartman, Saidiya V. 1997. *Scenes of Subjection: Terror, Slavery, and Self-Making in Nineteenth-Century America*. New York: Oxford University Press.
Henry, Sonny. 1959a. *Hogan Jazz Archive*. New Orleans, LA: Tulane University, 8 January.
Henry, Sonny. 1959b. *Hogan Jazz Archive*. New Orleans, LA: Tulane University, 21 October.
Higman, B.W. 2001. *Jamaica Surveyed: Plantation Maps and Plans of the Eighteenth and Nineteenth Centuries*. Kingston, Jamaica: University of West Indies Press.
Hite, Elizabeth Ross. n.d. "WPA Ex-Slave Narrative Collection, State Library of Louisiana." https://louisianadigitallibrary.org/islandora/object/state-lwp%3A8594
Ho, Fred. 2011. "How Does Music Free Us? 'Jazz' as Resistance to Commodification and the Embrace of the Eco-Logic Aesthetic." *Capitalism Nature Socialism* 22 (2): 52–61.
Johnson, Jerah. 1991. "New Orleans's Congo Square: An Urban Setting for Early Afro-American Culture Formation." *Louisiana History: The Journal of the Louisiana Historical Association* 32 (2): 117–157.
Julian, Jane. 1974. "Magnolia's Music." *The Mississippi Rag* 1 (9): 7–8.
Koenig, Karl. 1981. "The Plantation Belt and Musicians Part 1: Professor James B. Humphrey." *The Second Line* XXXIII: 24–49, Fall.
Koenig, Karl. 1982. "Professor James B. Humphrey – Part II." *The Second Line* XXXIV: 24–49, Winter.
Koenig, Karl. 1997. *Music in the Parishes Surrounding New Orleans: Ascension*. Running Springs, CA: Basin Street Press.
Kincaid, Jamaica. 2001. *My Garden*. New York: Farrar, Straus and Giroux.
Latrobe, Benjamin Henry. 1951. *Impressions Respecting New Orleans: Diary & Sketches, 1818–1820*. New York: Columbia University Press.
Lewis, Matthew Gregory. 1834. *Journal of a West India Proprietor*. London: John Murray.
Long, Edward. 1774. *The History of Jamaica*. London: T. Lowndes.
Mathieson, William Law. 1926. *British Slavery and Its Abolition 1823–1838*. London: Octagon Books.
McCusker, John. 2012. *Creole Trombone: Kid Ory and the Early Years of Jazz*. Jackson: University Press of Mississippi.
Métraux, Alfred. 2016. *Voodoo in Haiti*. Royal Oak, NZ: Pickle Partners Publishing.
Miller, Gene. 1994. "The Carpetbagger and the Professor." *New Orleans Music* 4 (6): 6–10.
Mintz, Sidney W. 1955. "The Jamaican Internal Marketing Pattern: Some Notes and Hypotheses." *Social and Economic Studies* 4 (1): 95–103.
New Orleans Picayune. 1874. (New Orleans, LA).
New Orleans Tribune. 1865. (New Orleans, LA).
Otto, Jesse J. 2011. "Dan Desdunes: New Orleans Civil Rights Activist and 'The Father of Negro Musicians of Omaha'." *Nebraska History* 92: 106–117.
Pearson, Elizabeth Ware. 1906. *Letters from Port Royal Written at the Time of the Civil War*. Boston: Book on Demand Ltd.
Robinson, Jim. 1958. *Hogan Jazz Archive*. New Orleans, LA: Tulane University, 10 December.
Rodrigue, John C. 2006. "Introduction." In *War, Politics, and Reconstruction: Stormy Days in Louisiana*, edited by Henry Clay Warmoth. Columbia: University of South Carolina Press.
Scott, Rebecca J. 2005. *Degrees of Freedom: Louisiana and Cuba After Slavery*. Cambridge, MA: The Belknap Press of Harvard University Press.
Scott, Rebecca J. 2007. "Public Rights and Private Commerce: A Nineteenth-Century Atlantic Creole Itinerary." *Current Anthropology* 48 (2): 237–256.
Senior, Bernard Martin. 1835. *Jamaica, as It Was, as It Is, and as It May Be*. London: T. Hurst.
Shugg, Roger Wallace. 1937. "Survival of the Plantation System in Louisiana." *The Journal of Southern History* 3 (3): 311–325.
Sitterson, Joseph Carlyle. 1953. *Sugar Country: The Cane Sugar Industry in the South, 1753–1950*. Lexington: University of Kentucky Press.
Somers, Robert. 1871. *The Southern States Since the War, 1870–1*. New York: Macmillan and Company.
Southern, Eileen. 1971. *The Music of Black Americans: A History*. New York: W.W. Norton.
Sugar Planter. 1866. (West Baton Rouge, LA).
Struge, Joseph, and Thomas Harvey. 1838. *The West Indies in 1837*. London: Hamilton, Adams and Co.
Thompson, Robert Farris. 1983. *Flash of the Spirit: African and Afro-American Art and Philosophy*. New York: Random House.

Tobin, Jacqueline L., and Raymond G. Dobard. 2011. *Hidden in Plain View: A Secret Story of Quilts and the Underground Railroad*. New York: Knopf Doubleday Publishing Group.

Tucker, Sherrie. 2016. "Jazz History Remix: Black Women from 'Enter' to 'Center'." In *Issues in African American Music: Power, Gender, Race, Representation*, edited by Portia K. Maultsby and Mellonee V. Burnim, 256–269. London: Taylor & Francis Group.

Vidal, Cécile. 2019. *Caribbean New Orleans: Empire, Race, and the Making of a Slave Society*. Chapel Hill: University of South Carolina Press.

Warmoth, Henry Clay. 2006. *War, Politics, and Reconstruction: Stormy Days in Louisiana*. Columbia: University of South Carolina Press.

West Baton Rouge Sugar Planter. 1866. (Baton Rouge, LA).

Wilson, Olly. 1992. "The Heterogeneous Sound Ideal in African American Music." In *New Perspectives on Music: Essays in Honour of Eileen Southern*, edited by Samuel Jr. Floyd and Josephine Wright, 327–340. Sterling Heights, MI: Harmonie Park Press.

Zoellner, Tom. 2020. *Island on Fire: The Revolt That Ended Slavery in the British Empire*. Cambridge, MA: Harvard University Press.

20

ECOSOCIALISM AND WORKPLACE DEMOCRACY

José Luis Haro García

Introduction

Can workplace democracy's proposals contribute to the articulation of an ecosocialist economic-productive system? In my opinion, the answer to this question is yes.

The ecosocialist alternative to the social and environmental ills caused by capitalist relations of production has been detailed in different chapters of this *Handbook*. Suffice it to point out here that the ecosocialist mode of production is characterized by, among other aspects, the promotion of an economy mainly on a local or regional scale centred on the generation of use value in which the activities of production, exchange, and consumption assume the conditioning factors and limitations that derive from social and environmental surroundings. Its economic project is not based on the continuous drive for economic growth and the concentration of capital, but on the redefinition of the links between labour, means of production, and goods produced. In short, ecosocialism means that economic activities adapt to local social and environmental conditioning, thus maintaining the bases of sustainability while promoting the emancipation of the working classes (Gorz 1995; Kovel and Löwy 2001; Polányi 2007).

The articulation of such a socio-economic model requires the democratization of the enterprise. Such a notion is based on the theoretical and practical experiences of workplace democracy over the last two centuries, as an alternative emanating from the world of work and not only to correct the worst dynamics of capitalist-based industrial societies (Mason 1982; Fernández Steinko 2002). This is in sharp contrast to the capitalist company, which is characterized by the fact that the agents with the power to manage it are shareholders who do not have to be physically present in the company for it to function properly. Shareholder ownership makes it possible for the same person to own different companies in distant locations and to exchange these property rights in the market, which is a much more complicated issue when linking ownership and the provision of labour (Dow 2003, 10–11). The alternative of workplace democracy entails practices and organizational forms that give workers the capacity to organize the tasks of their workplace as well as to intervene in the management of the company, either directly or through the election of representatives, but always under a criterion of equal participation based on the criterion of one person, one vote.[1]

This chapter argues that workplace democracy is necessary to reinforce ecosocialism and to promote a first-order strategy to achieve ecosocialist political objectives. First, the normative

10.4324/9780429341427-24

and conceptual affinities between the two will be discussed. Then a brief outline is given on what the literature shows about the environmental performance of existing democratically managed enterprises. The chapter concludes with some considerations of the strategy of the ecosocialist movement and the world of work in relation to enterprise democratization.

Ecosocialism, workplace democracy, and work employment

In different historical periods, labour movements have pushed, with greater or lesser success, for the democratization of enterprises and the economy in general, although the historical process that led to industrial societies led to the marginalization of those movements (Mason 1982, 136). In the twentieth century, in fact, the consolidation of industrialized societies implied the hegemony of Taylorist industrial organization, which grants the owners of capital (whether private or public) all managerial authority. This authority takes the form of an autocratic and hierarchical model of management by agents, whether owners or shareholders, who are not directly involved in production activities and in the framework of which workers are subjected to the principle of tutelage (Dahl 2002, 101).

However, although proposals for workplace democracy have been in the minority, it is also true that, in specific and unfortunately fleeting periods, they have been defended by broad social groups, even transcending Left-wing parties. These periods are known as democratization cycles (Fernández Steinko 2002). The first two cycles were a consequence of the two world wars and the public's consideration that one of the fundamental causes of the collapse of democratic regimes was the concentration of economic power in a few hands (Fernández Steinko 2002, 3). The third cycle begins at the end of the 1960s and coincides with the birth of contemporary environmental activism in the industrialized countries. In this third period, social movements driven by the new middle classes no longer demanded only incremental improvements but also a qualitative transformation of their societies. They demanded new levels of autonomy and personal self-realization; they did not demand higher incomes to consume more, but more leisure time and a critical re-evaluation of the indiscriminate growth of the economy and its personal, social, and environmental effects (Castoriadis and Cohn-Bendit 1982, 37). In this context, demands re-emerged for the democratization of the economy and enterprises through active and egalitarian participation and the establishment of meaningful forms of work (Gorz 1995, 155). These democratizing dynamics were cut short by the mid-1970s economic crisis and early 1980s neoliberal reaction.

At the same time, the third democratization cycle had special significance for incipient political environmentalism because it had affinities with its own normative, conceptual, and programmatic framework, including a critique of wage labour (Bahro 1986; Castoriadis and Cohn-Bendit 1982; Commoner 1971; Schumacher 2011). In fact, ecologism has been considered, along with feminism, as the first ideology that challenged the form of labour characteristic of industrial societies: labour employment. This is the institutional form that work takes in industrial societies, whether capitalist or not, and is characterized by, among other things, the fact that it is a commercial activity carried out in exchange for a wage. In these societies, the organization of time, citizenship rights, and the relationship of the individual with other central institutions of society – the family, the community – are conditioned on the relationship that one has with work employment, thus becoming an institution that shapes one's psychology and status (Köhler and Martín 2010, 17).

Classical socialism and liberalism conceive of work employment as an intrinsically positive institution, a gateway to full citizenship and the integration of the individual into society. In contrast, ecologism takes a more cautious approach: work employment is the form that certain productive activities take within the economic system of the industrial order and is related to the evils that beset contemporary societies, such as the loss of autonomy, the commodification and erosion of the institutions of social reproduction and of life, the invisibilization and subordination of women's work, the continuous expansion of the scale of production and consumption, and, ultimately, environmental deterioration. It is conceived as an ambivalent activity that, while creating the world, contributes to its destruction (Riechmann 1998, 34). Hence, it is argued that "the green economy is above all a new way of working" (Lipietz 1993, 50).

Inextricably linked to work employment and workers' gradual loss of autonomy are technologies that make possible the existence of both this form of work and the capitalist enterprise as we know it. E. F. Schumacher (2011, 156) argued in the 1970s that the technologies employed were at the origin of both alienation in the workplace and the deterioration of the environment and of human nature itself, an idea widely shared by environmentalists (Goldsmith et al. 1972; Bahro 1986; Bookchin 1999; Kemp and Wall 1990).

We must understand technology here in its broader sense, including the organizational technologies that make possible the existence of the capitalist enterprise (Chandler 1988). A technology, in short, that enables the organization of production goods, raw materials, energy, information, and people under autocratic and hierarchical criteria in order to make possible the appropriation of value and the avoidance of the social and environmental costs generated by the owners of capital (Kapp 1950). Such a perspective on organizational technology helps make workplace democracy proposals adaptable to different realities, from the large, capital-intensive, industrial-type shareholding firm to small, labour-intensive production units operating in very different economic environments.

On the other hand, as ecofeminist critique teaches us, the institution of work employment has a strong gender bias that does not include all productive activities carried out by human beings. Work employment, and the economic system itself, are based on structures of patriarchal domination that allow a whole series of care and life reproduction activities mostly carried out by women in the private sphere. These social reproduction activities are neither recognized nor valued as work that is necessary for the reproduction of life and societies and, therefore, for the very existence of the economic system itself (Herrero 2010; Salleh 1994).

In the same sense, the conditions of tutelage and alienation that workers suffer are not the same in different jobs within the same company, nor between companies, nor between companies operating in different sectors or territories. In the latter case, there is a hierarchization between capitalist core and periphery through which the core expel the productive activities involving greater environmental degradation and the articulation of more unhealthy and dangerous jobs. In this way, people in the peripheral territories of the capitalist system assume more difficult working and living conditions, as environmental justice activists have denounced (Shiva 2010).

Hence, the striving for workplace democracy does not exhaust the ecofeminist analysis of the necessary reconceptualization of the activities socially considered as work, nor the analysis of the geographical or cultural inequalities caused by capitalism within the working class itself, but focus on the conditions that working people endure within the production units we know as companies and the implications that such conditions generate in the personal, socio-economic, and environmental spheres. In short, calls for workplace democracy consider the company as a political space of the first order. And it is in this sense that such ideas are necessary for the development of a comprehensive proposal of ecosocialist transformation, adaptable to environmental, economic, and social contexts far removed from the industrialized core capitalist countries (Boillat, Gerber, and Funes-Monzote 2012).

The main dilemma is the degree to which work can give rise to fully autonomous, creative, and meaningful activity. Ecosocialism assumes that socially necessary work will always involve some alienation and loss of autonomy on the part of working people. So, a two-prong strategy to reduce it as much as possible is by: (a) democratizing enterprises, extending cooperative forms of organization, and encouraging self-organization in the workplace (Carter 1996) and (b) promoting other self-managed social spaces of production and reproduction, beyond the enterprise and economic system itself, replacing them with spheres of convivial logic (Illich 1974; Gorz 1995). Convivial logic entails productive activities more related to creativity and self-consumption in such a way as to pre-empt the commercialization of activities necessary for the reproduction of life. The aim is to restrict the social space and the hours dedicated to work employment to those goods that cannot be produced in any other way and to generate new working spaces where worker autonomy and creativity can be expanded (Gorz 1988). In short, from an ecosocialist perspective, the articulation of the economic system requires a transformation of its main institutions, including work employment and the business institutions, through the incorporation of forms of workplace democracy that allow workers and other agents or subjects affected by their activity (neighbours, suppliers, clients, etc.) to participate, to varying degrees, in determining an enterprise's course of action, with the aim of environmental sustainability and increasing worker autonomy.

This framework is based on a characterisation of sustainability as open or normative, as opposed to the closed or technocratic model of sustainability (Arias 2008, 179). The latter conceives sustainability as an absolute good that is knowable with exactitude while the open conception assumes that, given limited rationality, interdependency, and complexity (as all social and environmental systems are), such exact knowledge is not possible. Consequently, an open-ended version of sustainability cannot be ensured by technical cadres who enjoy technocratic legitimacy; the environmental goods and repercussions at stake require the active participation of all those affected so as to be able to make, in terms of a deliberative process, decisions that are more in line with the interests and knowledge of the citizenry as a whole (Greenwood 2010). In short, the ecosocialist perspective provides a reflexive and discursive approach that advocates for the democratization of companies in order to guarantee the autonomy of their workers and the sustainability of their activity.[2] This implies, as the workers' movements have long articulated, that democracy cannot remain at the gates of the company as this is a key arena in which the (un)sustainability of socio-environmental relations is decided.

Workplace democracy contributes to the sustainability of the economy

The few studies that analyze the environmental performance of democratically managed companies indicate that some of their characteristics may contribute to their being better adapted to the requirements derived from their social and environmental surroundings (see also Haro García 2018). This embeddedness (Polányi 2007) in their social and physical environment is mainly due to the link that such companies establish between the provision of work and access to its democratic management. A paradigmatic example of an enterprise that links democratic access to management with the provision of actual work is the worker or producer cooperative.

But why should linking work and the right to participate in management contribute to a better fit between company and environment? This consequence is due to a characteristic of work: that is, its inalienability. The provision and performance of work in a company cannot be abstracted from the social, psychological, and material conditions of the person who provides it. The participation of the worker ensures that his or her needs and perceptions, those of his or her family, and those of the community, are incorporated into decision-making (Dow 2003, 39). This linkage assumes that the incentive structure operating in a democratic firm differs from that in the capitalist shareholding firm. And, indeed, such a difference helps account for some of the advantages that different theoretical approaches and case studies have identified in the functioning of democratically managed enterprises, generally in contrast to the shareholding capitalist enterprise equivalent:

- A predisposition to opt for more prudent and less risky business strategies (Vanek 1971; Schweickart 2009)
- Differences in the technical organization of production related to a greater willingness to increase the efficiency of non-labour inputs: namely, raw materials and energy (Booth 1995; Fakhfakh, Pérotin, and Gago 2012) and to promote the physical and mental health of workers (Blauner 1964; Knudsen, Busck, and Lind 2011)
- Increased willingness to invest in reducing negative externalities and promoting sustainable management of raw materials (Askildsen et al. 2006; Booth 1995)
- Reduced likelihood of environmental dumping or non-compliance with environmental regulations (Miller 1991; Rocheleau 1999)
- Greater willingness to orient business strategies towards the long term and to incorporate the vision and interests of young workers (Dean 2014; Pérotin 2016)
- A propensity to stabilize firm size once the optimal size has been reached, rather than a tendency to scale up continuously (Booth 1995; Vanek 1971)
- Compatibility with steady-state or declining economic systems and greater ease of implementation of labour policies that lead to reduction and greater sharing of working hours (Rocheleau 1999; Schweickart 2009)

As we can see, there is sufficient evidence to argue that the democratization of production units can be a good strategy within an ecosocialist framework and programme. However, more empirical research is needed to better understand the impacts generated by democratic enterprise and its interaction with the basic institutions of the economic system. The firm, democratic or not, operates within an incentive environment that conditions the possible courses of action available to it (Carter 1996, 68–70). Consequently, if proposals for the democratization of the firm are to be effective, they must be accompanied by a consistent reform of the other institutions of the economic system so that democratic firms can develop their potential without being penalized for doing so. In short, the democratization of enterprises must be accompanied by the democratization of the economic system as a whole, in line with proposals for economic democracy. Analyses already exist of the interdependence between democratic enterprise and the economic system from an ecological perspective. These should allow for further research and a consequent programmatic delineation (Gorz 1995; Kelly 2012; Schweickart 2000, 2009; Söderbaum 2008; Turnbull 1994, 2011).

Workplace democracy and ecosocialist strategy and programmes

Proposals for workplace democracy can contribute to enabling the convergence of the world of work and environmental movements and, in general, of all civil and political movements advocating for the deepening of democracy. Although the relationship between the labour and early environmental movements was not at first straightforward (Fernández Buey 1998; Obach 2002), ecosocialism has always maintained the existence of a compatibility between the two, considering that the degradation of workers' living conditions and environmental degradation have a common origin in capitalist relations of production. Thus, the maturation and consolidation of the ecosocialist framework have contributed to overcoming the mutual misunderstandings of the 1970s and 1980s, giving way to more conciliatory positions like those characterizing the world of work and trade unionism since the 1990s.

In this sense, workplace democracy approaches reinforce the alliance of environmental and labour movements to articulate a new sustainable socio-productive model. This compatibility is based on a rereading of the values of autonomy and sustainability and their interrelation, from an environmentalist perspective, as well as on a revision of the institution of work employment and of the business institution itself, which in fact entails a critical revision of the productivist assumptions so frequently inhering the world of work.

In an ecosocialist agenda, workplace democracy can no longer be considered solely as a normative issue to be subordinated to the technical requirements of maximizing social production. It must also be understood as a necessary instrument for the articulation of enterprises more inclined to opt for sustainable courses of action. And this requires the preservation of a sphere of autonomy and discursive participation that allows for the reflexive determination of the paths to sustainability by all the agents involved (Haro García 2018, 99). Otherwise put, ecosocialism must contribute to the elaboration of a new theorization of work employment and the company within the framework of its economic system. A new company model that should orient its organizational techniques towards democratic, horizontalist, and open systems with mechanisms for accounting and validation of performance of a dialogical and reflexive nature (i.e., that allows all agents to participate and assert their needs and preferences in relation to the activity of the company itself) (Söderbaum 2006): a democratic company. In this way, these organizations will become more compatible with the conditioning factors derived from social and environmental reproduction, orienting themselves first and foremost towards satisfying local needs, preserving the autonomy and self-development of the workers and the communities in which they are based and that they serve, and orienting their activity towards long-term goals in ways that are compatible with the needs of future generations.

In this context, ecologically informed trade union action should give greater importance to workplace democracy approaches. These approaches are not opposed to the interests of the world of work as a collective or social class and should be conceived as an expression of the free development of all entities: of ecosystems, societies, communities, and associations, including companies, and of workers in the workplace. The remaking of business along democratic lines is, however, only one piece of the articulation of an ecosocialist economic system. For such a transformation to have the desired effects, the whole of the economic institutions must be transformed to give rise to responsible production and consumption practices. It will have to be open to deliberation of which needs should be satisfied through the production of goods and services within the economic system and which should be kept away from the narrowly economic logic. Depending on how these questions are resolved, one can envisage social models of production that are based on the degree of preponderance that the company, as a specialized organization, and work employment would maintain, as the form that work takes within the framework of the economic system. Logically, this scheme will require translation and contextual adaptation to the conditions and needs of each context.

In my opinion, the ecosocialist framework assumes the maintenance of a specialized economic system as it presents certain advantages for the correct satisfaction of some of the basic needs of the population as a whole. This also implies maintaining the existence of the institution of work employment and enterprises. However, the ecosocialist framework closely delimits this sphere by restricting the goods and services that must be produced under its logic – curbing the growing commodification of different social spaces – and through the democratization of the main institutions of the economic system and, among them, of companies. On the other hand, the ecosocialist framework promotes the development of other spaces of production and consumption with a convivial logic beyond the economic system, in which work lacks commercial logic and is identified with creativity, with activities that are an end in themselves and that are oriented towards the reproduction of personal, social, and ecological life. A plural economy (Söderbaum 2008) is thus drawn, articulated under the democratic principle, in which different sensibilities and purposes operate and different forms of production and consumption and, therefore, different forms of work are fully recognized beyond work employment. In this sense, a workplace democracy approach becomes a necessary element for the programmatic articulation of the ecosocialist model insofar as it contributes to increasing both the degree of autonomy enjoyed by people in the workplace and the environmental sustainability of business activity. From a social and environmental justice point of view, therefore, the time has come to advocate for the recovery and reworking of such an approach, given the new planetary scale of risks facing humanity today and in future.

Notes

1 In this chapter, the concept of workplace democracy is used as a synonym for a democratically managed company. See Haro García (2018, 13–20) for a discussion and analysis of the differing concepts used relative to the democratization of companies.

2 This perspective implies putting in crisis the performance evaluation tools of the conventional capitalist company and opting for reflexive evaluation models, which incorporate the interests, skills, and perspectives of the different agents, highlighting in this aspect the proposals of dialogical accounting (Brown 2009) or the positional analysis proposed by Söderbaum (2006).

References

Arias Maldonado, Manuel. 2008. *Sueño y Mentira del Ecologismo*. Madrid: Siglo XXI.

Askildsen, Jan Erik, Uwe Jirjahn, and Stephen C. Smith. 2006. "Works Councils and Environmental Investment: Theory and Evidence from German Panel Data." *Journal of Economic Behavior and Organization* 60 (3): 346–372.

Bahro, Rudolph. 1986. *Building the Green Movement*. London: GMP Publishers.

Blauner, Robert. 1964. *Alienation and Freedom, the Factory Worker and His Industry*. Chicago: The University of Chicago Press.

Boillat, Sébastien, Julien-François Gerber, and Fernando R. Funes-Monzote. 2012. "What Economic Democracy for Degrowth? Some Comments on the Contribution of Ecosocialist Models and Cuban Agroecology." *Futures* 44 (6): 600–607.

Bookchin, Murray. 1999 [1982]. *La Ecología de la Libertad, la Emergencia y Disolución de las Jerarquías*. Translated by M.G. Burello. Móstoles: Nossa y Jara Editores.

Booth, Douglas E. 1995 "Economic Democracy as an Environmental Measure." *Ecological Economics* 12 (3): 225–236.

Brown, Judy. 2009. "Democracy, Sustainability and Dialogic Accounting Technologies: Taking Pluralism Seriously." *Critical Perspectives on Accounting* 20 (3): 313–342.

Carter, Neil. 1996. "Worker Co-Operatives and Green Political Thought." In *Democracy and Green Political Thought: Sustainability, Rights and Citizenship*, edited by Brian Doherty and Marius de Geus, 56–75. London: Routledge.

Castoriadis, Cornelius, and Daniel Cohn-Bendit. 1982. *De la Ecología a la Autonomía*. Barcelona: Mascarón.

Chandler, Alfred D. 1988 [1977]. *La Mano Visible, la Revolución de la Dirección de la Empresa Norteamericana*. Translated by Ángeles Conde. Madrid: Ministerio de Trabajo y Seguridad Social.

Commoner, Barry. 1971. *The Closing Circle*. London: Jonathan Cape.

Dahl, Robert Alan. 2002 [1985]. *La Democracia Económica: Una Aproximación*. Translated by Mireia Bofill. Barcelona: Hacer.

Dean, Andrés. 2014. "Do Worker Cooperatives Degenerate?" http://vox.lacea.org/?q=abstract/cooperativas-de-trabajadores (accessed 15 December 2019).

Dow, Gregory K. 2003. *Governing the Firm, Worker's Control in Theory and Practice*. Cambridge: Cambridge University Press.

Fakhfakh, Fathi, Virginie Pérotin, and Mónica Gago. 2012. "Productivity, Capital, and Labor in Labor-Managed and Conventional Firms: An Investigation on French Data." *Industrial & Labor Relations Review* 65 (4): 847–879.

Fernández Buey, Francisco. 1998. "Programas Sindicales, Intereses Obreros y Reivindicaciones Ecologistas en la Lucha por un Mundo Habitable" In *Trabajar sin Destruir: Trabajadores, Sindicatos y Ecologismo*, edited by Francisco Fernández Buey and Jorge Riechmann, 133–163. Madrid: HOAC.

Fernández Steinko, Armando. 2002. *Experiencias Participativas en Economía y Empresa: Tres Ciclos para Domesticar un Siglo*. Madrid: Siglo XXI.

Goldsmith, Edward, Robert Allen, Michael Allaby, John Davoll, and Sam Lawrence. 1972. *Manifiesto para la Supervivencia*. Translated by Miguel Paredes. Madrid: Alianza Editorial.

Gorz, André. 1995 [1988]. *Metamorfosis del Trabajo*. Translated by Mari-Carmen Ruiz de Elvira. Madrid: Sistema.

Greenwood, Dan. 2010. "Facing Complexity: Democracy, Expertise and the Discovery Process." *Political Studies* 58 (4): 769–788.

Haro García, José Luis. 2018. "Ecologismo y Democracia en la Unidad de Producción: Una Reconstrucción Crítica desde la Teoría Política Verde." Doctoral thesis, Department of Constitutional Law and Political Science, Law School, Universitat de Barcelona, Barcelona.

Herrero, Yayo. 2010. "Hacia una Necesaria Reformulación del Trabajo." *Ecología Política* 40: 17–20.

Illich, Ivan. 1974. *La convivencialidad*. Barcelona: Barral Editores.

Kapp, Karl William. 1950. *The Social Costs of Private Enterprise*. Cambridge: Cambridge University Press.

Kemp, Penny, and Derek Wall. 1990. *A Green Manifesto for the 1990s*. London: Penguin.

Kelly, Marjorie. 2012. *Owning Our Future*. San Francisco: Berret-Koehler Publishers.

Knudsen, Herman, Ole Busck, and Jens Lind. 2011. "Work Environment Quality: The Role of Workplace Participation and Democracy." *Work, Employment & Society* 25 (3): 379–396.

Köhler, Hölm-Detlev, and Antonio Martín. 2010. *Manual de la Sociología del Trabajo y de las Relaciones Laborales*. 3rd ed. Madrid: Delta.

Kovel, Joel, and Michael Löwy. 2001. "Manifiesto Ecosocialista." http://rcci.net/globalizacion/2002/fg259.htm (accessed 15 December 2019).

Lipietz, Alain. 1993. *Vert Espérance, l'Avenir de l'Écologie Politique*. Paris: La Découverte.

Mason, Ronald M. 1982. *Participatory and Workplace Democracy*. Carbondale and Edwardsville: Southern Illinois University Press.

Miller, David. 1991. "A Vision of Market Socialism." *Dissent* 406–414, Summer.

Obach, Brian K. 2002. "Labor-Environmental Relations: An Analysis of the Relationship Between Labor Unions and Environmentalists." *Social Science Quarterly* 83 (1): 82–100.

Pérotin, Virginie. 2016. "Democratic Firms: Assets for the Long Term." In *Institutions for Future Generations*, edited by Iñigo González-Ricoy and Axel Gosseries, 331–352. Cambridge: Cambridge University Press.

Polányi, Károly. 2007 [1944]. *La Gran Transformación, los Orígenes Políticos y Económicos de Nuestro Tiempo*. Translated by Eduardo Suarez. Buenos Aires: Fondo de Cultura Económica.

Riechmann, Jorge. 1998. "Trabajo y Medio Ambiente en la Era de la Crisis Ecológica." In *Trabajar sin Destruir: Trabajadores, Sindicatos y Ecologismo*, edited by Francisco Fernández Buey and Jorge Riechmann, 13–34. Madrid: HOAC.

Rocheleau, Jordy. 1999 "Democracy and Ecological Soundness." *Ethics & the Environment* 4 (1): 39–56.

Salleh, Ariel. 1994. "Nature, Women, Labour, Capital: The Deepest Contradiction." *Ecologia Política* 7: 35–47.

Schumacher, Ernst F. 2011 [1973]. *Lo Pequeño Es Hermoso*. Translated by Oscar Margente. Madrid: Akal.

Schweickart, David. 2000. "Not with a Band but a Whimper: Capitalism/Ecology/Economic Democracy." *Theoria* 95: 1–33.

Schweickart, David. 2009. "Is Sustainable Capitalism an Oxymoron?" *Perspectives on Global Development and Technology* 8 (2–3): 559–580.

Shiva, Vandana. 2010. "Earth Democracy: Beyond Dead Democracy and Killing Economies." *Capitalism Nature Socialism* 21 (1): 83–95.

Söderbaum, Peter. 2006. "Democracy and Sustainable Development: What Is the Alternative to Cost-Benefit Analysis?" *Integrated Environmental Assessment and Management* 2 (2): 182–190.

Söderbaum, Peter. 2008. *Understanding Sustainability Economics: Towards Pluralism in Economics*. London: Earthscan.

Turnbull, Shann. 1994. "Stakeholder Democracy: Redesigning the Governance of Firms and Bureaucracies." *Journal of Socio-Economics* 23 (3): 321–360.

Turnbull, Shann. 2011. "Achieving Environmentally Sustainable Prosperity." https://papers.ssrn.com/sol3/papers.cfm?abstract_id=1769349 (accessed 15 December 2019).

Vanek, Jaroslav. 1971. *The Participatory Economy: An Evolutionary Hypothesis and Strategy for Development.* Ithaca, NY: Cornell University Press.

21
PEOPLE'S RESISTANCE AGAINST GEOENGINEERING

Silvia Ribeiro

Introduction

One of the most dangerous proposals to confront climate change is geoengineering, which is the technological and large-scale manipulation of the global climate that its proponents claim will counteract symptoms of climate chaos. It is particularly dangerous due to its impacts on people and the environment. The technologies are being deployed to continue business-as-usual emissions of greenhouse gases (GHG), thereby worsening the already dire climate crisis.[1]

For the industrial world, with high GHG emissions and home to polluting transnational companies, geoengineering presents a convenient "technological solution" that might allow them to continue emitting greenhouse gases without having to make radical changes in production and consumption patterns, or to the capitalist social relations that underpin them. It would also mean that new businesses could be developed, selling technologies that are supposed to remove and store carbon or manipulate the environment to lower global temperatures. Geoengineering is a key area of concern for those fighting for a "just transition" towards ecosocialism and other Indigenous and grassroots democratic political economies and cosmovisions. Behind geoengineering, which in many cases has military origins, there is a confluence of economic interests involving powerful industries and geopolitical actors (Fleming 2010).

Climate justice movements, especially Indigenous peoples and peasant organizations, along with environmental organizations and many scientists, have been strongly critical of the idea of geoengineering, arguing that it is a dangerous "false solution" to climate change.

Geoengineering and some proposed technologies

Geoengineering refers to a series of proposals for large-scale intervention and alteration of ecosystems as a "technofix" to climate change. Proponents often classify them under two main umbrella concepts: solar radiation management (SRM) and carbon dioxide removal (CDR). These proposals may involve technological interventions on land, on the sea, or in the atmosphere. None of them attempt to address the causes of climate change; instead, they aim to manage some of its symptoms: e.g., to lower the temperature or remove carbon from the atmosphere once it is emitted.

10.4324/9780429341427-25

There are some 20 to 25 geoengineering proposals, including injecting sulfates or other chemicals into the stratosphere in order to block sunlight, aiming to produce a dimming effect; brightening marine clouds to reflect more sunlight back into space; installing facilities to absorb carbon dioxide from the atmosphere and then burying it in marine or geological reservoirs; fertilizing the ocean with iron or urea to stimulate plankton blooms, hoping that they absorb carbon dioxide and sink it to the ocean floor; altering the chemistry of the sea to make it more alkaline; and mega-plantations of transgenic crops that would supposedly absorb more carbon or reflect more sunlight. All these proposals, deployed at the scale necessary to influence climate change, carry huge environmental, biodiversity, social, and economic risks, as well as unpredictable and often synergistic negative effects involving transboundary impacts.

While each proposed geoengineering scheme has specific risks and potential effects, all share a number of negative impacts. To start, they all function as excuses to avoid or delay real reduction of greenhouse gas emissions, thus perpetuating the use of fossil fuels. As the causes for climate change are not addressed, the deployment of geoengineering would create captive markets. The fact that powerful countries and private actors control the technologies will increase geopolitical tensions and global inequities.

The workings of the climate are not sufficiently known, and neither are dynamic ecosystems essential for life on the planet. Its disruption with geoengineering implies large risks and uncertainties. The mega scale needed to influence the global climate means that potential impacts would also be huge. For instance, solar geoengineering techniques (e.g., stratospheric aerosol injection) would mask the temperature increase while the causes for global warming (gas emissions) continue. This means that once deployed, it cannot be stopped, and it will require steadily larger amounts of aerosols. If interrupted – intentionally, for political changes, or through sabotage or human error – it would cause a "termination shock" with a sudden temperature rise that would be worse than if the technologies had never been used (Trisos et al. 2018).

In brief, geoengineering would create "winners and losers" – some regions may have a better temperature, but others will suffer more extreme droughts and floods. Furthermore, some technologies can also be weaponized. The next section poses the question: Who would control the global thermostat?

Oil industry and billionaires behind geoengineering

The fossil fuel industry has been systematically researching climate change since 1940, including ways to modify the climate, and companies like ExxonMobil, Shell, BP, Total, and Chevron have patented different geoengineering techniques. The industry has even ventured into the research and development of solar radiation management (SRM) techniques, ocean alkalinization, and other areas.

As the Center for International Environmental Law (CIEL) reports, "Industry advocates openly believe Carbon Capture and Storage (CCS) and Carbon Dioxide Removal (CDR) are essential to save coal, ensure the future of oil and gas, and 'unlock' unburnable carbon" (CIEL, ETC, and HBF 2019). The Centre shows that 85 percent of US subsidies for CCS and direct air capture (DAC) would flow to what is known as "enhanced oil recovery," thus increasing fossil fuel exploitation (CIEL 2019a).

The technique for enhanced oil recovery, developed decades ago, injects CO_2 into almost-depleted oil wells to access the deepest reserves. Rebranded under the banner of carbon capture and storage, the technique is being presented as a way to reduce climate change. Because it is said to have social benefit, proponents claim that it should be financed using public subsidies.

However, if one takes into account energy consumption over the full life cycle of this technique, it is not proven that it really "sequesters" carbon (CIEL, ETC, and HBF 2019).

More recently, Chevron, Occidental Petroleum, and BHP have invested over $70 million in the DAC company Carbon Engineering, founded by Harvard researcher David Keith. Tar sands millionaire Murray Edwards and Bill Gates are among its initial investors. Keith is one of the most outspoken promoters of geoengineering and is also working to deploy a controversial open-air SRM experiment called SCoPEx (Krauss 2019; Barnard 2019).

Despite the high-level interest from transnational companies and some development of CCS and DAC technologies, most geoengineering proposals are in the realm of speculative theories. Currently, the majority of geoengineering research is being carried in the US, with other research projects in the UK, China, and other high-greenhouse-gas-emitting countries such as Japan, Australia, Korea, and the European Union. Some of these research teams also attempt to carry out open-air experiments. For a list of geoengineering research and investment projects, see the geoengineering interactive map (HBF and ETC Group 2021).

Some of the richest billionaires have become increasingly involved in funding geoengineering research and experiments. This is an extremely concerning trend, because the world of philanthro-capitalism is highly untransparent and undemocratic, with strong lobby capacity to promote particular scientific interests (Ribeiro 2020). Next, I look at opposition to geoengineering at United Nations (UN) conventions.

Civil society resisting geoengineering at United Nations conventions

In 2007, ETC Group and Greenpeace were among the first organizations to bring the discussion on geoengineering inside several United Nations (UN) conventions, to seek a moratorium. Their concerns were quickly echoed by many other civil society organizations and social movements at the Convention on Biological Diversity (CBD) and the London Convention and London Protocol on the Prevention of Marine Pollution (LC/LP). Thanks to initiatives introduced by a number of governments (spanning all continents), both bodies echoed the need to apply extreme precaution to geoengineering, which was expressed in the decisions on three *de facto* moratoria on geoengineering and ocean fertilization.

The discussions at CBD included those centred on the publication of several technical reports on geoengineering that were peer reviewed by all parties and stakeholders. These led to three other reports that were the basis for the discussions at several CBD Conferences of the Parties, resulting in two consensus decisions on *de facto* moratoria against ocean fertilization and geoengineering in general. All documents and decisions are compiled at a dedicated site at the CBD (2021).

The London Convention/London Protocol (LC/LP) on the Prevention of Marine Pollution by Dumping Wastes and Other Matter has been discussing ocean marine geoengineering techniques for over a decade. LC/LP parties unanimously adopted a resolution in 2008 calling on parties to not allow ocean fertilization activities other than those for "legitimate scientific research." This was followed by another resolution in 2010 that set out strict conditions defining what constitutes "legitimate scientific research." In 2013, the London Protocol parties adopted (again unanimously) an amendment to give the prohibition and assessment framework legal force and to open the way for other marine geoengineering activities to be discussed in the future.

The UNFCCC has not discussed geoengineering as a whole but has approved carbon capture and storage to be included as part of the carbon removal projects in the clean development mechanism. Geoengineers were therefore enthusiastic about the possibility of advancing other geoengineering technologies in the Paris Agreement, but this does not mean there is any consensus about this. UNEA 4 has also discussed a proposal from Switzerland and other governments to conduct further studies on geoengineering techniques with a view towards considering further regulations, but the proposal was withdrawn after hard opposition from the United States and other high-emitting countries (CIEL, ETC, and HBF 2019).

In all of these fora, several civil society constituencies have taken consensus positions against geoengineering, either in general or with respect to particular technologies. For instance, since 2008, civil society organizations that are registered with the CBD Alliance have expressed opposition to geoengineering in all Conference on Biological Diversity COPs (CBD Alliance 2016).

At the UNFCCC, the Women and Gender Constituency and the environmental NGOs, demanding climate justice, have also expressed opposition to geoengineering, including in the latest COP 25 in Madrid in 2019. The Women and Gender Constituency has demanded that all governments participating in climate negotiations declare geoengineering and bioenergy with carbon capture and storage (BECCS) a "no go." According to the Constituency:

> Geoengineering, consisting of large-scale manipulation of the Earth's system using a wide range of technologies, is an unreliable and untested technofix that would create more problems than what it would solve. These types of false solutions serve to uphold business as usual rather than challenge and dismantle the root causes of climate chaos. The side-effects of geoengineering could be disastrous, globally and intergenerationally unjust, and potentially irreversible. BECCS for example, would require vast amounts of land, likely leading to the displacement of communities and conflicts, jeopardizing communities and women's rights.
>
> *(WGC 2019)*

Furthermore, Climate Action Network International, which includes over 1,300 NGOs, adopted a common position on solar radiation management (SRM) in September 2019 (CAN 2019). It includes four points:

1 Robust adaptation and mitigation actions are the first-line solutions to climate change. SRM is not a substitute for either and should not be seen as climate action.
2 Recognition of the inherent transboundary nature and significant and unknown risks (geopolitical, social, environmental, and ethical).
3 Strong opposition to deployment of SRM.
4 Strong opposition to real-world experiments.

In sum, the work of civil society has focused on three areas: producing critical research and information documents oriented to inform and build capacity within civil society; generating local resistance to specific projects, often with international support; and participating in research and information for governments and governmental delegates at the United Nations (UN). Next, I address some of the organizations and movements resisting geoengineering.

Resistance to geoengineering

The Hands Off Mother Earth! (HOME) Campaign

In 2010 at the World People's Conference on Climate Change and the Rights of Mother Earth in Cochabamba, Bolivia, the widest global network of organizations and movements collaborated to resist geoengineering, and the Hands Off Mother Earth! (HOME) Campaign was established. The World People's Conference, hosted by the government of Bolivia, gathered 35,000 participants from popular movements, civil society, and governments. It was a global reaction to the lack of will of the main polluters to take real action in the climate negotiations at the 2019 UNFCCC COP15 in Copenhagen. The conference included a rejection of geoengineering and all other false solutions to climate change in its Final Declaration (GARN 2010).

Among the founders of HOME were La Via Campesina, the Indigenous Environmental Network, Friends of the Earth International, the ETC group, and others. Since the establishment of the HOME campaign, there has been local resistance to various projects and numerous debates and seminars at different UN fora. However, due to the lack of real action on climate change by the highest-emitting-country governments, the idea of advancing geoengineering has gained traction, and the number of research and project proposals has increased. The introduction of the deceptive concept of "net zero emissions" (instead of real zero emissions) underpins the illusion that technofixes such as geoengineering could compensate for a lack of real reductions and even a continued increase in CO_2 emissions.

In October 2018, 110 civil society organizations and popular movements from five continents signed and made public the HOME Manifesto against geoengineering. It was issued the same week that the IPCC met in Korea to negotiate the conclusions of its new report on how to limit global warming to 1.5 degrees Celsius (CJA et al. 2018). The Manifesto demands the following:

- A ban on all geoengineering field experiments and deployment
- An end to all open-air experiments, including SCoPEx in the US, which proposes to inject sulphate particles and other materials into the stratosphere; the Ice911 project in Alaska, which would scatter millions of tiny glass bubbles over Arctic ice; the Marine Cloud Brightening project in Monterrey Bay, California, which would inject salt water into marine clouds to whiten them in order to reflect sunlight; and the oceans fertilization project in Chile and Peru
- An end to carbon capture and storage and direct air capture projects because they perpetuate fossil fuel extraction and combustion and an end to bioenergy with carbon capture and storage projects, which, besides being unproven and not technically feasible, would have grave consequences for land use, food sovereignty, the environment, and biodiversity
- Support for the diversity of alternatives to confront climate change that are already proven and less risky but are being sidelined in climate change deliberations

The HOME Manifesto has been signed by over 200 organizations from 45 countries, among them 31 international networks, including La Via Campesina, the Climate Justice Alliance, Friends of the Earth International, the World March of Women, Oilwatch, Oil Change, Alianza por la Biodiversidad en América Latina, the Center for International Environmental Law, the Heinrich *Böll* Foundation, Biofuelwatch, Corporate Accountability International, Focus on the Global South, the Third World Network, the World Indigenous Women's Alliance, Development Alternatives with Women for a New Era (DAWN), the ETC Group, the Global Forest Coalition, the World Rainforest Movement, the Grassroots Global Justice Alliance, the Indigenous Environmental Network (IEN), and others (HOME 2018).

Indigenous peoples' resistance to geoengineering

> As Indigenous Peoples, we are unified in our opposition to all forms of geoengineering. As human beings, we are entirely dependent upon our respectful

> relationship with the natural world. We are now faced with the consequences of the exploitation of the natural world that threaten the future existence of all life on Mother Earth. The Indigenous traditional teachings, lifestyles, spirituality, cultures and leadership of our people has sustained us for millennia and will do so for countless future generations but only if the world adheres to the Natural Laws of Creation and the Precautionary Principle. Geoengineering acts against all of those.
>
> *Tom BK Goldtooth, ETC Group (2020)*

Indigenous peoples' organizations have been very active in the opposition to geoengineering. They quickly realized that geoengineering proposals have the potential to unbalance vital global ecosystems, and that they violate all the values of respect and care for nature that are embedded in native cultures. They have identified these proposals as a new form of colonialism, a new way of invading their territories and communities, from oceans to sky, from sky to earth. Indeed, all proposals to conduct field geoengineering experiments in North America that have been announced are scheduled to take place on Indigenous territories (just the same as in the past, with nuclear testing and other experiments). This is the case for Harvard University's project SCoPex (aerosol injection) in Southern US; the Artic Ice Project to disseminate micro glass beads in Alaska and other Indigenous territories in Canada; the Marine Cloud Brightening Project in Ohlone territory in Monterey Bay, California; and the defeated ocean fertilization project in Haida Gwaii, Canada.

The Indigenous Environmental Network (IEN) is among the early whistleblowers on geoengineering and actively participated in the founding of the HOME Campaign in 2010 in Bolivia and in the global manifesto against geoengineering that was launched in 2018. At the launch of the HOME Campaign, Ben Powless of the Mohawk Nation said:

> For too long our peoples' bodies and lands have been used to test new technologies. Now, in response to climate change, these same people want to put Mother Earth at risk with geoengineering technologies. We can't afford to threaten our planet in this way, especially when simple, just and proven solutions are at hand.
>
> *(Powless in ETC Group 2010)*

Tom BK Goldtooth explains that many Indigenous communities in North America have been enclosed in reservations and forced and/or coopted to allow polluting industries, such as oil and coal, onto their territories. In some cases, geoengineers are using this situation to legitimize their projects while continuing to carry out experiments and geoengineering developments on Indigenous lands. In this way, they try to evade the obligation to ensure the free, prior, and informed Consent (FPIC) outlined in the UN Declaration on the Rights of Indigenous Peoples (UNDRIP) (ETC Group 2020).

Indeed, many Indigenous organizations at IEN and on other continents have expressed a clear opposition to geoengineering and its mechanisms, including the introduction of the concept of "net zero" emissions. They identify it as a colonization of the sky through geoengineering and carbon markets, and they have also built an international coalition to oppose these called Sky Protectors. IEN is also a member of the Climate Justice Alliance (CJA) and the international movement It Takes Roots, which both have a clear stand against all forms of geoengineering (CJA 2019).

The Climate Justice Alliance has, in turn, carried out very important work informing others involved in grassroots organization in North America, communicating about how they are or could be affected by proposed geoengineering experiments and projects (CJA 2020).

Next I list some examples of Indigenous peoples' and peasants' actions to resist geoengineering:

Ocean fertilization in Haida Gwaii and beyond

In 2012, a company founded by, among others, geoengineer Russ George (who had previously attempted ocean fertilization experiments in the Galapagos Islands with his company Planktos) conducted an extensive illegal ocean fertilization experiment off the Pacific coast of Canada in Haida Guaii, territory of the Haida Indigenous nation. George founded a new company called the Haida Salmon Restoration Corporation (HSRC), whose staff was almost entirely composed of non-Indigenous scientists. However, the project was presented to the world as supported by the Haida Community of Old Masset in the Haida Gwaii Archipelago. When it later became clear that the company had not complied with Canadian law, company officials appealed, using the Haida people's claims of sovereignty to defend themselves, and even raised a Haida flag (instead of a Canadian one) while illegally dumping iron at sea.

This geoengineering project created conflicts within the Haida Indigenous community, many of whom had been misled into believing that it was a fish stock recovery project. Eventually, the Council of Chiefs and the Haida Nation issued a rejection of geoengineering, signed by Haida Nation President Guujaw, making it clear that the HSRC's actions did not reflect Haida Nation values:

> The consequences of interfering with nature on this scale are not predictable and pose unacceptable risks to the marine environment. Our people, along with the rest of humanity, depend on the oceans and cannot leave their fate to the whim of a few.
>
> *(Guujaw 2012)*

The experiment was revealed and made public to the Indigenous community by the ETC Group. At the same time, it was denounced by members of the HOME Campaign and the civil society CBD Alliance at the Convention on Biological Diversity as violating the *de facto* moratorium on ocean fertilization and geoengineering. The project was later suspended, and the Canadian government took legal action against the founders of the HSRC. Nevertheless, the project organizers had already implemented the largest illegal open-air experiment on ocean fertilization.

After the suspension of this project, some of the same people who were part of the project on Haida territory started another company, Oceaneos, in Chile and Peru. This was later renamed as a research organization for "ocean seeding," in an attempt to circumvent the CBD and London Convention/Protocol moratoria on ocean fertilization. Scientists and organizations in Chile denounced this attempt and have been successful in stopping these open-air experiments in Chile so far (Leiva 2019).

SCoPEx: Colonizing the sky

Civil society and Indigenous people also opposed Harvard University's project stratospheric controlled perturbation experiment (SCoPEx), an aim to develop solar radiation management, when leaders announced that the experiment would take place in Arizona. The location of the experiment has now been changed, and the new location has not yet been announced, but it could well be in New Mexico. This would be a small experiment, but its meaning is much larger. The deployment of solar radiation management (SRM) techniques will have global implications, and allowing SCoPEx would be considered more a political signal, opening the race for open-air experiments on this set of dangerous proposals, than a technical experiment. Small experiments on SRM will not uncover anything about the technology's effect on climate, precisely because they are small, but they are still steps that would set a precedent with respect to large-scale implementation (GEM 2021b).

As the University of Oxford, Professor Raymond Pierrehumbert noted in *The Bulletin for Atomic Scientists*:

> [SCoPEx] is mostly a stunt to break the ice and get people used to the idea of field trials of albedo modification. The experiment would not address any of the really major issues raised in the US National Research Council climate intervention report, which indeed are hard to experiment with short of something that amounts to full-scale deployment.
>
> *(Pierrehumbert 2017)*

Tupac Enrique Acosta, a member of the organization Tonatierra Nahuacalli in Arizona and the Indigenous Environmental Network (IEN), spoke against the commodification of the sky:

> With today's geoengineering flagship projects such as SCoPEx in Tucson, we see the sails on the horizon of yet another flotilla of invasions just like the Niña, the Pinta, the Santa Maria – or the Mayflower. This time it is the sky itself that is being commodified and marketed. We stand in solidarity and commitment to the Cochabamba Protocols in defense of the Territorial Integrity of Mother Earth and we say NO to Geoengineering! We deny consent!
>
> *(CJA et al. 2018)*

David Keith, founder of SCoPEx and Harvard´s Solar Geoengineering Research Program, has been announcing the realization of an open-air SRM experiment since 2012, and by now, he may have acquired a critical amount of funding for the project from billionaires and foundations in the US. To manage the opposition to the project, a supposedly independent SCoPEx advisory committee, mostly composed of US-based academics, was formed. This committee was denounced by the HOME Campaign as an attempt to substitute for real societal participation. An open letter denouncing this action was sent to the committee with signatures from 80 national and international organizations (HOME 2020b). As a result, the executive director of the California Strategic Growth Council, who was the chair of the SCoPEx advisory committee, had to resign (HOME 2020a).

Artic Ice Project: Ice911

Another proposed SRM experiment is the Artic Ice Project, formerly called Ice911. This project aims to spread millions of tiny glass bubbles over Artic ice in the attempt to make it more reflective and slow down the melting process in summer months. Leslie Field, the founder and technology director of the project, is a former Chevron employee (GEM 2018).

The project claims to work in collaboration with the Ukpeaġvik Iñupiat Corporation in Ukpeaġvik (also known as Barrow) in Alaska, where there are 13 communities in the bay area.

The project intended to deploy the experiment without the free, prior informed consent procedure, a right that the communities have according to the United Nations Declaration on the Rights of Indigenous Peoples (UNDRIP).

Adrienne Titus, an Inupiaq organizer with the Native Movement in Fairbanks, says that rather than a solution to climate change, geoengineering experiments like Ice 911 are just another risk factor for communities in the Arctic. In a fragile ecology, many things could go wrong with the introduction of millions of tiny glass "microspheres." Titus asked,

> What happens when this stuff gets in the water, and what effect will it have on the krill and all the other microorganisms that are in the top surface of the ocean? Then the fish that eat them, and all of the other sea mammals and animals that rely on this ecosystem, this area, this water?
>
> *(Titus in GEM 2019)*

They are also very worried about the effects on people inhaling the dust that would be spread and eating the fish and other marine animals that would swallow the glass (GEM 2019).

Farmers contesting geoengineering

La Via Campesina (LVC), the largest peasant organization at the global level with over 200 million members on all continents, including 182 organizations in 81 countries, has long opposed geoengineering and was one of the founders of the HOME Campaign. La Via Campesina has reaffirmed its opposition to geoengineering as a "false solution" to climate crisis in several of its demonstrations parallel to UNFCCC climate negotiations and at its own global conferences, most recently in the Basque country in 2017. In 2018, LVC was among the initiators of the HOME Manifesto. La Via Campesina affirms that peasant agriculture and agroecology are among the most important real solutions that can prevent climate change and feed the majority of the world's population, on the basis of social and climate justice instead of dangerous proposals like geoengineering (HBF 2018).

Intentionally altering the climate at global level or even the weather at the local or regional level will have effects on rain and wind patterns, with potential negative effects on agriculture and fisheries. Also, projects such as ocean fertilization will have direct negative impacts on the marine food chain and algae cultivation, and bioenergy with carbon capture and storage (BECCS) would compete for land, water, and nutrients used for food production. All these particularly affect peasant and family farmers, as well as pastoralists, artisanal fisheries, and algae producers. Because of this, in 2007, organizations defending over 10,000 artisanal algae cultivators in the Philippines successfully defeated a urea ocean fertilization project that the company ONC attempted to deploy in the Sulu Sea. This example was taken as one of the precautionary tales that led to the UN moratoria on ocean fertilization.

Concluding remarks

The fight over geoengineering is a high-risk battle that has only just begun. While the seriousness of the impacts of climate change clearly demonstrates the need for radical changes to capitalist production and consumption, powerful actors are putting forward extreme technologies as proposals to "solve" the climate crisis to block those that might affect their bottom lines.

Billionaires and fossil fuel industry proponents increasingly invest in and push geoengineering techniques as false so-called "solutions" to climate change, in the hope that their unproven promises function as alibis for the continuation of their high-GHG-emitting activities. Geoengineering would provide them with an additional new climate business.

To advance these technologies, they need to convince governments and United Nations agencies to rescind the precautionary approach and not implement independent overview that would show the risks of geoengineering or expose how it aggravates climate and social injustices. The deployment of geoengineering at a large scale needs to be preceded by small-scale experiments, which are now overwhelmingly proposed to be carried out on Indigenous territories and the lands of peasant communities.

This chapter has addressed prominent examples of the growing resistance to geoengineering by civil society organizations and popular movements, at both global and local levels. These movements are fighting in a variety of ways to defend their territories, cultures, and livelihoods and to advance grassroots democratic control over the means of life and decision-making processes, thus strengthening the path towards ecosocialism. They are organizing for a "just transition" to a culturally diverse and economically and socially just world in harmony with all life. These struggles have up until now stopped the rapid advancement of these dangerous projects. More support and broader commitment from movements, communities, and organizations will be key to stopping the mounting threats of geoengineering.

Note

1 Two points of critical sources of information on geoengineering: (a) the HOME network, the ETC Group, the Heinrich Böll Foundation, and Biofuelwatch, outline the history of geoengineering, the economic and political actors involved, the proposals and their impacts, the debates and decisions at the UN, and the alternatives put forward by the movements. This book is a useful primer for understanding the issue (ETC Group, HBF, and BFW 2018). (b) These organizations collaborate to maintain the electronic portal Geoengineering Monitor that provides updates about geoengineering developments and reflects the actions and reports of the HOME Campaign. Another important reference is an interactive map that shows where different geoengineering projects and research and experiments sites are being planned or have been carried out, with some descriptions about their current status (HBF and ETC Group 2021). For a more detailed description of the techniques and impacts of geoengineering, see the technology briefings at Geoengineering Monitor (GEM 2021a).

References

Barnard, Michael. 2019. "Chevron's Fig Leaf: Carbon Engineering Burns Natural Gas to Capture Carbon from the Air." *Cleantechnica*, 12 May. https://cleantechnica.com/2019/04/12/chevrons-fig-leaf-part-1-carbon-engineering-burns-natural-gas-to-capture-carbon-from-the-air/

Center for International Environmental Law (CIEL). 2019a. "Press Room: Fuel to the Fire: How Geoengineering Threatens to Entrench Fossil Fuels and Accelerate the Climate Crisis." *Center for International Environmental Law*, 13 February. www.ciel.org/news/fuel-to-the-fire-how-geoengineering-threatens-to-entrench-fossil-fuels-and-accelerate-the-climate-crisis/

Center for International Environmental Law (CIEL). 2019b. "Fuel to the Fire: How Geoengineering Threatens to Entrench Fossil Fuels and Accelerate the Climate Crisis." *Center for International Environmental Law*. www.ciel.org/reports/fuel-to-the-fire-how-geoengineering-threatens-to-entrench-fossil-fuels-and-accelerate-the-climate-crisis-feb-2019/ (accessed 10 April 2021).

Center for International Environmental Law (CIEL), ETC Group, and Heinrich Böll Foundation. 2019. "The UN Must Take on Geoengineering Governance." *Center for International Environmental Law, ETC Group, Heinrich Böll Foundation*, 11 March. www.geoengineeringmonitor.org/2019/03/the-un-must-take-on-geoengineering-governance/

Climate Action Network (CAN). 2019. "Position on Solar Radiation Management (SRM), September 2019." https://climatenetwork.org/wp-content/uploads/2019/09/CAN-SRM-position.pdf (accessed 1 April 2021).

Climate Justice Alliance (CJA). 2019. "It Takes Roots." https://climatejusticealliance.org/workgroup/it-takes-roots/ (accessed 1 April 2021).

Climate Justice Alliance (CJA). 2020. "Geoengineering 101: Hacking the Planet." https://climatejusticealliance.org/geoengineering/ (accessed 1 April 2021).

Climate Justice Alliance (CJA), Biofuelwatch, Indigenous Environmental Network, Friends of the Earth International, La Via Campesina, and ETC Group. "110 Civil Society Organizations and Popular Movements Demand an Immediate Stop to Climate Geoengineering." 4 October. www.geoengineeringmonitor.org/2018/10/news-release-hands-off-mother-earth-110-civil-society-organizations-and-popular-movements-demand-an-immediate-stop-to-climate-geoengineering/

Convention on Biological Diversity (CBD). 2021. "Climate Related Geoengineering and Biodiversity." www.cbd.int/climate/geoengineering/ (accessed 25 April 2021).

Convention on Biological Diversity Alliance. 2016. "Geoengineering is a Distraction from the Real Priorities Emission Reductions." *Convention on Biological Diversity Alliance*, 16 November. www.cbd-alliance.org/en/open-letter/2016/geoengineering-distraction-real-priorities-emission-reductions

ETC Group. 2010. "Hands Off Mother Earth!" 21 April. www.etcgroup.org/content/hands-mother-earth

ETC Group. 2020. "Geoengineering and Decolonization: Podcast #7 Interview with Tom Goldtooth." 11 November. www.etcgroup.org/content/7-geoengineering-and-decolonization

ETC Group, Heinrich Böll Foundation, and Biofuelwatch. 2018. "The Big Bad Fix: The Case against Climate Geoengineering." www.etcgroup.org/content/big-bad-fix (accessed 10 April 2021).

Fleming, James R. 2010. *Fixing the Sky: The Checkered History of Weather and Climate Control*. New York: Columbia University Press.

Geoengineering Monitor (GEM). 2018. "The Ice 911 Project: Geoengineering Experiment Briefing." 6 April. www.geoengineeringmonitor.org/2018/04/ice-911-geoengineering-experiment-briefing/

Geoengineering Monitor (GEM). 2019. "Arctic Geoengineering Experiment is Dangerous, Lacks Community Consent: Inupiaq Organizer." 14 February. www.geoengineeringmonitor.org/2019/02/arctic-geoengineering-experiment-is-dangerous-lacks-community-consent-inupiaq-organizer/

Geoengineering Monitor (GEM). 2021a. "Technology Briefings." www.geoengineeringmonitor.org/technologies/ (accessed 15 April 2021).

Geoengineering Monitor (GEM). 2021b. "Current Geoengineering Attempts Briefing: SCoPEx." 8 February. www.geoengineeringmonitor.org/2021/02/current-geoengineering-attempts-briefing-scopex-2021/

Global Alliance for the Rights of Nature (GARN). 2010. "People's Conference on Climate Change and the Rights of Mother Earth." https://therightsofnature.org/cochabama-rights/ (accessed 10 April 2021).

Guujaw, President of the Haida Nation. 2012. "Let It Be Known." *Council of the Haida Nation*, 18 October. http://aptn.ca/news/wp-content/uploads/sites/4/2012/10/Haida.pdf

Hands Off Mother Earth (HOME). 2018. "Manifesto against Geoengineering." www.handsoffmotherearth.org/ (accessed 10 April 2021).

Hands Off Mother Earth (HOME). 2020a. "A California Government Backs Away from Solar Geoengineering Project – but Doesn't Withdraw." 17 June. www.geoengineeringmonitor.org/2020/06/california-government-backs-away-from-solar-geoengineering-project-but-doesnt-withdraw/

Hands Off Mother Earth (HOME). 2020b. "HOME Campaign Statement on the SCoPEx Advisory Committee and Its Legitimation of Solar Geoengineering." 31 July. www.geoengineeringmonitor.org/2020/07/no-to-solar-geoengineering-and-meaningless-participation/

Heinrich Böll Foundation (HBF). 2018. "Radical Realism for Climate Justice." *La Via Campesina*. www.boell.de/en/radicalrealism?dimension1=ds_geoengineering

Heinrich Böll Foundation (HBF) and ETC Group. 2021. "Interactive Map on Geoengineering." https://map.geoengineeringmonitor.org/ (accessed 12 April 2021).

Krauss, Clifford. 2019. "Blamed for Climate Change, Oil Companies Invest in Carbon Removal." *New York Times*, 7 April. www.nytimes.com/2019/04/07/business/energy-environment/climate-change-carbon-engineering.html

Leiva, Samuel. 2019. "Geoingeniería en Chile." *Revista América Latina en Movimiento* 543: 26–28. www.alainet.org/es/articulo/202429

Pierrehumbert, Raymond T. 2017. "The Trouble with Geoengineers 'Hacking the Planet'." *Bulletin of the Atomic Scientists*, 23 June. https://thebulletin.org/2017/06/the-trouble-with-geoengineers-hacking-the-planet/

Ribeiro, Silvia. 2020. "Billionaires and Geoengineering." *Agencia Latinoamericana de Información (ALAI)*, 10 October. www.alainet.org/en/articulo/209701?language=es

Trisos, Christopher H., Giueseppe Amatulli, Jessica Gurevitch, Alan Robock, Lili Xia, and Brian Zambri. 2018. "Potentially Dangerous Consequences for Biodiversity of Solar Geoengineering Implementation and Termination." *Nature Ecology & Evolution* 2: 475–482. www.nature.com/articles/s41559-017-0431-0

Women Gender Constituency. 2019. "WGC Feminist Demands at COP25." 2 December. https://womengenderclimate.org/wgc-feminist-demands-at-cop25/

22

GREEN REFORMS AND INDIVIDUAL INTERVENTIONS IN THE GREEN NEW DEAL TRANSITION TO ECOSOCIALISM

Pritam Singh

Introduction

We wish to focus on two aspects of the political economy of the transition to ecosocialism which either have not received adequate attention in the literature on the Green New Deal transition to ecosocialism or, in fact, have been criticized as antithetical to the project of making a transition to ecosocialism. The first deals with the importance of green reforms under capitalism on the transition path to ecosocialism, and the second deals with the role of individuals in crafting political, social, and economic interventions for the move towards ecosocialism.

Green New Deal

Three points are central to the argument of the Green New Deal transition to ecosocialism. The first is that, in the current context of global capitalism, there is no prospect for a sudden or revolutionary rupture with capitalism to usher in a socialist regulation of economy and society. The second is that, given the environmental catastrophe humanity is currently faced with, it is historically necessary to make a break with the environmentally destructive capitalist mode of production and usher in an ecologically oriented socialist era. The third is that there is an urgent need for a multitude of macro but also micro changes relating to production, exchange, distribution, and consumption within the confines of capitalism, which collectively and cumulatively have the potential to become a powerful mass in charting the transition path to socialism that is compatible with ecological sustainability. The third point can be considered the key one, incorporating the first and the second.

It is in the context of this third point that we examine the role of green reforms under capitalism and that of individual economic, social, political, and cultural initiatives in making a transition to ecosocialism.

10.4324/9780429341427-26

Green reforms under capitalism

The ecosocialist position on averting global climate change that poses an existential threat to planet earth rightly emphasizes the necessity of the historical task of replacing capitalism with ecosocialism. This necessity can be posed in terms of ecosocialism versus barbarism in the same way Rosa Luxemburg posed the question of socialism versus barbarism, first articulated by Karl Kautsky (Angus 2014), in the context of averting the human costs of the First World War that was driven by inter-imperialist rivalry (Luxemburg 2010). However, the sharp polarity posed in this manner between socialism/ecosocialism and capitalism calls upon the urgency of the political project of revolutionary overthrow of capitalism. This urgency was viewed by Rosa Luxemburg as historically necessary due to the feared end of civilization caused by the impending First World War, while the urgency now might be seen as due to the impending environmental catastrophe if the global heating and mass extinction of species taking place are not stopped. During the time of Rosa Luxemburg, the advocacy of a political project of undertaking reforms to bring about socialist transformation of a capitalist society would have been dismissed by revolutionary socialists as the politics of reformism, social democratic compromise with capitalism, and not recognizing the scale of danger presented to humanity by the impending First World War. In a somewhat similar manner, the defense of green reforms under capitalism as a transitional path to ecosocialism can be construed and dismissed as supportive of green capitalism and green accommodation to capitalism and not recognizing the severity of the existential crisis humanity faces now due to the fast-accelerating global climate change. Such a dismissal of the green reforms under capitalism would be a mistake because of the failure to recognize the critical distinction between two sharply opposing positionings of such reforms. One position on green reforms can be viewed as a part of defending the project of green capitalism to reject the necessity for socialism. The opposite position on green reforms is to critique the green capitalism perspective and to underscore the necessity for socialism and the role of green reforms in the project of making a transition to ecosocialism. Understanding the fundamental differences between these two contrasting perspectives on green reforms under capitalism is of critical importance, both analytically and politically.

The green capitalism perspective can be understood to be driven by three considerations: one, the realization that the current neoliberal global capitalism is environmentally destructive; two, that capitalism needs to show flexibility and adaptability by incorporating many green initiatives; and three, that to avert global environmental catastrophe resulting from the current global capitalism, it is not necessary to replace capitalism with socialism; instead, it is possible to introduce green reforms to capitalism to make it compatible with ecological sustainability.

The ecosocialist perspective, in sharp contrast with the green capitalism perspective, is based on recognizing that the conflict between capital and nature is fundamental and cannot be resolved by merely making green accommodations under capitalism. The centrality of the profit-maximizing imperative of capital leads to capital viewing nature from the objective of extracting surplus from it. This imparts a short-termism character to the operation of capital and, therefore, leads to capital being inherently incompatible with sustainability, which is essentially long term in character. Additionally, a capitalist economy is characterized by many capitals competing with each other. Each individual capital is in pursuit of profitability in its own domain and is indifferent to the overall macro consequences, social or ecological, of its profitability pursuit. This individual profitability pursuit with its micro-capitalist rationality leads to macro-ecological irrationality manifested through overproduction, waste, pollution, and global climate change. It is possible that an individual firm governed by the imperative of profitability may incorporate environmentally efficient technology such as renewable energy to outcompete its rivals. Such a business initiative by a single firm is likely to generate the illusion of the possibility of compatibility between profits and sustainability – a possibility that is central to the argument in favour of green capitalism.

To puncture that illusion, it is important to grasp that another central dimension of the logic of capital accumulation is the need for continuous expansion of consumption to valorize or realize the surplus value generated in the production and labour process through technologically induced methods to increase the relative rate of surplus value. A critical implication of this dimension of capital accumulation is that in the short term, a single firm's success in combining profitability objectives with ecologically sustainable goals would clash in the long term with the macro needs of capitalism to continuously expand consumption, which is environmentally destructive. What seems possible, therefore, in the short term at the level of a single capitalist business in combining profits with ecology is not possible in the long term at the level of the macro-capitalist economy. Heuristically, we can say that if there were to be one capital governing global capitalism that was free from the external pressure of competition, it could transcend the short termism that is characteristic of many capitals competing and could view the sustainability of nature as necessary for the long-term survival of that sole capital. Such a single capital governing global capitalism would then need globally centralized planning to coordinate the continuous availability of nature for extraction. Such a centrally planned capitalism operating on a global scale and governed by a single capital is a logical impossibility. This means that many micro green capitalist initiatives are possible, but greening capitalism as a whole at a macro systemic level is not possible.

Having recognized that green capitalism at a macro aggregative level is not possible raises the need to interrogate further the potentialities of micro green initiatives towards greening capitalism and the role of this greening capitalism in creating the material foundations for the ecosocialist economy of the future. At a general level, socialist support for green reforms under capitalism could be considered akin to socialist support for democratic reforms under capitalism. Socialists support democratic reforms without entertaining any illusion that capitalism, with its huge and structural class inequalities, can ever be fully democratic. Despite this structural limitation, socialists still support democratic reforms under capitalism. This is for two main reasons: one, to expand the space for the organization of collectives of workers, peasants, service-sector employees, women, racial minorities, linguistic and national minorities, sexual minorities, disabled, homeless and other discriminated social groups and identities so that such collective struggles can exert their social and political weight to reduce multiple forms of inequalities under capitalism and two, to limit the centralized power structures through increasing the countervailing power of democratically elected assemblies, free media, independent judiciary, and academia. Both these main reasons for socialists supporting democratic reforms under capitalism have another underlying logic: that is, to create cultural and material foundations for socialism that would also be democratic. If a capitalist society is characterized by highly embedded authoritarian structures, bureaucratic institutions, and hierarchical social practices, the socialism that would emerge as an alternative to such a capitalist society would also be marred by these authoritarian, bureaucratic, and hierarchical deformities. The painful historical experiences of the post-1917 Soviet Union and subsequent "socialist" alternatives in Eastern Europe, China, Southeast Asia, and Cuba provide undeniable evidence of how an alternative is influenced very critically by what it is an alternative to.

Green reforms within the confines of a capitalist economy can be viewed at different levels: global, national, regional, and local. The 2015 Paris Accord is the best example of a set of intended green reforms at a global level without questioning or altering the fundamental logic or institutions of capitalism. Another global green reforms initiative is the ENO (Environment

Online) Green Cities Network programme launched in 2000, which is a global virtual school and network linking thousands of schools in 150 countries. Tree planting is the most popular activity of the programme, and the linked schools aimed to plant 100 million trees by 2017. The scheme, launched first by Bangladesh in 2002, to ban thinner plastic bags altogether after they were found to have choked local drainage systems during floods is an example of a national-level green reform measure that has led to other countries either adopting or modifying the scheme, such as shops charging for plastic carrier bags. Similarly, green industrial policies suited to the geographical and ecological conditions of different regions reflect green reform initiatives at a regional level (Grillitsch and Hansen 2019). An excellent example of a local-/city-/village-level green reform initiative is the urban green spaces initiative in Melbourne (Bush 2020).

Even if the green reforms, whether at the global, national, regional, or local level, are strictly within the logic of adaptability to a capitalist economy, these reforms must be viewed and supported as part of the transition path to ecosocialism. Green capitalism is better than dirty capitalism in laying the material foundations for green socialism in the same way as democratic capitalism is better than authoritarian capitalism in creating the politico-cultural conditions for democratic socialism. Even if the green reforms under capitalism are pursued and supported by those who subjectively view these reforms as aimed at merely greening capitalism, such reforms are objectively subversive to capitalism insofar as these reforms weaken capital's power to subjugate nature.

The role of individual green actions under capitalism in the transition to ecosocialism

There is a strong and persistent tradition in Marxist historical writings that values large social actions in historical change. This is primarily due to two reasons. First, it is due to any larger social change being viewed as the result of collective action. A revolution is the epitome of celebration of collective action, but even changes that are significant but far less than a revolution are considered the result of a collective action or a series of interconnected collective actions. Second, it is due to a reaction against and critique of what is often referred to as bourgeois individualism. Even when an individual role or action is valued as, for example, Lenin's role during the October–November turmoil in Russia in 1917 in giving a call for a revolutionary overthrow of the provisional government by the Soviet workers, it is mainly seen in the context of the larger social movements of workers, peasants, soldiers, and the Soviets. An ordinary micro activity of a citizen while going about their daily life is not considered worthy of any serious attention.

In the environmental movement, too, attributing importance to individual actions is largely viewed with suspicion or as promoting complacency about the anti-environmental nature of the capitalist system as a whole. Heather Rogers, while highlighting the role of the corporate world in creating waste, dismisses the individual's environmental actions: "People are induced to accept individual, personal responsibility for cleaning up the environment and are lulled into a sense of complacency by the idea that they are actually doing something effective" (2006, 249). Smith (1998, 107) expresses a similar view that seeking personal solutions to problems created by capitalism leads to distraction from confronting the inequalities of the economic system. The argument being advanced here in highlighting the importance of individual actions is not that individual action on its own can deal with the inequities of capitalism or the environmental devastation caused by capitalism, but that such actions can have the cumulative effect of activating individuals into making a contribution to an equitable and sustainable economic system. If the individual activity is trashed as meaningless, or even counterproductive for lulling them into complacency, it can also lead to a feeling of helplessness in facing the gigantic task of confronting capitalism and can demotivate individuals.

In appreciating the role of individual action in the transition to ecosocialism, these actions can be viewed in the interconnected domains of economy, politics, social practices, culture, and art.

In the economic domain, the most significant green transitional intervention an individual can make is in the sphere of consumption. The sphere of consumption embraces a wide variety of interconnected activities, including the consumption of food, clothes, and footwear; the consumption of energy in cooking, travel, and lighting the house; and the consumption of time and materials in reading, writing, leisure enjoyment, physical exercise, and sexual pleasure. This list is certainly not exhaustive, but even this list indicates the multiple forms of environmental impact of each one of these activities in consumption of resources, the waste that emerges from that consumption, and the impact of that waste in straining the absorptive capacity of the biosphere. An individual aware, even if only partially, of the environmental impact of any of these activities can proactively make a change in the forms and scale of consumption in their daily life.

Marxist political economy has a tendency to overplay the primacy of production and to view consumption as a mere derivative of production (Singh 2021). This tendency to overemphasize production over consumption has been strengthened as a reaction against the neo-classical marginalist economics which fetishize utility from consumption as a utility theory of value against the Marxist labour theory of value (Bharadwaj 1986). Undoubtedly, there has been a tradition even in Marxist theory of giving due weight to consumption in the analysis of capitalism as, for example, in Rosa Luxemburg's work on imperialism (1951), in which she argued that the limit to capitalist expansion was determined by the limits to consumption caused by the ultimate exhaustion of markets for commodities. Marx's *Capital* and *Grundrisse* do not view consumption as a mere passive derivative of production as it is seen in the productivist tradition of Marxian political economy, and one of the key contributions of ecosocialist and green political economy has been to bring back the importance of consumption in the analysis and critique of capitalism, especially its nature-destroying character (Singh 2021).

The significance of individual interventions in the economic sphere in contributing to the transition to ecosocialism must be understood and appreciated in the context of the increasing ecological attention given to the role of consumption in capitalism's nature-destroying character.

Regarding an individual's intervention in the domain of politics to make a contribution to the ecosocialist transition, it is important to emphasize the distinction between two types of individual interventions: namely, one as a member of a collective, political party, political group. or trade union and the other as not being actively involved in any collective organization. What is being discussed here is not the role of an individual as a member of a collective because that involves a consideration of a role at another level: i.e., the role of the collective and not that of an individual per se. We wish to reflect on the role of an individual who is *not* a part of any collective. Such an individual who keeps away from any such collective, either due to their temperament or the compulsions of such a person's circumstances, also has a potential contribution to the ecosocialist transition. For the sake of clarity, we may characterize such an individual as a passive participant in ecosocialist transition to differentiate such a person from an active participant who is involved in some form of collective political activity. The minimum contribution such a passive individual can make is by voting for a socialist/green candidate in a local, regional, or national election. A slightly higher level of individual contribution could be making financial contributions, however small, to socialist/green political parties. To exercise one's right to vote in a capitalist society is, perhaps, the most passive micro-political activity anyone eligible to vote can undertake. The contribution of such a micro-political activity can vary from enabling socialist/green parties to get visibility in the public space to such parties scoring electoral victories at different levels of governance in a country from local councils to a national parliament.

In any country, the vast majority of people, whether of voting age or below, are not active participants in politics. The traditional socialist politics overvalues active political participation. Meetings, rallies, demonstrations, trade-union militancy, public speaking, and contesting elections are part of the valued traditions of socialist politics. This activist view of political participation has an inherent gender bias. Due to women's more complex role in household management, this activist bias favours men's participation and dominance in politics. In imagining the political path to ecosocialist transition, it is of critical importance that the significance of the role and contributions of non-activists should be understood, recognized, and incorporated in socialist strategies.

In the domain of an individual's role in social practices and culture towards promoting ecosocialist transition, the most significant contribution an individual can make is shunning consumerism in one's lifestyle. Capitalist economy, through the institution of commercial advertisements, is based on promoting consumerism and commodity fetishism. From the classic work of Vance Packard (1957) to more modern commentaries on consumer capitalism (Panayotakis 2011), the role of the advertising industry in creating wants, needs, and consumer culture has become widely known. The traditional macroeconomics textbooks have popularized the concept of conspicuous consumption as if the phenomenon of conspicuous consumption is a part of eternal human nature. As opposed to conspicuous consumption, an individual, through a daily lifestyle of simplicity, can have a counter-demonstrative effect against conspicuous consumption.

The overarching importance of individual green actions in the transition to ecosocialism is that the individuals participating in such actions remain mentally engaged with the larger project of greening the economy, even when they are not able or not inclined to participate in more collectivist activities such as demonstrations and strikes.

In arguing in defense of the importance of individual actions, the ecosocialist perspective is adding a new dimension to the praxis of socialism in contrast with the dominant view in socialist orthodoxy of arguing only for macro or large actions as legitimate for making a transition to ecosocialism.

Conclusion

The orthodox socialist politics has eulogized revolutionary overthrow of capitalism and the action of large social classes (working class or peasantry) in bringing about the historic break with capitalism. We have argued for an ecosocialist politics that recognizes the historic necessity of making a break with capitalism to save planet earth from the environmental catastrophe resulting from the current phase of global climate change. Such a break does demand significant mobilization of those who are affected by the inequalities and environmental devastation caused by capitalism. As a further development of that ecosocialist politics, we have argued that there is a space for green reforms under capitalism and individual actions to contribute to ecosocialist transition. We have argued that if green reforms lead to making capitalism green, even if that is partial and cannot be complete, that greening creates the material foundations for socialist alternatives that would also be green. Dirty capitalism's socialist alternative is also likely to be dirty. Understanding this perspective on green reforms also enables us to shed any possible illusion that capitalism as a whole can be greened. The fundamental macro conflict between capital and nature leaves no scope for such an illusion.

Since every individual action has an environmental implication in terms of resource use and waste creation, the awareness of this implication raises the importance of individual action which could not be captured in the orthodox pre-environmental phase of socialist politics when, to take an example, an individual worker going on a strike against a capitalist employer was not only meaningless and ineffective; it was also, in fact, counter-productive to the worker individually and to his trade union collectively. Ecosocialist politics has expanded the scope of a variety of individual actions in the sphere of economy, politics, and culture aimed at contributing to ecosocialist transition.

References

Angus, Ian. 2014. "The Origin of Rosa Luxemburg's Slogan 'Socialism or Barbarism'." *Climate and Capitalism*, 22 October. https://climateandcapitalism.com/2014/10/22/origin-rosa-luxemburgs-slogan-socialism-barbarism/ (accessed 25 February 2021).

Bharadwaj, Krishna. 1986. *Classical Political Economy and Rise to Dominance of Supply and Demand Theories.* New Delhi: Sangam.

Bush, Judy. 2020. "The Role of Local Government Greening Policies in the Transition Towards Nature-Based Cities." *Environmental Innovation and Societal Transitions* 35: 35–44, June.

Grillitsch, Markus, and Teis Hansen. 2019. "Green Industry Development in Different Types of Regions." *European Planning Studies* 27 (11): 2163–2183.

Luxemburg, Rosa. 1915 [2010]. *Rosa Luxemburg: Socialism or Barbarism, Selected Writings*, edited by Helen C. Scott and Paul Le Blanc. London: Pluto.

Luxemburg, Rosa. 1951. *The Accumulation of Capital.* New Haven, CT: Yale University Press.

Packard, Vance. 1957 [1958]. *The Hidden Persuaders.* Cardinal ed. New York: David McKay, Pocket Books.

Panayotakis, Costas. 2011. *Remaking Scarcity: From Capitalist Inefficiency to Economic Democracy.* London: Pluto.

Rogers, Heather. 2006. "Garbage Capitalism's Green Commerce." In *Socialist Register 2007*, edited by Leo Panitch and Colin Leys with Barbara Harriss-White, Elmar Altvater, and Greg Albo, 231–253. London: The Merlin Press, New York: Monthly Review Press, and Halifax: Fernwood Publishing (Special issue on "Coming to Terms With Nature.").

Singh, Pritam. 2021. "Capitalism, Ecology and Eco-Socialism." In *Global Political Economy: A Critique of Contemporary Capitalism*, edited by V. Upadhyay and Paramjit Singh, 353–375. New Delhi: Aakar.

Smith, Toby M. 1998. *The Myth of Green Marketing: Tending Our Goats at the Edge of Apocalypse.* Toronto: Toronto University Press.

23

EXTINCTION REBELLION

Crisis, inaction, and the question of civil disobedience as ecosocialist strategy

Samuel Alexander and Peter Burdon

Introduction

In this chapter, we assess the political strategy of "civil disobedience" – the practice of non-violently breaking the law to advance social, political, or environmental causes. Our contention is that while civil disobedience can make people feel uncomfortable, it is important for a society to understand the motivations for civil disobedience and evaluate the reasons given for practicing this radical and disruptive strategy for societal change. Might civil disobedience be needed to advance the vision of an ecosocialist society? We consider this question in relation to the nascent Extinction Rebellion movement, which has brought activists together around the world.

To address this issue, we first provide a brief introduction to Extinction Rebellion and their demands and tactics. Following this we unpack the meaning of civil disobedience and consider whether (and in what circumstances) it can be justified within a liberal democracy. Here we argue that civil disobedience mediates the boundary between the need for change and the need for stability. The law both gains and retains its legitimacy through contestation, antagonism, and the ability of a people to dissent. Finally, we conclude our discussion by considering how Extinction Rebellion can learn from socialist debates about the tension between crowds and structures. More specifically, we contend that Extinction Rebellion should be more expressly political in its identity and build structures that are accountable and able to support a long-term movement.

Extinction Rebellion: Principles and practices

The Extinction Rebellion (XR) has emerged as one of the most active and prominent faces of the environmental movement around the world today. While XR is a decentralized movement, it is important to note from the outset that its most prominent spokespeople have not aligned themselves with ecosocialism or class-based struggle. XR describes itself as "apolitical" or "beyond politics" and seeks to incorporate every person "regardless of ethnicity, race, class, gender, gender identity" (Extinction Rebellion 2019).[1] This is a pragmatic position that seeks to incorporate the widest possible subset of civil societies in acts of non-violent civil disobedience. However, as we argue later in this chapter, it is also something that ought to be challenged as the movement grows and seeks to make broader demands.

10.4324/9780429341427-27

While a "protest" on a particular issue may come and go, a "rebellion" defines itself by the breadth of its opposition and the refusal to fade away, even in the face of slow progress or backlash from the state. Whether the Extinction Rebellion can live up to its name remains to be seen, but the forces of resistance do seem to be on the rise (Read 2019).

The Extinction Rebellion (or XR) has three principles or demands:

1. Government must tell the truth by declaring a climate and ecological emergency, working with other institutions to communicate the urgency for change;
2. Government must act now to halt biodiversity loss and reduce greenhouse gas emissions to net zero by 2025; and
3. Government must create and be led by the decisions of a citizens' assembly on climate and ecological justice.

All these principles deserve critical consideration and ongoing debate (see Farrell et al. 2019), and reasonable people can accept them, challenge them, or disagree with aspects of them. Indeed, XR itself views these demands as part of an ongoing process of discussion and refinement, and the way the movement and its key issues are framed has not been free from criticism, even by sympathetic voices (see, e.g., Resilience 2019; Knights 2019). What is clear is that achieving the goals of XR will raise all sorts of deep complexities and thorny challenges, which may only be resolvable – again, if resolvable at all – through the messy process of lived experience and experimentation.

A swift decarbonization of the global economy is an intimidating task, supported by the science but utterly unprecedented in human history. Certainly, there is no detailed blueprint to tell us how to do it. But there is a clear distinction between XR and most other forms of thinking and practice in the environmental movement today. In the attempt to respond appropriately to climate breakdown and the broader environmental crisis (see Steffan et al. 2015), XR is explicitly holding up non-violent civil disobedience as an important and perhaps necessary part of the socio-political strategy for achieving a just and sustainable world (Extinction Rebellion 2019a; Hallam 2019).

Some commentators will be tempted to dismiss XR activists as mere "trouble makers" or even "criminals," but such reactions, though understandable, risk mischaracterizing these ethically motivated actions that are designed to be confronting, inconvenient, and disruptive. Even though most people probably have reservations and concerns about civil disobedience, we must nevertheless appreciate that many of the most significant social and political advances over the last century owe much to social movements that engaged in civil disobedience as a primary strategy (Chenoweth and Stephan 2012). One might think especially of Gandhi and the independence movement from British rule, the suffragettes movement, and the civil rights movement. These esteemed traditions raise the disconcerting question: Might future advances in society also demand civil disobedience?

What is civil disobedience, and is it justified?

We are of the view that Extinction Rebellion and related movements are likely to grow over coming years. We have written previously about the centrality of capitalism to the environment crisis and socialist organizing (Alexander and Burdon 2017). Rather than retracing this material, we think it is critical to examine a defining feature of XR – that is, an openness to civil disobedience as a strategy for change. What is civil disobedience? And when, if ever, can it be justified?

In essence, civil disobedience can be defined as "a public, non-violent and conscientious breach of law undertaken with the aim of bringing about change in laws or government policies" (Brownlee 2007, np). For present purposes, we will assume that for disobedience to be "civil," it has to be non-violent, and indeed this accords with the explicit and unconditional commitment XR has to non-violence (Farrell et al. 2019). In an important aside, empirical studies show that movements committed to non-violent disobedience tend to be twice as successful in achieving their aims as violent demonstrations (Chenoweth and Stephan 2012; Chenoweth 2017);[2] thus, XR's principled commitment to non-violence is also pragmatic. One might add that it is also a diverse strategy, with Gene Sharp famously listing 198 ways to practice non-violent resistance (Sharp 1973). Before engaging in such acts, however, individuals and groups should ask themselves: Can civil disobedience ever be justified in a democracy?

It can be helpful to begin assessing civil disobedience in relation to basic democratic theory. Imperfect though it is, XR has taken root most strongly in countries that have democratic systems of governance. Among other things, this means that citizens and (in most cases) permanent residents get to vote on who will represent them in government, and government includes a legislative branch that creates law and an executive branch that enforces it. Since we all have, in theory, an equal opportunity to influence the law-making process through the ballot box, it is generally assumed that we should obey the law because the democratic process is the best way to organize and structure society and develop public policy that serves the common good.

From this perspective, an opponent of civil disobedience might argue as follows: "We can't all break the law every time we disagree with it. Imagine how unstable society would be if that happened. If we don't like what is happening, we can campaign for change like everyone else, and if we succeed, we can vote the existing government out of power through the electoral process and vote in a new government." In this way, democratic societies are said to have created the institutions and processes needed for their own peaceful improvement.

So the main objection to civil disobedience is this: "If you disagree with a law or policy, don't break that law or policy; instead, campaign to get it changed through the democratic process. If you are permitted to break the law just because you disagree with it, then why can't anyone break a law they disagree with?" At first instance, perhaps, this objection seems quite powerful. Indeed, Immanuel Kant argued that "[a]ll resistance against the supreme legislative power . . . is the greatest and most punishable crime in the commonwealth, for it destroys its very foundations" (Kant 1970, 81). In the twentieth century, T. H. Green (1907, 111) made this statement about the relationship between obligation and democracy:

> Supposing then the individual to have decided that some command of a "political superior" is not for the common good, how ought he to act in regard to it? In a country like ours, with a popular government and settled methods of enacting and repealing laws, the answer of common sense is simple and sufficient. He should do all he can by legal methods to get the command cancelled, but till it is cancelled, he should conform to it.

Underlying these views is the idea that if people only abide by laws they agree with, then the rule of law will break down. To some extent, then, we might all have sympathy with the political assumption that we ought to obey laws – even laws we don't agree with.

But it is one thing to make that broad and pragmatic concession. It is quite another to suggest that all laws, always, ought to be obeyed. If obedience to law were unconditional and absolute by virtue of the democratic process, it would follow that civil disobedience is always unjustified.

How might acts of civil disobedience be interpreted within the contested disciplines of legal and political theory?

First of all, one might argue that civil disobedience is potentially justifiable when the mechanisms of democracy are not working properly, such that laws do not represent the will of the people. This can occur when laws and policies are shaped by the undemocratic influence of foreign governments, billionaires, mass media conglomerates, or other corporate lobby groups (see, e.g., Mayer 2016; Tham 2010). In such cases, one might suggest that laws produced by undemocratic processes do not demand our political allegiance since they were not produced through fair, robust, and representative democratic processes. There is also a second way in which it might be argued that civil disobedience is justified: that is, to recognize that there is a distinction between law and morality or a distinction between what is law and what is just. Often, we might admit, there is much overlap between law and justice. The more overlap the better. But any thinking person knows that often in history, and no doubt still today, there are times when we see a clear difference between what is "law" and what is "just" – even if justice is essentially a contested term. In other words, democracy may be the best form of government, but this does not mean that a democracy always gets things right. Rather, democracy, when it is functioning properly, reflects culture, and there is no reason to think that cultural norms and expectations are always just. Put more directly, a functioning democracy can produce unjust laws when a citizenry knowingly and voluntarily votes for policies that are unjust (even if they are not considered unjust by those voting for them).

For example, we know that democracies have historically declared it illegal to engage in same-sex relationships, and today most members of liberal democracies recognize that such laws were and are in breach of basic human rights. In the past, laws produced in democracies have institutionalized slavery, ratified unjust wars, legally entrenched racial segregation, criminalized homosexuality or particular religious practices, prohibited women and people of colour from voting, and so forth. Again, what is law does not automatically overlap with what is just. Nobody can deny that unless you still believe in the divine right of kings. At such times when a law or policy is clearly unjust (e.g., recognizing ownership of persons as slaves), a case can be made that there is a place for civil disobedience in democratic societies, on the grounds that we must accept that even democratically produced laws sometimes get it wrong – sometimes *really* wrong.

There is a rich and revered tradition in legal and political theory that recognizes and accepts these broad lines of argument (see review in Brownlee 2007). In other words, it is widely accepted that there is a proper place for civil disobedience in liberal democratic societies. In fact, as we look back on social movements in history – whether it is Gandhi's campaign for independence, Martin Luther King, Jr. and the civil rights movement, or Emmaline Pankhurst and the suffragette movement – some of the greatest leaps forward in social and political progress have been the result of acts of civil disobedience. It would show a gross lack of historical understanding to dismiss civil disobedience as a regressive social practice. The powerful but uncomfortable inference is that future acts of civil disobedience may also be required to advance our state of society.

Civil disobedience and the Extinction Rebellion

So how does this apply to the Extinction Rebellion? There are, as we have just implied, two main ways to evaluate civil disobedience. On the one hand, an argument could be made that we live in democracies that are at least partially broken, such that the laws and policies that are produced are sometimes undemocratic because of the undue influence corporate interests have had on the legislative process – for example, the fossil fuel industry, the Murdoch media, or other powerful economic forces (see, e.g., Market Forces 2019; Tham 2010; Cooke 2019; Knaus 2018). Roger Hallam (2019, 18), one of the founders of XR in the UK, has expressed this view:

> The political culture of Western democracies has changed from a reformist to a revolutionary structure. It is no longer possible to save our society through small incremental steps. Mass political disruption is now required.

This suggests that even if our culture wanted a strong climate response, vested interests would interfere with any such response and ensure that law and policy kept things more or less as they are. To some extent, this may be part of the reason many countries continue to promote regressive climate policy. In words often attributed to anarchist writer, Emma Goldman: 'If voting changed anything, it would be made illegal."

Perhaps the more powerful argument for civil disobedience, however, is that overall, most countries have not come to terms with the magnitude of climate breakdown and the broader environmental crisis. After all, as noted earlier, it is still quite easy to distance ourselves from the impacts of these crises, and recent elections in the UK, the US, and Australia have resulted in governments that celebrate coal and essentially deny that climate change is a serious problem. For these reasons, among others, governments around the world are losing their legitimacy and social base (Mair 2009).

So we might draw an analogy here with the anti-slavery or civil rights movements in the US. Where once the state sanctioned and supported the moral wrongs of slavery and segregation, today the state sanctions and supports the moral wrong of climate breakdown. Activists who engaged in civil disobedience during the civil rights movement might accept that white people were, in fact, voting for racist laws and public policy but justify their disobedience on the grounds that racist laws and policies were wrong and deserved to be disobeyed. We cannot say that the anti-slavery activists or civil rights activists were wrong to break the law and engage in non-violent acts of civil disobedience. Those racist laws were grossly immoral, and they deserved to be disobeyed. Rosa Parks was right not to give up her seat on the bus on that fateful day in 1955 even though it violated the laws and regulations. According to Henry David Thoreau (1982), who published his famous essay on civil disobedience in 1849, this strategy is not just a right but at times a duty. It is no surprise, then, that Gandhi, Martin Luther King, Jr., Emmaline Pankhurst, and countless other social activists have been inspired to engage in such acts and are now revered for their bravery.

Let us ask with Thoreau: Are we expected to resign our conscience to the legislator? Why have a conscience, then, if we are simply expected to uncritically affirm all acts of government? We must be human beings first and subjects of the state afterward. As Thoreau (1982, 111) argued, "it is not desirable to cultivate a respect for law, so much as for the right," and indeed, Thoreau argued that respect for law can, at times, make us daily agents of injustice. In relation to his own time, he argued that one could not be associated with the US government without disgrace for he could not recognize as *his* government what was also the *slave's* government. He concluded that if a government's law is of such a nature that it requires you to be the agent of injustice to another, then break the law. "Let your life be a counter friction to stop the machine," he declared (Thoreau 1982, 120). "Cast your whole vote, not a strip of paper merely, but your whole influence" (Thoreau 1982, 122).

In the UK, after merely eight days of Extinction Rebellion activities, more than 1,000 people had been arrested for their civil disobedience. While no one should fetishize "being arrested" as the only way to participate in XR, and the movement should recognize also that people have different "biographical availabilities" for being arrested (Beyerlein and Bergstrand 2013), the fact is that all acts of civil disobedience raise the possibility of being arrested and perhaps imprisoned. No doubt acts of civil disobedience will be perceived by many as annoying and inconvenient and unnecessarily disruptive, but that calculus always has to be weighed against the moral wrong that is motivating the disobedience (see Monbiot 2019). Slavery and segregation were also "inconvenient" . . . for those who suffered under racist laws! In that light, the inconvenience caused by sit-ins and bus boycotts pales in comparison. Similarly, when environmentalists engage in acts of civil disobedience to resist ecocide, the extinction of species, and the unfolding climate emergency, some sectors of society will no doubt be appalled and dismiss the activists as "troublemakers" and "criminals." Civil disobedience may indeed be inconvenient to many people. But to evaluate the legitimacy of the civil disobedience, one has to resist superficial analyses and ask how that inconvenience compares to future suffering and, indeed, the suffering already being caused by environmental breakdown (Spratt and Dunlop 2019; Nixon 2013).

We can now bring the analysis to a head. Just imagine, for example, that in 10 or 20 or 30 years – it doesn't really matter when – we discover that today's high-impact modes of production and consumption have led to even more alarming ecosystemic breakdown, a future that has mountains of scientific support (see, e.g., Steffan et al. 2015; Spratt and Dunlop 2017). Suppose the climate reaches its tipping point; many nations enter indeterminate and intensifying drought; food production drops even as population grows, leading to mass famine and increased geopolitical tension and war; suppose in 10 or 20 years, the arctic ice cap disappears, and the methane release from the permafrost induces a swift jump in global temperatures. Suppose any number of such things happen, and people begin to die. When we look back on today, we will ask ourselves: Did we do enough? Were we complicit in a broken system? Should we have been so obedient given that we knew our gutless governments were leading us down a dead end?

These questions are not for us to answer – we are still struggling with them ourselves. We will remain sympathetic critics and revise our views as new evidence and insight emerge. None of us can condemn or condone the actions of XR in advance of their particular, context-dependent manifestations. One might sympathize with XR in general while disagreeing with specifics, or vice versa. These are very personal questions (with social effects) which we must meditate on with due diligence. But our point is that if the future turns out the way the best scientists are predicting it will turn out if business as usual continues (for reviews, see Steffan et al. 2015; Spratt and Dunlop 2017, 2019), then the younger generation might well ask us what we did to resist the foreseeable collapse of ecosystems and the humanitarian catastrophes such breakdowns are already inducing.

Conclusion: From crowds to structures

Having presented a strong theoretical justification for non violent civil disobedience, we conclude now with a reflection on how XR might be influenced by socialist theory in the future. While the movement has, to date, described itself as apolitical, we don't think that this can or should be sustained in the future. One of the most hopeful outcomes of XR has been that it has provided a home to a generation of young activists whose work has also been an education in organizing and movement building. But if there is one additional lesson that XR should learn from past movements, it is that they need to develop structures and processes that are accountable, give the moment a coherent voice, and enable themselves to transcend the three demands outlined at the beginning of this chapter.

This, of course, is an old debate in socialist literature (Holloway and Picciotto 1979). One recent statement is from Jodi Dean (2016, 3), who critiqued the Occupy movement for its insistence that "everyone is an autonomous individual." This individualism contrasted with more collectivist statements such as "We are the 99 percent" and was a symptom of the way Left movements can unconsciously adopt the logic of communicative capitalism. As Dean (2016, 31) contends: "the era of communicative capitalism is the era of commanded individuality." Of course, as XR has shown, there is something powerful in the promise of a crowd, and it can give participants a sense of "intense belonging" (Dean 2016, 161). However, only formal structures are capable of "remembering, learning, and responding" to the environmental crisis and the barbarism of capitalism. Only collective structures that advocate an explicit political worldview "[enable] the crowd to endure as a rupture with capitalism . . . hold[ing] open the gap for the people as the collective subject of politics" (Dean 2016, 260–261).

As noted earlier, XR proposes political revolution through a series of grassroots citizen assemblies. At the moment, there is very little detail about what this might look like (Hallam 2019, 21–22), but those involved in XR should learn from past socialist struggles whose histories are filled with lessons learned. In this context, being ahistorical is not a virtue. And while an apolitical posture might help XR appeal to the largest portion of the population, that is not a neutral position either. As Marx (1978, 3) understood long ago, indifference to the powers that organize society – powers of wealth, education, race, gender, class, etc. – is to side with the status quo. For example, if extractive capitalism is a site of social power and something that frames political demands, then for the state, to feign neutrality is to side with existing privilege and foreclose certain kinds of arguments from gaining traction.

Notes

1 The apolitical nature of XR is further exemplified in Lucas (2019) and Raworth (2019).

2 Extinction Rebellion's strategy is grounded in the idea that only 3.5 percent of the population needs to be engaged in active disobedience for political change. For a critique, see Ahmed 2019.

References

Ahmed, Nafeez. 2019. "The Flawed Social Science Behind Extinction Rebellion's Change Strategy." *Medium*. https://medium.com/insurge-intelligence/the-flawed-science-behind-extinction-rebellions-change-strategy-af077b9abb4d (accessed 17 December 2019).

Alexander, Samuel, and Peter Burdon. 2017. "Wild Democracy: A Biodiversity of Resistance and Renewal." *The Ecological Citizen* 1 (1): 45–54.

Beyerlein, Kraig, and Kelly Bergstrand. 2013. "Biographical Availability." *The Wiley-Blackwell Encyclopedia of Social and Political Movements*. https://onlinelibrary.wiley.com/doi/abs/10.1002/9780470674871.wbespm012 (accessed 5 September 2019).

Brownlee, Kimberley. 2007. "Civil Disobedience." *Stanford Encyclopedia of Philosophy*. https://plato.stanford.edu/entries/civil-disobedience/ (accessed 5 September 2019).

Chenoweth, Erica. 2017. "It May Only Take 3.5% of the Population to Topple a Dictator – with Civil Resistance." *The Guardian*, 2 February (accessed 5 September 2019).

Chenoweth, Erica, and Maria Stephan. 2012. *Why Civil Resistance Works: The Strategic Logic of Non-Violent Conflict*. New York: Columbia University Press.

Cooke, Richard, 2019. "News Corp: Democracy's Greatest Threat." *The Monthly*, May.

Dean, Jodi. 2016. *Crowds and Party*. New York: Verso.

Extinction Rebellion. 2019a. http://rebellion.earth (accessed 5 September 2019).

Farrell, Clare, Alison Green, Sam Knights, and William Skeaping, eds. 2019. *This Is Not a Drill: An Extinction Rebellion Handbook*. London: Penguin.

Green, Thomas Hill. 1907. *Lectures on the Principles of Political Obligation*. London: Longmans.

Hallam, Roger. 2019. "Common Sense for the 21st Century: Only Non-Violent Rebellion Can Now Stop Climate Breakdown and Social Collapse." Pre-Publication Working Draft, version 0.3.

Holloway, John, and Sol Picciotto. 1979. *State and Capital: A Marxist Debate*. London: Edward Arnold.

Kant, Immanuel. 1970. *Kant's Political Writings*, edited by H.S. Reiss. Cambridge: Cambridge University Press.

Knaus, Christopher. 2018. "Australia's Political Parties Got $62m in 'Dark Money' Donations Last Year." *The Guardian*, 3 September.

Knights, Sam. 2019. "Extinction Rebellion: We Need to Talk About the Future." *Medium*, October. https://medium.com/@sam.j.knights/extinction-rebellion-we-need-to-talk-about-the-future-95459aa4d4e0 (accessed 7 November 2019).

Lucas, Caroline. 2019. "A Political View." In *This Is Not a Drill: An Extinction Rebellion Handbook*, edited by Clare Farrell, Alison Green, Sam Knights, and William Skeaping. London: Penguin.

Mair, Peter. 2009. *Ruling the Void: The Hollowing of Western Democracy*. New York: Verso.

Market Forces. 2019. "Friends in High Places: Fossil Fuel Political Donations." *Market Forces*, February. www.marketforces.org.au/politicaldonations2019/ (accessed 7 September 2019).

Marx, Karl. 1978. "On the Jewish Question." In *The Marx Engels Reader*, edited by R.C. Tucker. New York: W.W. Norton & Company.

Mayer, Jane. 2016. *Dark Money: How a Secretive Group of Billionaires Is Trying to Buy Political Control in the US*. London: Scribe.

Monbiot, George. 2019. "History Will Be Kind to Heathrow Climate Protesters Who Stop Us Flying." *The Guardian*, 4 September.

Nixon, Rob, 2013. *Slow Violence and the Environmentalism of the Poor*. Cambridge, MA: Harvard University Press.

Raworth, Kate. 2019. "A New Economics." In *This Is Not a Drill: An Extinction Rebellion Handbook*, edited by Clare Farrell, Alison Green, Sam Knights, and William Skeaping. London: Penguin.

Read, Rupert. 2019. "How a Movement of Movements Can Win: Taking XR to the Next Level." *Rupert Read's Website*, 13 August. https://rupertread.net/writings/2019/how-movement-movements-can-win-taking-xr-next-level (accessed 5 September 2019).

Resilience. 2019. "Articles on Extinction Rebellion." www.resilience.org/?type=Filter+by&s=%22Extinction+Rebellion%22 (accessed 5 September 2019).

Sharp, Gene. 1973. "198 Methods of Non-Violent Resistance." www.aeinstein.org/wp-content/uploads/2014/12/198-Methods.pdf (accessed 5 September 2019).

Spratt, David, and Ian Dunlop. 2017. *What Lies Beneath: The Scientific Understatement of Climate Risks*. San Diego, CA: Breakthrough Institute, September.

Spratt, David, and Ian Dunlop. 2019. "Australia's Climate Stance Is Inflicting Criminal Damage on Humanity." *The Guardian*, 3 August.

Steffan, Will, Wendy Broadgate, Lisa Deutsch, Owen Gaffney, and Cornelia Ludwig. 2015. "The Trajectory of the Anthropocene: The Great Acceleration." *The Anthropocene Review* 2 (1): 81–98.

Tham, Joo-Cheong. 2010. *Money and Politics: The Democracy We Can't Afford*. Sydney: UNSW Press.

Thoreau, Henry. 1982 [1849]. "Civil Disobedience." In *The Portable Thoreau*, edited by Carl Bode, 109–137. New York: Penguin.

PART IV

Power struggles on institutional terrains

24

CONCEPTUALIZING DEMOCRATIC ECOSOCIALISM

A personal journey

Hans A. Baer

Introduction

Politically, I became radicalized during the 1960s, not so much within the bowels of the university but rather in the aircraft industry, where I worked as an engineer between September 1966 and January 1970. My growing alienation with work, particularly at the Boeing Company in the Seattle area, and disenchantment with the state of the world prompted me to change careers at age 25 and become an anthropologist in order to understand the state of the world better and perhaps in some small ways contribute to making the world a better place. Shortly after completion of my PhD studies in 1976 and while teaching, I began to define myself as socialist and identified myself as such to my students, as I have ever since.

During the 1980s I did a lot of reading about the reasons for the discrepancies between the ideals of socialism and the efforts to create it in places such as the Soviet Union, Eastern Europe, China, Cuba, etc. As an anthropologist, I thought it would be far better for me to spend time in a socialist-oriented or post-revolutionary society than to read about it. Bearing this in mind, I applied to the Council of International Exchange of Scholars (CIES) for a Fulbright Lectureship in the Soviet Union in 1986 and again in 1987. While I was one of 35 nominees in both years, I was not one of the 15 accepted by the Soviet Ministry of Education for reasons unknown to me. When I called my CIES contact asking his advice on what to do, he noted that the Council had established a new exchange programme with the German Democratic Republic (GDR), adding that the GDR was seeking someone to teach an American studies course. I revised and submitted my proposal to teach "The Anthropology and Sociology of the United States." I was offered a Fulbright Lectureship to teach *Landeskunde der USA* (Area Studies of the USA) in the Department of British and American Studies at Humboldt University in East Berlin. I spent the better part of the period from July 1988 to February 1989 teaching and conducting ethnographic fieldwork on the side in the GDR, a period that proved to be its last days, before hardly anyone envisioned its collapse and absorption in October 1990 into the Federal Republic of Germany. I returned to the new German state for ten weeks on a sabbatical in early 1991, for a couple of weeks in the summer of 1991, and yet again for eight weeks from May to July of 1995.

Based on my reading of and first-hand experiences in the GDR, I concluded that the GDR, along with other post-revolutionary societies, constituted a transitional authoritarian society

10.4324/9780429341427-29

between capitalism and socialism that would have to go through a process of democratization and be part and parcel of a "permanent revolution" around the world (Baer 1992, 1998). I also witnessed the environmental devastation in the GDR due to the use of brown coal for generating power and the pollution emanating from the chemical industry, particularly in the vicinity of Halle and Leipzig. In large part, however, such environmental devastation in the Soviet-bloc countries was related to efforts to catch up with developed capitalist countries, which were able to achieve a higher material standard of living overall due to their reliance on cheap resources and labour in the peripheral and semi-peripheral capitalist countries.

The failure to achieve an authentically socialist world system was ultimately related to both internal forces specific to each post-revolutionary society and external forces that created a hostile environment for equitable development. The contradictory nature of Leninist regimes imploded first in Eastern Europe in 1989, particularly highlighted by the opening of the Berlin Wall, and in the collapse of the Soviet Union in 1991. In the case of China, its Communist Party leaders embraced capitalist structures as a means of rapid economic development to the point that some scholars argue that it now constitutes a state capitalist society, entailing tremendous social inequalities and environmental devastation.

Conversely, global capitalism is increasingly manifesting glaring contradictions. But there are voices bucking this massive global system from many quarters, including the anti-corporate globalization movement, the most progressive segments of the labour and environmental movements, the peace movement, Indigenous and ethnic rights movements, and radical feminists, as well as socialists, anarchists, and even Left-leaning liberals and social democrats. Indeed, over the course of the past 15 years or so, a distinct climate movement has emerged in both the global North and the global South, one that has built upon warnings about the dangers of climate change and global warming emanating from climate scientists, environmental groups, and Indigenous groups, particularly in the Arctic and South Pacific.

The emergence of the concept of democratic ecosocialism

In 1996, Merrill Singer, Ida Susser, and I employed the term *democratic ecosocialism* as a vision for creating a healthy world in our medical anthropology textbook, which went through two subsequent editions, 2003 and 2013 (Baer, Singer, and Susser 2013). We simply conflated the two previously existing terms *democratic socialism* and *ecosocialism*, which I summarize later.

Democratic socialism

Numerous Marxists have asserted that socialism is inherently more democratic than capitalist societies could ever be, and thus democracy is an inherent component of authentic socialism. Miliband (1994, 51) delineates three core propositions that define socialism: (a) democracy, (b) egalitarianism, and (c) socialization or public ownership of a predominant portion of the economy. Socialist democracy would not be identical to total state ownership and centralized planning but could include collective, cooperative, and even individual property. Tariq Ali (2009, 88) argues that twenty-first-century socialism should include political pluralism, freedom of speech, access to the media, the right to form trade unions, and cultural liberty. In recent years, various Leftists have been revisiting the concept not only of socialism but also of communism. Alan Badiou argues that "more than ever, we can, we must, and we will reactivate the communist hypothesis" (Badiou and Gauchet 2016, 48). Jodi Dean (2016, 1), another proponent of the communist hypothesis, argues that an alternative goal to capitalism is needed in "light of the planetary climate disaster and ever intensifying class war as states distribute wealth to the rich in the name of austerity." Most scholars who have embraced the communist hypothesis have not discussed the global crisis and how communism, let alone socialism, might serve to solve it. For example, Badiou and Žižek have dismissively asserted that "ecology has become the new opium for the masses" (quoted in Foster and Clark 2016, 9).

Ecosocialism

In the past, Marxian political economy has tended to give, at best, passing consideration to environmental factors. However, various Marxist theorists, including Herbert Marcuse, Erich Fromm, E. P. Thompson, Raymond Williams, Andre Gorz, Barry Commoner, and Rudolf Bahro served as precursors to present-day ecosocialism (Wall 2010, 82–89). A growing number of neo-Marxian scholars, as well as other radical scholars, have been attempting to integrate ecological considerations into their analyses of various types of social formations and societies. Such endeavours have been referred to as the *political economy of ecology*, *green socialism*, *ecosocialism*, *radical ecology*, *socialist ecology*, *social ecology*, *eco-anarchism*, and *earth democracy*. Foster maintains:

> Although contributions to ecological thought within the Marxist tradition have existed since the beginning – going to Marx himself – ecosocialism, as a distinct tradition of inquiry, arose primarily in the late 1980s and early'90s under the hegemony of green theory (and in the context of the crisis of Marxism following the downfall of Soviet-societies). The general approach adopted was one of grafting Marxian conceptions onto already existing green theory – or, in some cases, grafting green theory onto Marxism.
>
> *(Foster 2014, 57)*

However, the earliest use of the term *ecosocialism* may harken back to a pamphlet entitled *Ecosocialism in a Nutshell* published in 1980 by the Socialist Environment and Resources Association (Albritton 2019). Albritton (2019, 5) argues that "since the publication of this pamphlet, 'ecosocialism' has come to be seen by large numbers of people as the theoretical and action concept most appropriate for mobilizing against capitalism in the twenty-first century."

Ecological Marxism or ecosocialism has made some headway among Chinese Marxist scholars. Wang, He, and Fan (2014, 49) believe that ecological Marxism "can help remind the CCP of its ecological responsibility, and because environmental issues involve the vital interests of the people, it can remind the Party of its social responsibility as well." However, they admit that the "mainstream in China still tends to rely on the modernistic, technologically determinist, developmentalist way of thinking to address the problems facing China," despite official CCP declaration of commitment to ecological civilization.

Ecofeminism emerged out of the feminist, peace, and environmental movements in the late 1970s and 1980s. It seeks to eradicate not only injustice against women and the environment, but all forms of social injustice. Ecofeminism assumes various genres, including liberal ecosocialism, cultural ecofeminism, social ecofeminism, and socialist ecofeminism (Merchant 2005, 200–211). Socialist ecofeminism makes the category of reproduction as opposed to production central to the notion of achieving a socially just and environmentally sustainable world system. Ariel Salleh (1997, 12–13) argues that the ecological crisis stems from a Eurocentric capitalist patriarchal culture built on the domination of nature and domination of women "as nature." She argues that global justice and sustainability demand that the "North will have to review its high-tech consumption in favour of more species-egalitarian models by the South provisions itself (Salleh 1997, 142).

Democratic ecosocialism

Democratic ecosocialism embraces the following principles:

- An economy oriented to meeting basic social needs – namely food, clothing, shelter, education, health, and dignified work
- A high degree of social equality
- Public ownership of the means of production
- Representative and participatory democracy
- Environmental sustainability

Democratic ecosocialism rejects a statist, growth-oriented, productivist ethic and recognizes that humans live on an ecologically fragile planet with limited resources that must be sustained and renewed as much as possible for future generation.

The vision of democratic ecosocialism closely resembles what world system theorist Terry Boswell and Christopher Chase-Dunn (2000) term *global democracy*, a concept that entails the following components:

- An increasing movement towards public ownership of productive forces at local, regional, national, and international levels
- The development of an economy oriented towards meeting social needs, such as basic food, clothing, shelter, and healthcare and environmental sustainability rather than profit making
- The eradication of health and social disparities and the redistribution of human resources between developed and developing societies and within societies in general
- The curtailment of population growth that in large part would follow from the previously mentioned conditions
- The conservation of finite resources and the development of renewable energy resources.
- The redesign of settlement and transport systems to reduce energy demands and greenhouse gas emissions
- The reduction of wastes through recycling and transcending the reigning culture of consumption

Democratic ecosocialism constitutes what Erik Olin Wright (2010) terms a *real utopia*, a vision that is achievable but only through much theorizing and social experimentation. As the existing capitalist world system continues to self-destruct due to its socially unjust and environmentally unsustainable practices, many of which result in greenhouse gas emissions that contribute to anthropogenic climate change, democratic ecosocialism seeks to provide a vision to mobilize human beings around the world, albeit in different ways, to prevent ongoing human socio-economic and environmental degradation, including catastrophic climate change.

How to go from A to B

The literature on ecosocialism has become very extensive, as have visions of what it should constitute. For instance, Fred Magdoff and Chris Williams (2017, 285), drawing on the Indigenous notion of *buen vivir* (living better), referring to people living a good life communally in harmony with nature, delineate the following characteristics of how a future ecological society might operate: (a) equality and democratic decision-making, (b) full participation of everyone on an equal basis; (c) provision for a full life; (d) democratic provision of economic, political, and social rights and responsibilities; (e) an economy oriented to meeting people's needs and respecting the environment; (f) rotation of jobs and duties within workplaces; and (g) the replacement of nation-states and national borders with regional associations in which water, energy, and food are shared according to need.

However, with a few exceptions, most conceptions of ecosocialism are vague on how humanity should proceed from the capitalist world system to an ecosocialist world system, one that would be democratic as well. Derek Wall proposes ten transitional policies that he hopes will promote discussion:

- Supporting Indigenous control of rainforests and other vital ecosystems
- Empowering workers to assume control of bankrupt businesses
- Utilizing government bailouts to support mutualistic resources
- Converting arms and SUV production for public programmes
- Promoting open-source patenting
- Land reform
- Large-scale support for libraries and other social services
- A tax and welfare system to support common ownership
- Competition reform to transform ownership
- Public ownership of pharmaceuticals and healthcare (Wall 2010, 68)

Sanjeev Ghotge (2018, 13) delineates 13 "building blocks of a green socialist political economy," too lengthy to list here.

In my case, while not seeking to create a blueprint *per se* for creating an alternative world system that will be manifested in different ways in the many countries around the world, in *Democratic Ecosocialism as Real Utopia*, I delineate several system-challenging reforms that potentially could facilitate a transition from the existing capitalist world system to a democratic ecosocialist world system (Baer 2018). These include:

- The creation of new Left, anti-capitalist parties designed to capture the state in countries around the world
- The implementation of greenhouse gas emissions taxes at the sites of production that include measures to protect low-income people
- Social planning and the increasing public and social ownership in various means of production
- Increasing social equality within nation-states and between nation-states and achieving a sustainable global population size
- Implementation of workers' democracy
- The creation of meaningful work and shortening the work week
- The creation of a new global steady-state or net zero–growth economy that will entail growth for some populations but degrowth for other populations, depending on the level of abundance or lack thereof
- The adoption of energy efficiency, renewable energy sources, appropriate technology, and green jobs
- The development of sustainable public transportation and travel
- The implementation of environmentally sustainable food production and forestry
- Resisting the culture of consumption and adopting sustainable and meaningful consumption patterns

- The implementation of sustainable trade
- The implementation of sustainable settlement patterns and local communities

Achieving social equality within settler colonial states may require huge reparations to Indigenous peoples and even the creation of partially or even fully autonomous micro-states where Indigenous peoples can achieve sovereignty.

The creation of new Left parties that capture the state could pave the way for implementing many of the other radical transitional reforms. These transitional steps constitute loose guidelines for shifting human societies or countries towards democratic ecosocialism and a safe climate, but it is important to note that these developments will require a global effort, including the creation of progressive global climate governance regimes. Unfortunately, the existing United Nations Framework on Climate Change (UNFCCC), which has had 25 Congresses of the Parties (COPs) since 1995, despite some incremental steps forward, operates by and large within the dictates of the capitalist world system and the parameters of its growth paradigm, often framed under the rubric of *sustainable development*, and generally entails voluntary commitments to reducing greenhouse gas emissions, including under the provisions of the highly touted 2015 Paris Agreement. Despite perhaps the best of intentions, the emissions rose from 280 ppm of CO_2 at the time of Industrial Revolution to 415 ppm at the present time, and they are still rising.

Social movements should play a crucial role in initiating the radical transitional reforms that I have delineated. Also, radically changing the world will require a multiplicity of interstitial movements situated within mainstream society in which people develop alternative social relations and communities (Holloway 2010). Social movements in the global North for the most part have historically focused on relatively limited objectives, whether better wages and working conditions in the case of the labour movement, voting rights and economic and educational opportunities in the cases of the women's and civil rights movements, regulation in the case of the environmental movement, and so on. Conversely, national liberation and anti-racism movements, such as Indigenous sovereignty movements (particularly ones emanating from the global South), tend to be more comprehensive and anti-systemic in their objectives. Much of the climate movement in the global North has tended to stress climate change mitigation strategies that come under the rubric of *ecological modernization* – that is, an emphasis on renewable sources of energy, energy efficiency, electric cars, and mass transportation – but to deemphasize social justice issues. In contrast, the climate movements emanating from the global South have given much more attention to issues of social parity, such as the fact that the developed countries have historically contributed and continue to contribute much more on a per capita basis to greenhouse gas emissions than have or do the developing countries, even China (the leading total emitter in the world, even superseding the United States, but not on a per capita basis). Climate justice activists as opposed to climate activists *per se* often say, "Not climate change, system change." While they are not always clear on what sort of next system they envision, democratic ecosocialism constitutes for me an alternative to both the existing capitalist world system and some form of green capitalism, a vision which has been highly touted, ranging from Al Gore's version to the Green New Deal being proposed in the United States and elsewhere (Gore 2009; Pettifor 2019).

The challenge anti-systemic movements now face is uniting in their struggle against a common foe: namely, global capitalism. Ethnic, national, religious, and cultural differences too often have divided subalterns around the world and prevented their forming a united revolutionary movement. Furthermore, Leftist governments have warred with each other or splintered along national lines, as was the case with the debacle of the Second International.

A viable anti-capitalist movement will have to address the material impoverishment of much of the world's population. Many parties, ranging from the World Bank to entertainment celebrities, make appeals to "eradicate extreme poverty" or "make poverty history." However, "make wealth, particularly extreme wealth, history," and the eradication of poverty will inevitably follow. Personally, I hope that the democratic ecosocialist vision will serve as an integrative focus for anti-systemic movements, including the climate justice movement, within nation-states and transnationally, although I recognize how daunting this task will be.

Ecosocialism in action

At the organizational level, ecosocialism is situated "amongst green parties, social movements, socialist groups and indigenous networks" (Wall 2010, 2). *Green Left* is an ecosocialist network within the Green Party of the UK. The Ecosocialist International Network was created in 2007 and drew its inspiration in large part from the 2001 Ecosocialist Manifesto written by Joel Kovel and Michael Löwy. Some European parties define themselves as ecosocialist, including the Red-Green Alliance in Denmark, the Left Bloc in Portugal, the Socialist Left Party in Norway, and the Parti de Gauche in France. The US Green Party has adopted an ecosocialist platform and developed a Green New Deal years prior to the highly diluted version from part of the US Democratic Party. To a large extent, at the organizational level, ecosocialism remains an internet phenomenon. Ecosocialist Horizon states on its website that it "seeks to advance ecosocialism as a world-view and as a movement capable of offering real answers to the crises caused by capitalism" (http://ecosocialisthorizons.com/). *Climate & Capitalism* is an online ecosocialist journal edited by Canadian Ian Angus (http://climateandcapitalism.com). *Monthly Review*, due largely to the influence of its editor John Bellamy Foster, exhibits a strong ecosocialist stance, as does the journal *Capitalism Nature Socialism*.

In Australia, the Socialist Alliance, a political party that runs candidates in selected federal, state, and city council electorates, has a strong ecosocialist orientation as manifested in many of the articles it publishes in *Green Left*, promoting ecosocialist analyses and climate justice campaigns. Solidarity, a smaller Australian socialist group, also has a strong ecosocialist orientation. The Australian Greens have ecosocialist members, but for the most part, they do not have a strong identifiable presence. I personally have spoken about democratic ecosocialism as an important climate mitigation strategy at conferences and meetings of the Socialist Alliance, Solidarity, and local Greens branches as well as in classes, academic seminars and conferences, the New International Bookshop in Victorian Trades Hall, the Melbourne Peace Memorial Unitarian Church, and the University of the Third Age.

Conclusion

As humanity proceeds into the twenty-first century, its survival as a species appears to be more and more precarious, particularly given the impact of climate change that, in a multiplicity of ways, looms on the horizon. Global, regional, and local temperature records are repeatedly being broken around the world, accompanied by extreme climatic disruptions, ranging from wildfires and droughts, melting glaciers, hurricanes and storm surges, and rising sea levels to the melting of the Arctic icecap and polar vortexes that can create sudden cold spells in the Northern hemisphere. While the capitalist world system appears to be well entrenched on numerous fronts, there are numerous cracks in the system, manifested in phenomena such as Brexit in the UK, the election of Donald Trump as president of the United States in 2016 (fortunately followed by his defeat in 2020), and the rise of neo-fascist governments and parties around the world, tensions within the European Union, the threat of nuclear war, ongoing conflicts in many parts of the world, the growing concentration of wealth, climate crises of many sorts, the acidification of the oceans, and the extinction of many creatures ranging from birds and animals to marine life.

My own sense is that overall things over the course of the next few decades will get worse before they get better, and there is no guarantee that they will get better. Thus, in terms of the foreseeable future, I am very much in agreement with Wallerstein (2007, 382), who maintains:

> I do not believe that our historical system is going last much longer, for I consider it to be in terminal structural crisis, a chaotic transition to some other system (or systems), a transition that will last twenty-five to fifty years. I therefore believe that it could be possible to overcome the self-destructive patterns of global environmental change into which the world has fallen and establish alternative patterns. I emphasize however my firm assessment that the outcome of this transition is inherently uncertain and unpredictable.

During 2020, much of the world's attention on the ravages of climate change turned to a more immediate crisis – namely, the COVID-19 pandemic – which probably started in the open markets of Wuhan, China, in late 2019. Wallace, Liebman, Chaves, and Wallace observe,

> The overlapping economic geography extends back from the Wuhan market to the hinterlands where exotic and traditional foods are raised by operations bordering the edge of a contracting wilderness. As industrial production encroaches on the last of the forest, wild food operations must cut farther in to raise their delicacies or raid the last stands. As a result, the most exotic of pathogens, in this case bat-hosted SARS-2 find their way on a truck, whether in food animals or the labor tending them, shotgun from one end of a lengthening periurban circuit to the other before hitting the world stages.
>
> *(Wallace et al. 2020, 5)*

While a growing number of critical scholars acknowledge that global capitalism constitutes the overarching driver of anthropogenic climate change, one of the smaller elephants in the room has been the growing number of airplane flights worldwide, at least until the coronavirus (COVID-19) forced reluctant governments around the world to greatly restrict the number of flights, which played a key role in turning a localized into a global epidemic. Airplanes of many sorts (commercial, military, and private) have become an integral component of modern cultural and social life and a source of tremendous profit making and thus an integral component of the capitalist world system. Airplanes and international shipping serve to transport both human actors and commodities in order keep the world system functioning: however, with dire environmental, climatic, and health consequences (Baer 2020; Campling and Colas 2021).

More so than has ever been the case, it is essential for critical scholars and activists to envision future scenarios and strategies for achieving an alternative world system. More important is developing strategies to shift from the existing system of globalized capitalism to an alternative that transcends its numerous contradictions and limitations. While presently and for the foreseeable future, the notion that democratic ecosocialism may be eventually implemented in any society or in several linked societies may appear absurd, history tells us that social changes can occur very quickly once certain social structural and environmental conditions have reached a tipping point, a term that has become popular in climate science.

References

Albritton, Robert. 2019. *Ecosocialism for Now and the Future: Practical Utopias and Rational Action*. London: Resistance Books.

Ali, Tariq. 2009. *The Idea of Communism*. London: Seagull Books.

Badiou, Alain, and Marcel Gauchet. 2016. *What Is to Be Done? A Dialogue on Communism, Capitalism, and the Future of Democracy*. London: Polity.

Baer, Hans A. 1992. "The Rise and Fall of the German Democratic Republic." *Monthly Review* 43 (11): 30–40.

Baer, Hans A. 1998. *Fallen Walls and Tarnished Ideals: An Ethnography of East Germany Before and After Unification*. Lanham, MD: University Press of America.

Baer, Hans A. 2018. *Democratic Ecosocialism as a Real Utopia: Transitioning to an Alternative World System*. New York: Berghahn.

Baer, Hans A. 2020. *Airplanes, the Environment, and the Human Condition*. London: Routledge.

Baer, Hans A., Merrill Singer, and Ida Susser. 2013. *Medical Anthropology and the World System: A Critical Perspective*. 3rd ed. Westport, CT: Praeger.

Boswell, Terry, and Christopher Chase-Dunn. 2000. *The Spiral of Capitalism and Socialism*. Boulder, CO: Lynne Rienner.

Campling, Liam, and Alejandro Colas. 2021. *Capitalism and the Sea: The Maritime Factor in the Making of the Modern World*. London: Verso.

Dean, Jodi. 2016. *Crowds and Polity*. London: Verso.

Foster, John B. 2014. "Paul Burkett's *Marx and Nature* Fifteen Years After." *Monthly Review* 66 (7): 56–62.

Foster, John B., and Brett Clark. 2016. "Marxism and the Dialectics of Ecology." *Monthly Review* 68 (5): 1–17.

Ghotge, Sanjeev. 2018. "Climate Change and Marx in the Twenty-First Century." *Capitalism Nature Socialism* 29 (3): 11–20.

Gore, A. 2009. *Our Choice: A Plan to Solve the Climate Crisis*. London: Bloomsbury.

Holloway, John. 2010. *Crack Capitalism*. London: Pluto Press.

Magdoff, Fred, and Chris Williams. 2017. *Creating an Ecological Society: Toward a Revolutionary Transformation*. New York: Monthly Review Press.

Merchant, Carolyn. 2005. *Radical Ecology: The Search for a Liveable World*. 2nd ed. New York: Routledge.

Miliband, Ralph. 1994. *Socialism for a Sceptical Age*. London: Verso.

Pettifor, Ann. 2019. *The Case for the Green New Deal*. London: Verso.

Salleh, Ariel. 1997. *Ecofeminism as Politics: Nature, Marx, and the Postmodern*. London: Zed Books.

Wall, Derek. 2010. *The Rise of the Worldwide Ecosocialist Movement*. London: Pluto Press.

Wallace, Rob, Alex Liebman, Luis F. Chaves, and Rodrick Wallace. 2020. "COVID-19 and Circuits of Capital: New York to China and Back." *Monthly Review* 72 (1): 1–15.

Wallerstein, Immanuel. 2007. "Commentary 205: Climate Disasters: Three Obstacles to Doing Anything." Fernand Braudel Center, 15 March. www.resilience.org/stories/2007-03-15/climate-disasters-three-obstacles-doing-anything/ (accessed 29 March 2021).

Wang, Zhihe, Huili He, and Meijun Fan. 2014. "The Ecological Civilization Debate in China: The Role of Ecological Marxism and Constructive Postmodernism – Beyond the Predicament of Legislation." *Monthly Review* 66 (6): 37–59.

Wright, Erik O. 2010. *Envisioning Real Utopias*. London: Verso.

25

MONEY-FREE ECONOMIES AND ECOSOCIALISM

Anitra Nelson

Only some theorists in the political stream that Rubel and Crump (1987) have identified as nonmarket socialists are also ecosocialists in that their visions of a free association of producers operating without money encompass the challenges of ecological limits. Yet an efficient and effective mode of production for ecosocialism would seem impossible without moving beyond the market, money, and capital as nonmarket socialists suggest. Indeed, we need to establish money-free economies that make ecological, social, and humane values central within direct and participatory decision-making over production for people and planet on the basis of real needs.

This chapter briefly outlines our conjunctural challenges and a personal intellectual journey; identifies fatal weaknesses of money economies that appear in fully developed form within capitalism; sketches a nonmonetary post-capitalist local-global commons, Yenomon; identifies some examples oriented towards money-free societies; and proposes the future constitution of this as yet underdeveloped area of thought and study as "real value studies."

To be in the belly of the beast

Even if ecosocialism only arose as an identifiable movement in the 1980s, like degrowth and many current environmental, social, and anti-capitalist movements, its intellectual wellsprings, in the form of books, protests, and discourses, go back at least as far as the 1960s. Following ecological footprint calculations, early in the 1970s, we as a species started to overexploit the planet – our use of ecological resources and services surpassed earth's regenerative capacity. Since then, we have increasingly degraded earth's potential to support all kinds of species. Impacts of rising carbon emissions from human activities are only the tip of the iceberg of our environmental crises, but they alone threaten the future of the human species.

Simultaneously, over the past half century, generations of youth have increasingly identified as citizens of the world with the state becoming contentious and irrelevant as deeper forms of direct and participatory democracy evolved; paid work has become suspect as stable employment diminished and unemployment and the precariat developed; late-stage capitalism has inserted property rights into post-socialist, previously non-capitalist regions and new sectors – for instance, patenting nature's seeds and people's ideas; and we have become more unequal in terms of material wealth, income, and power over managing production.

10.4324/9780429341427-30

My strategic interrogations of money, capital, and markets developed during various activities and intense discourses in the women's, environmental, and socialist movements in the 1970s. I became increasingly aware that capitalists hid and enacted their power in monetary processes and market (il)logic that the state supported and internalized. Within all these movements, fractious divisions emerged between those working to adapt to and reform capitalism and more fundamental anti-capitalists. Nevertheless, even many anti-capitalists imagine post-capitalism maintaining states and markets, albeit modified. As streams in such movements responded to neoliberalism by engaging in the market rather than with the state, I became exquisitely aware of the commanding power of money, not simply the power of those with money but of the language and logic of money and the ways in which certain women, socialists, and environmentalists treated money as a tool and not a barrier to change, to revolution.

I had the opportunity to explore the nature of money when I successfully proposed an interdisciplinary critique of Marx's concept of money as the Revolutionary Area of Studies topic for my doctoral thesis, conducted at La Trobe University (Melbourne, Australia). Consequently, I discovered the work of Rubel and Crump (1987), Cleaver (1979) – and later, in the same vein, Holloway (2015) and ecofeminists such as Habermann (2018) – who, in contrast to mainstream interpretations of Marx, highlight the political, nonsensical, and omnipotent power of money – supporting my instinct that moving beyond money was significant and central to imagining truly existing ecosocialism.

Mainstream interpretations of Karl Marx's work developed a dogma that analytically separated trade from production for trade, devaluing money as superficial and prioritizing "value" as central. This is curious if only because a close reading of Marx's work and appreciation of his dialectical approach make such an interpretation a travesty of both his method and his intent. Exchange value cannot exist without money, and production for trade cannot exist without trade; they coexist as content and form. In fact, this very compaction and amalgamation of the two implies that socialism means moving beyond money, as indicated by Marx, particularly in his earliest works (Nelson 1999, vii–ix, 1–22).

If Che Guevara envied North Americans' ideal revolutionary location in "the belly of the beast" (Rubin 1976), how might he consider this tinderbox time we live through now? We seem to face a material and global conjuncture in which ecosocialism is the perfect answer. An ecosocialist struggle is so logical, immediately necessary, and constructive that questions only remain about how to achieve it.

To be in the belly of the beast, twenty-first-century global capitalism, is to experience the full force of both massive economic and political inequalities and imbalances and tensions between human practices and planetary ecosystem well-being. Monetary value dominates social and environmental values structurally, so dualism proliferates in market activities as a practical fact. We cannot address ecological sustainability without minimizing inequalities between us, inequalities of both power and access to resources to sustain us. We need to act with urgency as all humanity might be lost even as Earth continues without us. We need to make decisions about production and ways of living based on social and environmental values. What is so problematic about money? Why would a transition to a money-free mode of production be strategically advantageous?

Money

In capitalism, money is the supreme value, akin to a god. Accounting, credits, and debts rule. Money is not "a thing" but a social force arising as a claim to marketed goods and services in a specific set of social relations and rituals. Money's four functions – as a unit of account, medium of exchange, means of payment, and store of value – are co-dependent dimensions of capital that amalgamate with totalizing, isolating, and marginalizing effects. Money is intrinsic to those capitalist dynamics that are responsible for social inequality, environmental unsustainability, and growing carbon emissions. Using money and markets as organizing principles of society limits and contorts relations between us and with nature, embedding dualism as material practices. Impacts of neoliberalism and climate change show how such developments have reached an historic crisis point.

Trading is the conflictual monetary practice on which mainstream economic efficiencies are argued. There will always be winners and losers, poverty and plenty, those with money and those without. Moreover, these social dynamics ignore or prevent ecologically sound production and exchange. An elite decides what is produced and how it is produced. Secretive and competitive privatized production means that supply rarely meets demand, resulting in unnecessary waste. In contrast, you can only buy need satisfiers if you have the money. Making an irony of capitalist "efficiency," productive, trading, financial, and commercial activities all use materials, human effort, and fossil-fuel energy expansively, to grow.

Indeed, we cannot define or conceive of either capitalism or capitalist growth without money. Growth boils down to money invested to make more money. Degrowth – minimizing material and energy used in production and meeting everyone's needs (as distinct from wants) – seems impossible without going well beyond capitalism. The degrowth movement highlights that this simplistic capitalist dynamic drives the market and that environments and peoples are increasingly incorporated into the pursuit of unending growth with disastrous social and environmental consequences (Exner 2014). Capitalist enterprises and capitalist societies have no operating principles for degrowing. Indeed, shrinking one's firm or the gross domestic product within capitalist processes is seen conventionally as collapse, bankruptcy, recession, or depression.

Profit is not a component of costs that exists simply to provide investors with a surplus but, rather, is a necessary buffer for any operator in private production for trade to stay liquid, avoid bankruptcy, and conduct legal operations. For instance, "not-for-profit" firms actually do earn a profit but just distribute it in unconventional, mainly community-oriented and ethical ways. In short, not only is degrowth anathema to capitalist operations, but also private firms in a competitive market cannot rationally operate except on the principle of growth.

Ideas of addressing social inequity, say with guaranteed minimum incomes (GMI), and environmental problems with environmental prices, accounting, offsetting, and trading proliferate. Yet GMI schemes cannot account for people's various needs because necessarily distinctive ways of satisfying them do not amount to equivalent sums of money per capita. Moreover, the market will not necessarily offer their specific needs. Similarly, environmental values, uses, and risks implicated in producing and trading any object or service can neither be accurately accessed monetarily nor reduced to a singular indicator in a price. Thus, we need to go beyond money, not just capital, to establish a fair and sustainable world.

This means production and distribution for use, decided and arranged locally within global commoning based on principles of fulfilling the basic needs of all people and the regenerative needs of the rest of nature. There are strategic benefits to this nonmonetary route as an establishing principle of ecosocialism. The imaginary sketched in the next section incorporates everyone as both a decision maker and as a worker and focuses on satisfying the basic needs of everyone, minimizing inequities at the same time as creating a mode of production that respects earth's limits.

Furthermore, money is the strategic organizing principle of capitalism and its Achilles heel: no money, no capitalists. The end of money eradicates a critical base on which counter-revolutions arise. The national communist experiments of the twentieth century failed in terms of environmentally unsustainable practices and political elitism – instituting a managed top-down socialism in which modified markets and money, as well as inequities, persisted (Nelson 2011). Surrounded by capitalism and maintaining trading relationships, communist states faced internal and external contradictions and destabilization. In contrast, nonmonetary production and exchange – collective planning, commoning, and sharing – undercut capitalists' seat of power absolutely. The strategy of citizens gaining direct power over production means, arguably, that they value this power and are likely to fight for it. Significantly, without money and trade, capitalist practices cannot exist.

But what might such a society look and feel like? How might it operate?

Nonmonetary economies

This sketch of a nonmonetary post-capitalist local-global commons, Yenomon, draws on earlier work (Nelson 2016) and a chapter from a forthcoming book (Nelson Forthcoming). In Yenomon, we approach the world using "real values," an expanded concept of "use values" that incorporates all those environmental and social qualities and quantities that comprise or meet the needs of both people and the planet. Real values draw on biological, ecological, and other social and natural scientific knowledge. Production and exchange are based on real values. Caring and decision-making are based on real values. We live for, of, and by real values.

There is no private property but, rather, intricate, humane, and ecologically sensible principles for use rights, which entail responsibilities to use human and non-human nature in ways which perpetuate our and other species coexistence on earth. We engage in nurturing, gardening, and caring – healing earth and people from the vestiges of the Anthropocene.

We live glocally, meaning we live by universal global principles, processes, and practices that guide us in our decision-making and activities in communities that are as collectively sufficient as possible. Our ecological footprints have shrunk below one planet footprints yet we live healthily. Our communities are settled in "ecotats," which meet our needs with sheltered areas for settlement and easily accessible water, and the land to which we have use rights supports food self-provisioning using permaculture and agroecology approaches.

Borders of ecotats are somewhat fluid – adjacent communities share use rights, say, to a co-located lake and woodlands. Indeed, managing our land and water sources demands that we work with neighbouring and regional communities. Our community has rotating representatives on numerous regional river, eco-corridor, forest, and ridges working groups.

What we need and cannot provide for ourselves locally through collective sufficiency is provided for through temporary, semi-permanent, or permanent exchange arrangements – "compacts" rather than monetary contracts. We make compacts with communities as immediate and as local as we can. We have compacts to give, compacts to take, and compacts that include arrangements to give and to take, often involving more than one community. Compacts allow us access to a surplus elsewhere, whether that surplus occurs by design or accident. Compacts focus on needs, not wants. Arrangements around wants do not have either the status or name of a compact. Wants are secondary.

Our community supports, say, around 100 people of all ages. We live by strong consensual decision-making and are skilful resolvers of conflict. We freely share our intricate knowledge of the land to which we have use rights. We have expansive gardens and orchards around buildings where we live and make and store things. We grow indigenous and productive plants, a balance guided by our needs and earth's regenerative needs. We have woodlands; water storage systems; animals that provide, for instance, eggs, milk, feathers, and skins; we build soil and use rocks. Much of our gardening, food self-provisioning, and design is inspired by permaculture. Our gardens produce vegetables and herbs, and our orchard trees bear fruit.

We evaluate the ways we produce things using holistic "real values," balancing human needs with ecosystem potential and limits. We use and make both simple and sophisticated tools. Most ecotats fulfil tasks according to a modicum of division of labour within regions of ecotats. Typically, one regional community has a "tofuary" directly linked to a farm co-producing edamame, soy milk, soy flour, soy butter, tofu, tempeh, miso, and tamari. All ecotats have repair centres; all our devices, clothes, and buildings are designed and created to last; and ways of maintaining and prolonging their life are part of our daily lives.

We have re-inhabited and renovated some buildings remaining from the Anthropocene and created some "new" simple dwellings and other buildings from recycled materials. Some of us live alone, some in couples; other households are collectives whose members have a private space but share living, dining, kitchen, laundry, and home work spaces.

Our substantively autonomous, collectively sufficient community is a primary ecomaterial unit and ecopolitical cell, where we make decisions. Efficient and effective production starts with individuals' needs and proceeds as creation on demand. I'll use the example of food.

Every year, each household creates its annual food order. Familiar with what grows best and most easily locally, we focus on achievable orders that we know will be checked by other community members, assessing whether we expect too much or not enough. When all orders are verified, the food working group adds an extra order for the buffer store and takes account of existing compacts. They plan to create a bit more than has been ordered, especially of more easily stored nutritious foods or products which can be used in other ways than as human food. This surplus, beyond use in our community, can be made available to fulfil other communities' needs or ecosystem needs more generally.

The total community order is divided into types and amounts of food so that the orchard, vegetable, dairy, and other working groups can assess the feasibility of meeting the community's demand. They take account of a range of natural, social, and practical matters which impact production. After all these deliberations, revisions, and discussions, the general assembly hears from all groups and individuals on challenges and solutions. Decisions are made to proceed with what looks feasible. Everyone must feel confident that the food order is of a sufficient size and achievable by the community, given other workloads.

During the rest of the year, weekly assemblies get updates on production and solve challenges as they arise. Because we focus on satisfying our basic needs within the regenerative limits of our locale – and somewhat beyond – distribution is decided simultaneously with confirmed orders. The order is fed into our collective agreements on our productive goals, and, once achieved, there is no need for markets. When the goods are ready for collection, those who ordered them are notified and arrange to collect them. We have open stores where unwanted things are left and available for anyone one to collect.

Each person contributes an obligatory average 35 hours per week to collective production, including a lot of what was unpaid domestic work in the Anthropocene and, added to paid work, amounted to around 60 hours per week in those times. In return, we have our basic needs met and a right to continuous input in decision-making on every aspect of our lives. Our commons blends into other commons. Our communities engage with other communities.

In Yenomon we learn "on the job" all the duties and responsibilities of regenerating our community and ecotat. Everyone's basic needs must be met; satisfying wants is secondary. Still, we engage in lots of wants in our free time, wants contained by the glocal principles by which we live, principles of respecting and caring for people and the planet. We roam free; we talk; we sing and celebrate. We engage in all kinds of low-impact artistic endeavours. We engage globally via a modest use of digital communications and commons.

There is no marked duality between ecological and social worlds in Yenomon. We meld our own sustenance with regenerating nature. Sustainability is a continuous work in progress of developing knowledge and skills around real values. Every new technique, every new way of doing and creating things reveals new real values, real values of us, real values of earth.

Money-free communities

The Yenomon imaginary is not so far from – indeed, intentionally draws on – certain realities, including my experience of activist occupations and sharing communities (Nelson 2018b; Kinkade and the Twin Oaks Community 2011). Today there is resistance and protest on various continents in struggles for substantive democracy along the lines of a community mode of production centring on use values, participatory governance, and direct democracy advocated by Uruguayan writer Eduardo Galeano who, arguably, can be placed in the nonmarket socialist camp (Nelson 2018a). Galeano (1991) specifically called for local sustainability among Latin American peoples: "It's out of hope, not nostalgia, that we must recover a community-based mode of production and way of life, founded not on greed but on solidarity, age-old freedoms and identity between human beings and nature."

Examples of a community mode of production exist in embryo, suspend stillborn, and continuously evolve. In the self-governing community mode of production, people's assessments of nature's potential and limits and their integral social commitments extinguish any false consciousness on needs for money or state. Money appears opaquely as a tool of class and power-over, a system for appropriating their power over production, a dynamic demanding competition and exploitation (Cleaver 2017, 228–233). Some examples include Zapatista and other indigenous and peasant communities of Mexico (Barkin and Lemus 2016); the self-governing communities of Rojava's Kurds in Northern Syria; communities of protest which become utopian settlements, such as occupiers/squatters in "zads"; and experimental communities in Germany eschewing "exchange-logic," the universal equivalent money (Habermann 2018).

In his analysis of Zapatista political and economic autonomy, Fitzwater (2019, 134) reports, "The first step that Zapatista autonomous government has taken to prevent the accumulation of economic power is to eliminate monetary compensation for those who govern." Members of council assemblies are sustained by *trabajos colectivos*. Direct democracy means that autonomous municipalities have diverse processes for practicing democracy "constantly defined and redefined by the communities," but all abide by seven Zapatista principles (ibid., 2, 52). Significantly, trade with the capitalist market and landed property relations impact the *trabajos colectivos* in "fundamental" ways (ibid., 111ff). As distinct from conventional paid work "that creates a system of control and domination," their political project aims to redefine it as "work that is done by everyone and that is good for everyone": in other words, "a form of work that strengthens the community's collective potential (*ch'ulel*) to decide together how they want to live, work, and sustain themselves and their organization" (ibid., 123).

Occupiers of the area planned for a mega-airport, Zone À Défendre (la zad) fields in Notre-Dame-des-Landes north of Nantes (France) has been a site of resistance for decades. It has attracted hundreds of thousands of protesters and spawned zads in other locations. Even before achieving a reversal of the planned mega-airport, its squatter core decided to turn the land into a liberated, creative, diverse, and relatively collectively sufficient territory, often referred to as "a new commune for the twenty-first century" – beyond state and market.

They regard the land as commons, developing trust and mutual commitments, relying on sharing – "you can use tractors, tools, or books without ever reaching for your wallet" – and referring to money as "an easy reimbursement for a lack of involvement in common life, a way of exempting oneself" (CMDO 2018, 10). Mauvaise Troupe Collective (2014, 19–20) report:

> We come together every week around what you might believe is a market – although nothing is sold and everything is by donation, i.e. people giving what they can – which enables us to open up to sharing some of the produce of the land. The surplus is used to feed other struggles, street kitchens or migrant squats in the city of Nantes. . . .
>
> We are not interested in self-sufficiency for itself. What is happening here is political autonomy. What we are inventing, through trial and error, is the capacity to collectively decide our own rules.

In the "myriad of other experiences in autonomy . . . flowering, outside of the logic of management and the market," they have created centres of production and repair such as bike and textile workshops, a flour mill, and meeting and eating places, along with media, energy, and health schemes (ibid., 19).

Inspiration for what Zad Inhabitant (2018, 24) calls "utopias with teeth" is sourced in the commune, and they hold hands with the revolutionary Kurdish communities of Rojava in tenuously held *de facto* autonomous regions in northern and eastern Syria, establishing a society based on direct democracy, gender equality, and ecological sustainability – where economic autonomy, a community-based and cooperative economy, means local self-organization. Even if leaders such as Öcalan (2017, 129) do not outright reject money or market, the evolution of self-governing sharing economies makes such a step feasible in the future (Tax 2016, 173–175).

Praxis

In the introduction I purposely referred to thought about and study of money-free economies as "underdeveloped." A money-free post-capitalist community mode of production has been trivialized and marginalized, advocates to date simply being treated, as in Melanie Safka's 1970 song, as "candles in the rain." In general, the imaginaries of Left and green thinkers lack a political consciousness of monetary values and relations as absolute barriers to achieving socialism. Yet, from our point of view, moving beyond money is a necessary, even if insufficient, step towards creating the ecosocialist project, ecosocialist structures that satisfy the needs of planet and people.

Market-based societies are ruled by the universal and mystical abstraction of money – prompting a plethora of anti-social dynamics, managerial elitism, and ecological unsustainability. Instead, our current conjecture demands glocal socialism: i.e., that we make collective decisions on co-creating collective sufficiency in local areas using universal principles. This theory is illustrated in the contemporary practices of certain revolutionaries in sites of protest, experimentation, and occupation in various locations around the world:

people intent on realizing ecosocialist values, politico-economic change, and ecological sustainability.

However, discourse on and study of the potential and limitations of such money-free economies in realizing ecosocialism is lacking (Nelson 2016). Significant clusters of activist scholars exist, including demonetization activists (Exner 2014; Exner et al. 2020), autonomist Marxists such as Harry Cleaver (2017, 228–233), and ecofeminists such as Ariel Salleh (2011) and a German stream represented here by Frederieke Habermann (2018).

I suggest that relevant works, investigations, and discourses might be usefully identified and built on as a transdisciplinary field of "real value studies." Spheres of self-organizing glocal ecosocialism oriented to community sufficiency and ecologically regenerative activities require deep interrogation. Real value studies would systematically and critically explore empirical case studies, such as those discussed here. Real value studies could develop theoretical discourses – say, on how "real values" are conceived of, treated, and identified by members in a community mode of production, interrogating criteria used for balancing competing and complementary social and environmental needs satisfaction in practice.

Within real value studies and money-free realities, ecosocialism could live as praxis.

References

Barkin, David, and Blanca Lemus. 2016. "Third World Alternatives for Building Post-Capitalist Worlds." *Review of Radical Political Economics* 48 (4): 569–576. https://doi.org/10.1177/0486613416665828

Cleaver, Harry M. 1979. *Reading 'Capital' Politically*. Austin: University of Texas Press.

Cleaver, Harry M. 2017. *Rupturing the Dialectic: The Struggle Against Work, Money, and Financialization*. Chico, Oakland, Edinburgh, and Baltimore: AK Press.

CMDO. 2018. "The ZAD Will Survive." *Conseil Pour le Maintien des Occupations*. https://zadforever.files.wordpress.com/2018/03/zad-will-survive-en-pdf.pdf (accessed 16 April 2021).

Exner, Andreas. 2014. "Degrowth and Demonetization: On the Limits of a Non-Capitalist Market Economy." *Capitalism Nature Socialism* 25 (3): 9–27. https://doi.org/10.1080/10455752.2014.882963

Exner, Andreas, Justin Morgan, Franz Nahrada, Anitra Nelson, and Christian Siefkes. 2020. "Demonetize: The Problem Is Money." In *Degrowth in Movements: Exploring Pathways for Transformation*, edited by Corinna Bukhart, Matthias Schmelzer, and Nina Treu, 159–171. Winchester and Washington: Zero Books.

Fitzwater, Dylan Eldredge. 2019. *Autonomy Is in Our Hearts: Zapatista Autonomous Government Through the Lens of the Tsotsil Language*. Oakland, CA: PM Press.

Galeano, Eduardo. 1991. "The Blue Tiger and the Promised Land." *NACLA Report on the Americas* 24 (5). https://nacla.org/article/blue-tiger-and-promised-land-0 (accessed 16 April 2021).

Habermann, Friederike. 2018. *Ausgetauscht! Warum Gutes Leben für alle Tauschlogikfrei sein Muss*. Roßdorf bei Darmstadt: Hrsg. Stiftung Fraueninitiative, Ulrike Helmer-Verlag.

Holloway, John. 2015. "Read *Capital*: The First Sentence or *Capital* Starts with Wealth, Not with the Commodity." *Historical Materialism* 23 (3): 3–26.

Kinkade, Kat, and the Twin Oaks Community. 2011. "Labour Credit – Twin Oaks Community." In *Life Without Money: Building Fair and Sustainable Economies*, edited by Anitra Nelson and Frans Timmerman, 173–191. London: Pluto Press.

Mauvaise Troupe Collective. 2014. *Defending the Zad*. Paris: Éditions de l'éclat.

Nelson, Anitra. 1999. *Marx's Concept of Money: The God of Commodities*. London and New York: Routledge.

Nelson, Anitra. 2011. "Money Versus Socialism." In *Life Without Money: Building Fair and Sustainable Economies*, edited by Anitra Nelson and Frans Timmerman, 23–46. London: Pluto Press.

Nelson, Anitra. 2016. "'Your Money or Your Life': Money and Socialist Transformation." *Capitalism Nature Socialism* 27 (4): 40–60. https://doi.org/10.1080/10455752.2016.1204619

Nelson, Anitra. 2018a. "The Political Economy of Space and Time in Eduardo Galeano." *Progress in Political Economy*, 20 February. http://ppesydney.net/political-economy-space-time-eduardo-galeano/

Nelson, Anitra. 2018b. *Small Is Necessary: Shared Living on a Shared Planet*. London: Pluto Press.

Nelson, Anitra. Forthcoming. *Beyond Money: A Postcapitalist Strategy*. London: Pluto Press.
Öcalan, Abdullah. 2017. *The Political Thought of Abdullah Öcalan*. London: Pluto Press.
Rubel, Maximilien, and John Crump. 1987. *Non-Market Socialism in the Nineteenth and Twentieth Centuries*. Houndmills and London: Palgrave Macmillan.
Rubin, Jerry. 1976. *Growing (Up) at Thirty-Seven*. New York: M. Evans.
Salleh, Ariel. 2011. "The Value of a Synergistic Economy." In *Life Without Money: Building Fair and Sustainable Economies*, edited by Anitra Nelson and Frans Timmerman, 94–110. London: Pluto Press.
Tax, Meredith. 2016. *A Road Unforeseen: Women Fight the Islamic State*, New York: Bellevue Literary Press.
Zad Inhabitant. 2018. *The Revenge Against the Commons*. https://zadforever.files.wordpress.com/2018/04/revenge-of-the-commons-eng.pdf (accessed 16 April 2021).

26
ECOSOCIALIST ECONOMICS

Pat Devine

Introduction

The discipline of economics is conventionally ascribed to moral philosopher Adam Smith under the name of "political economy." The first theoretical framework came to be known as "classical political economy," with prices based on the objective cost of production: the labour theory of value. After Marx adapted this theory to demonstrate how exploitation occurred under capitalism, it was replaced by the marginalist neo-classical framework, with its subjective preference theory of value and no room for exploitation. The discipline dropped the "political" and became known as "economics," as it still is today. Both these frameworks were analyses of economic activity under capitalism, with the economy seen as a distinct sphere of social life, separate from the rest of society and subject to its own laws, its own *modus operandi*, leading to what became the catchphrase "You can't buck the market."

Ecosocialism, in contrast to capitalism, is envisaged as a society in which the separation of economic activity from the rest of social life is overcome, with the different aspects of social life consciously integrated into an organic whole. Economic activity is therefore no longer subject to laws of its own in particular market forces with outcomes no one willed, but instead is shaped and planned by democratically determined objectives and priorities. As there is no distinct economic sphere, with its own *modus operandi*, to be analyzed, the concept of an "ecosocialist economics" may be a category mistake. However, much work that goes under the name of ecosocialist economics is concerned with two broad areas of great importance: first, progressive – sometimes prefigurative – reforms under the existing capitalist system, such as a Green New Deal and a just transition, and second, the values and priorities that would inform an ecosocialist society – in particular, social planning rather than market forces and use values rather than exchange value and private profit. This chapter discusses the first of these areas, and the second area is addressed in many of the other chapters. The rest of this chapter discusses the possible structure, institutions, and processes of an ecosocialist society.

The red and the green

Socialist concerns in their modern form have built on the insights of many socialist thinkers from the 1830s onwards, pre-eminently those of Marx and Engels. Marx analyzed exploitation

 10.4324/9780429341427-31

under capitalism, the system's anarchy of production resulting from the operation of market forces, and the various forms of alienation it produces. Engels (1850 [1987]) documented the disastrous impact capitalist industrialization had on people in his 1845 *Condition of the Working Class in England*. They argued that these negative characteristics of capitalism – exploitation, alienation, and the adverse conditions of the working class – were due to the contradiction between the social relations of production, based on private ownership of the means of production, and the social forces of production, *viz*. the cooperative techniques and social processes of production. They also recognized the deleterious effect that capitalist agriculture had on the soil, and this insight was extended to non-human nature in general by subsequent green writers and the developing green movements. John Bellamy Foster (1999) has used the term "metabolic rift" to capture this rupture of the organic link between human society and non-human nature. Marx's "first" contradiction of capitalism was within capitalism itself, between the private social relations of production and the cooperative social forces of production. This first contradiction was then later added to by O'Connor's (1988) "second" contradiction of capitalism, between the capitalist mode of production as a whole and non-human nature (elsewhere in this volume). The first and second contradictions underpin the socialist and green movements, respectively, parts of which have now co-evolved into the ecosocialist movement.

A convenient theoretical framework for thinking about these issues is that developed by Karl Polanyi (Polanyi 1944 [2001]). Polanyi argued that what he called "the self-regulating market," by which he meant capitalism, was so destructive of both human and non-human nature that it gave rise to protective movements and measures within capitalism to counter these adverse effects. He also argued that such measures interfered with capitalism's *modus operandi*, and this produced new crises, followed by moves to reinstate the unregulated, self-regulating market mechanism. Historically and geographically there have been – and still are – many different varieties of capitalism, some more regulated and some less regulated, but all still capitalist. Polanyi argued that in capitalist societies, the economy is always embedded in the rest of society, but it can be more embedded, as in social democratic Keynesian welfare state societies, or less embedded, as in the more free-market United States model (leaving aside the pre–World War Two, Roosevelt New Deal period). In both cases, the economy was still instituted as a distinct sphere of activity, and the societies were capitalist (Adaman, Devine, and Ozkaynak 2007). Only when society had regained full control over economic activity by abolishing the markets for what Polanyi called the "fictitious commodities" – labour, land (non-human nature), and money (capital) – could this dynamic of oscillating between more or less self-regulation of the economy be overcome (Devine 2018).

Ecosocialist objectives

My organizing vision of ecosocialism is the organic reconnection between the polity, the economy, and non-human nature. Only this can create the conditions for the sustainable well-being of both human and non-human nature. Through representative assemblies and self-governing associations, civil society exercises control over the state and mediates society's relationship with non-human nature, directly and through the state. (Here, the term "state" is used to mean the administrative structure that replaces the capitalist state's class repressive role) (Devine 1988 [2010]). Climate change and biodiversity loss mean that there has to be a fundamental and qualitative change in our ways of life. This will involve and affect technology, energy, agriculture, transport, town and country planning, caring work, the division of labour, and consumerism, of all of which, more later.

The transition to new ways of life will involve major changes in society, and, given the ecosocialist values of equality and solidarity, there must be a just transition to the new situation. "Just transition" is a framework developed by the trade union movement to secure workers' rights and livelihoods when economies are shifting to sustainable production and new lifestyles. Social and climate justice globally will require that, in the course of such a transition, the gross inequalities *within* society and *between* societies be addressed. Although the adverse effects of climate change and biodiversity loss affect everyone, they affect the poor most. A just transition will inevitably involve redistribution within and between countries.

Two current approaches with some bearing on the issues of historical, intra-generational, and inter-generational equality are the degrowth and steady state movements. Degrowthers correctly argue that global output is already unsustainably high and needs to be reduced. However, it is generally recognized that while this is true for the rich countries, it is not the case for the poorer countries, which still need to grow, albeit in a different way from the rich capitalist countries. What they need is the development of their productive capacity to a level at which it can satisfy basic needs, when it is not used for wasteful military production and conspicuous consumption. Steady-staters argue that the fact that further growth is unsustainable does not mean stasis – no change – but that the overall level of output should not increase. These approaches each draw attention to important issues, and both have implications for the ways in which economic activity should be measured, with new indicators needed in place of gross domestic product (GDP) or GDP per head. (For discussion of these issues, see Seaton 2019; Sen, Stiglitz, and Fitoussi 2010).

Classical Marxism envisaged abundance as a precondition for communism, with abundance conceived as the development of the productive forces, including centrally, labour, to a level of productivity at which everyone's needs could be satisfied. More recently this emphasis on the supply side, sometimes known as "productivism," has been qualified by recognition that the demand side – what counts as needs – also has to be taken into account. Under capitalism, the absence of people having significant real control over their lives leads to various forms of compensatory activity: notably, in the developed capitalist countries, turbo-charged consumerism. Kate Soper has developed the concept of "alternative hedonism" (Soper 2008; and elsewhere in this volume) as a way of countering the mistaken view that taking climate change and biodiversity loss seriously involves deprivation. She argues instead that alternative hedonism suggests the possibility of a better and more satisfying way of life. This is not (mainly or only) a moral concept but has empirical support in that when the material conditions of people's lives give them the opportunity to do so, they increasingly opt for more fulfilling ways of life rather than more and more consumerism, more and more "stuff": quality over quantity.

Substantive societal changes

The organizing principle of an ecosocialist society would be one in which the satisfaction of the needs required for global human flourishing also enables non-human nature to flourish, which would probably best be achieved through a system of participatory democratic planning. Production for use rather than profit would include at least food, clothing, and housing; clean air and water; health and education; transport; and culture. The satisfaction of these needs, in turn, depends on renewable energy; technology for production using minimal resources, enabling reuse and recycling, and producing minimal waste; and town and country planning that minimizes the need for travel by locating homes, schools and colleges, health centres, work, leisure facilities, and green spaces close to one another in consciously designed organically integrated communities. While this approach would inform new developments, how to apply these principles to the existing material infrastructure is an important matter for deliberative democratic discussion based on prefigurative achievements in the process of transition.

Participatory democratic planning, combined with the reduction of labour time required for the production necessary for human flourishing as productivity increases, would be a decisive step towards what Marx called "the kingdom of freedom." This is because, in addition to free time for creative and human developmental activities, a significant increase in free time is also a necessary condition for the participation of working people in democratic discussion and decision-making in society and in the management of economic activity.

I have generalized the concept of the "kingdom of freedom" to the abolition of the *social*, but not the *functional*, division of labour. The social division of labour means people doing the same category of activities for most of their lives. These categories include running and managing different political, economic, and voluntary organizations and associations; professionals such as doctors and nurses, academics and teachers, architects and engineers, lawyers and civil servants; skilled manual labour, such as electricians and carpenters, bricklayers and roofers, firefighters and other rescue workers; and unskilled labour. Within each category of social activity, there are, of course, different functional activities, and there are also some activities, such as caring work, much of it currently unpaid in capitalist societies, involving skills usually associated with more than one social category (Devine 1988 [2010])

Sectoral implications

The changes needed for transition to a different way of life enabling different lifestyles will involve major changes in the structure, characteristics, and size of the sectors of the economy, as well as in the pattern of needs that have to be satisfied. This is likely to be a long process, perhaps best thought of as a continuing journey rather than an end state. What follows in this section are indications of the directions of development needed, many of which are currently being fought for and some of which may be partially achieved in a prefigurative form within capitalism. But while capitalism is still with us, they will meet relentless opposition from the capitalist corporations and states, which will prevent, delay, and distort them. That is why we need fundamental societal transformation to ecosocialism.

Energy

There is now incontrovertible scientific evidence that the human-induced emission of greenhouse gases contributes greatly to climate change and that this has to be substantially reduced if we are to keep global warming to 2.0 or preferably 1.5 degrees above pre–Industrial Revolution levels. This means that carbon-based energy sources have to be rapidly phased out and replaced by green energy sources – wind, water, and solar, with some (in my view, mistakenly) also arguing for nuclear. There is much discussion about the best way to achieve the necessary transition: on shore or off shore; tidal, river, or storage based; local or regional, national, international, or global grid networks; and the possible need for reserve back-up generating capacity. Each has implications for employment, ecosystems, impact on natural beauty, and local control. Decisions that are to be effective can only be made on the basis of the available scientific evidence and the participation of those groups likely to be affected by them. The outcome is likely to be a combination of decentralized electricity generation and large-scale green generating centres.

The amount of generating capacity needed, of course, depends on how much energy societies need. Current proposals for reducing energy needs include homes, workplaces, and public buildings to be fully insulated; road building and traffic to be greatly reduced, with green motor vehicles replacing petrol and diesel vehicles; railway systems to be extended; and production in general to be more localized, rather than, as at the moment, globalized, to minimise the need for long-distance shipping and air traffic. Together, these would significantly reduce the demand for energy, making the transition to green energy less challenging. These sorts of desirable developments are currently referred to as "integrated energy systems."

Agriculture

Modern agriculture has largely undermined and commodified subsistence agriculture, and the remaining parts of the world where subsistence agriculture still predominates are under increasing pressure and stress. Modern agriculture is largely industrial and monocrop, dependent on chemical pesticides and weed killers, together with artificial fertilizers. The countryside in developed countries is largely sterile while urban areas exhibit much greater biodiversity. Foodstuffs in the more developed world are now largely globally traded commodities, with all the implications that has for global transport systems. Consumers in the rich countries increasingly expect their food to be available throughout the year, irrespective of local seasonality. Ecosocialist agricultural production is likely to be predominantly small scale, locally based, seasonal, and organic: in other words, subsistence oriented.

The "green revolution" in the 1950s and 1960s was an attempt to transform and greatly improve agricultural productivity, primarily in India and other then-less-developed countries, through the use of high-yield crop varieties and artificial fertilizers and pesticides. However, it has been argued that the green revolution failed to lead to long-term improvements in productivity and led to greater concentration in agriculture at the expense of small-scale producers (Cullather 2010). Organic agricultural production, by contrast, has been found in some cases to increase yields per acre through more labour-intensive methods, is less polluting, and is more beneficial to biodiversity. The most recent evidence suggests that feeding the expected future global population may require a mosaic of organic, conventional, and semi-natural agriculture, with natural habitats retained in agricultural landscapes together with extensive re-wilding. Organic production has the added advantage of being potentially more satisfying, in that it brings people closer to non-human nature and helps them overcome their alienation from what they produce (Johnson 2015; Pretty and Pervez Bharucha 2018). The popularity of allotments and community-supported agriculture in many countries is a further indication that Soper's alternative hedonism is not just moral advocacy but has an empirical material foundation.

Transport

Motor vehicles, shipping, and flying are significant contributors to greenhouse gas emissions and climate change. Current national targets for a reduction in these emissions exclude shipping and flying. In the lead-up to the first Earth Summit in Rio in 1992 on how to move forward, US President George Bush the elder stated, "The American way of life is not up for negotiation" (Economist 2003). Emphasis is still very heavily on "greening" energy production and only to a very limited extent on reducing the demand for energy in various ways. An ecosocialist transport policy would be part of a move towards creating a very different way of life from that which has emerged in the US and other affluent, albeit grossly unequal, societies today. Twenty years ago, Chinese cities were full of bicycles; today they are full of cars. Road traffic not only produces greenhouse gases, but it also produces air pollution and the illnesses associated with it, congestion, and traffic jams, and it contributes to the degradation of city centres.

Road building to ease congestion in societies today only encourages more traffic, yet since the advent of motor transport, cities have been designed to favour road transport. Road building and traffic in the developed capitalist countries need to be greatly reduced, with green energy–driven cars used to replace petrol and diesel vehicles for what necessary road travel remains. In general, though not indiscriminately, railway systems need to be extended and production in general needs to be more localized, rather than globalized, to minimise the need for long-distance shipping and air traffic. Food and manufactured goods are now largely globally traded commodities. Much local road traffic, such as the school run, is unnecessary and needs to be replaced by cycling and walking, which require the provision of multiple safe and pleasant pedestrian and cycle routes, having the added benefit of healthier lifestyles.

All this would involve major changes in people's lifestyles and for the workers who currently produce and operate our existing transport systems. That such changes are possible is evident from the movement towards such systems in an increasing number of cities and urban areas. However, the scale and speed of the changes needed means that the just transition movement becomes ever more important. It also means that the often woefully inadequate and overcentralized town and country planning systems need fundamental reform.

Town and country planning

As argued earlier, rather than productivism or production for private profit, in an ecosocialist society, the satisfaction of the needs required for global human flourishing in a way that enables non-human nature also to flourish would be a central organizing principle. This could best be implemented through a system of participatory democratic planning, including town and country planning, to minimize the need for travel by locating homes, work, schools and colleges, health centres, leisure facilities, and green spaces close to one another in consciously designed, organically integrated communities, as argued earlier.

In the parts of the world where town and country planning systems exist, they are today dominated by the capitalist corporate interests that are a central part of any current stakeholder interests involved in the decision-making process. In the UK, there is a history of democratically elected, local area councils which have rejected planning applications for fracking in their areas being overruled by central government, although after years of campaigning, fracking has now been prohibited. The same rejection of local objections has happened in relation to the proposal for HS2, a new high-speed cross-country train line in the north of England whose route is planned to run through local areas of natural beauty and of great importance for conservation. Of course, there will be occasions when higher-level considerations have to be given precedence over local concerns, but in an ecosocialist society, this would be decided democratically through a layered system of participatory planning.

Ecosocialism or barbarism

If we are to avoid the barbarism or dystopias already making headway in today's world, we need a transition to ecosocialism, which requires both a vision of how such a system might work and a strategy for how we might get there. If ecosocialist economics is to mean anything, I believe it must be a model of a participatory democratic system, specifying how it might work, its institutions, and its processes. How to arrive at such a system is a political project beyond the scope of this chapter, but as a starting point, I would begin with Gramsci's concepts of wars of position, organic crises, and wars of manoeuvre (Simon 1982 [1991]).

Institutions and processes

An ecosocialist society would be a democratic, self-governing society based on generalized participation in political and economic decision-making and, where appropriate, in the carrying out of the decisions made. One possible way of achieving this might be a system in which at all but the most local, small-scale levels, direct democracy would be combined with representative democracy. In the political sphere, civil society could elect representative assemblies, and the self-organizing associations of people with similar interests could elect members of a second "chamber of interests." In this way, representatives of the general interest and of the particular interests of the self-organizing groups would negotiate a set of policies and proposals that reflected the interests of all concerned.

Of course, there would inevitably from time to time be tensions and contradictions between the self-managed associations themselves and between them and the representative assembly. However, given the societal values of equality and solidarity expected to prevail in an ecosocialist society and a process of deliberation in which the different concerns and interests of those involved are made explicit, negotiation is likely to resolve most differences in a coordinated outcome that all could live with. The process of negotiated coordination is not one of aggregating existing preferences but is a transformatory process in the course of which people's preferences change as they become aware of other people's concerns. Ultimately, however, any remaining differences would have to be resolved through majority voting – subject, of course, to a society-wide framework of rules protecting legitimate minority rights and prohibiting illegitimate discrimination.

Through this process, society would be able to decide democratically on the priorities and direction of development desired for economic activity: the allocation of resources to investment and to welfare, social services, education, and culture and the conservation and enhancement of non-human nature. Implementation would be by a planning board consisting of representatives of the general interest and elected representatives of the different sectors and interests affected. Economic activity would then be undertaken by socially owned enterprises.

Social ownership

Under capitalism socialists have long advocated common ownership, and greens have advocated stakeholder ownership. Drawing on these prefigurative traditions, I have developed the concept of social ownership – ownership by those groups affected by the activities of the enterprise. I have also argued for the concept of the market to be deconstructed through a distinction between market *exchange* and market *forces*. Under capitalism, enterprises exchange the goods and services they produce using their existing productive capacity. Changes in the pattern of productive capacity occur as a result of market forces brought about primarily through what Schumpeter called "creative destruction," the replacement of existing goods and services by new ones: mainly, but not always, through the replacement of existing enterprises by new ones, with an unplanned outcome no one willed (Devine 1988 [2010]; Schumpeter 1942 [2006]).

Applying this distinction to an ecosocialist society, enterprises would engage in market exchange, and the social owners at this level would be its workers; the users of its output, input-supplying, or competing enterprises; the locality in which it is based; the industry or sectoral body to which it belongs; and environmental or equal opportunities interest groups. When it comes to changes in productive capacity, market forces would be replaced by negotiated coordination as the process through which decisions on investment or disinvestment are made. The social owners at this level would be the industry or sectoral negotiated coordination body made up of representatives of those interests that would be affected by the changes in capacity – enterprises in the industry or sector, input-supplying industries and users of its output, localities likely to be affected, and relevant local and reginal planning boards, plus again any environmental or equal opportunities groups. It will be apparent that the social owners at the enterprise and industry levels will be different. This is an example of the principle of subsidiarity.

Subsidiarity

Subsidiarity is the principle that decisions should be made at the most local level consistent with all groups significantly affected by the decisions being able to be involved in making them. Many issues affect a wider set of people than just those locally involved. Fair and effective political, social, economic, and ecological decisions typically depend on, among other things, an input of individual and social tacit knowledge at all levels, which can only be provided by those involved with the relevant knowledge and experience. Thus subsidiarity involves a layered structure of interdependent institutions and processes – local, regional, national, international, and global. This is clearly the case in relation to ecological issues such as climate change and biodiversity, as these can only be dealt with effectively on a global scale, with implementation in turn devolved to the relevant lower levels. Think globally; act locally.

Prices

The overall plan would be focused on the enhancement of use value, with the allocation of resources for technological development shaped accordingly. Here, use value would be value not only for humans but also for non-human ecology and biodiversity. Primary input prices would be determined according to both social and ecological criteria and, together with the costs of the intermediate inputs derived from them and the state of existing technology, would give a set of prices for all produced goods and services. The plan's structure would then reflect judgements on the relative use values, human and non-human, of different alternatives in relation to their costs. Prices would have primarily an accounting function in the process of allocating output to the democratically determined alternative uses for it – personal and collective consumption and investment in different geographic areas and regions, in different sectors of the economy, and in non-human ecology and biodiversity. The rapid development of electronic money would facilitate this.

Conclusion

Ecosocialism is based on the reasonable expectation that in a society liberated from capitalism, without classes, having achieved full social control over economic activity, having repaired the metabolic rift, and having overcome alienation, there would be time for personal development through cultural, sportive, playful, scientific, erotic, artistic, and political activities: time to stand and stare. Such a classless society would not be without contradictions and conflicts, but these would mainly be dealt with through the process of negotiated coordination within the layered structure of participatory democratic planning based on subsidiarity. It would be a learning process through which a self-governing society is able to recognize and correct mistakes. Ecosocialism would be not only a new mode of production in which economic decision-making is fully integrated into social decision-making, but also a new way of life, a new ecosocialist civilization, founded on an ecologically sustainable conception of the "good life."

Acknowledgements

I have benefitted greatly from discussions of the UK-based Red Green Study Group and the Montreal-based Karl Polanyi Institute over many years. Many thanks are also due to Ted Benton, Richard Kuper, Elena Lieven, Simon Pirani, and David Purdy for their valuable comments on a preliminary draft.

References

Adaman, Fikret, Pat Devine, and Begum Ozkaynak. 2007. "Reinstituting the Economic Process: (Re) embedding the Economy in Society and Nature." In *Karl Polanyi: New Perspectives on the Place of the Economy in Society*, edited by Mark Harvey, Ronnie. Ramlogan, and Sally Randles, 93–112. Manchester: Manchester University Press.

Cullather, Nick. 2010. *The Hungry World: America's Cold War Battle Against Poverty in Asia*. Boston: Harvard University Press.

Devine, Pat. 1988 [2010]. *Democracy and Economic Planning: The Political Economy of a Self-Governing Society*. London: Polity Press.

Devine, Pat. 2018. "Planning for Freedom." In *Karl Polanyi's Vision of a Socialist Transformation*, edited by Michael Brie and Claus Thomasberger, 2009–2020. Montreal: Black Rose Books.

Economist. 2003. "Leader, a Greener Bush." 13 February. www.economist.com/leaders/2003/02/13/a-greener-bush

Engels, Friedrich. 1850 [1987]. *Condition of the Working Class in England*. London: Lawrence & Wishart.

Foster, John Bellamy. 1999. "Marx's Theory of Metabolic Rift." *American Journal of Sociology* 105 (2): 366–405.

Johnson, Nathaniel. 2015. "Do Industrial Agricultural Methods Actually Yield More Food per Acre Than Organic Ones?" *Grist*, 14 October. https://grist.org/food/do-industrial-agricultural-methods-actually-yield-more-food-per-acre-than-organic-ones/ (accessed 11 January 2020).

O'Connor, James. 1988. "Capitalism Nature Socialism: A Theoretical Introduction." *Capital Nature Socialism* 1 (1): 11–38.

Polanyi, Karl. 1944 [2001]. *The Great Transformation*. Boston: Beacon.

Pretty, Jules, and Zareen Pervez Bharucha. 2018. *Sustainable Intensification of Agriculture*. London: Earthscan Food and Agriculture.

Schumpeter, Joseph. 1942 [2006]. *Capitalism, Socialism and Democracy*. London: Routledge.

Seaton, Lola. 2019. "Green Questions." *New Left Review* 115: 105–129, January–February.

Sen, Amartya, Joseph Stiglitz, and Jean-Paul Fitoussi. 2010. *Mismeasuring Our Lives: Why GDP Doesn't Add Up: The Report*. New York: New Press.

Simon, Roger. 1982 [1991]. *Gramsci's Political Thought: An Introduction*. London: Lawrence & Wishart.

Soper, Kate. 2008. "Alternative Hedonism, Cultural Theory and the Role of Aesthetic Revisioning." *Cultural Studies* 22 (5): 567–584.

27
SOLAR COMMUNISM

David Schwartzman

Introduction

The concept and the name *solar communism* came to me in the aftermath of the collapse of the Soviet Union at the end of 1991, with the first record in my notebook dated 18 January 1992: "Solar Communism A Scenario for 2050." I was on a trip to the West Coast of the US with my family in the summer of 1992 when I wrote further notes, which turned into a 1992 article in which I stated:

> The second contradiction of capitalism forces a reexamination of the very principle proposed to guide a communist society: '*From each according to her ability, to each according to her needs*.' We must now define "each" and "her" as referring both to human beings and to nature (ecosystems).[12] Further, if socialism as a transition to communism is to be viable, this new principle must be progressively applied to this mixed social formation, between two modes of production. . . . I suggest we begin, as socialists, to project a vision of a radical reconstruction of society together with a restoration of the biosphere, namely – a goal of *planetary solar communism* that would harness the power of information technologies and renewable energy sources to realize the dream of utopian and scientific communism.
>
> *(Schwartzman 1992, 18, emphasis added)*

Footnote 12 in that article is my caveat:

> I may have come across this formulation in the green or ecologist literature. I just don't remember precisely where, so I apologise now for possible nonattribution. Identifying nature with the feminine gender has ancient roots. Some greens have used the Term "Gaia," the mother of earth in Greek mythology, for nature – this is also the term used for a new scientific theory of the biosphere elaborated by Lovelock and Margulis; see Scientists on Gaia, MIT Press, 1991. (In retrospect, Gaia is still not a theory in the scientific sense, rather a very fruitful set of hypotheses. As far as I know I have priority for this reconceptualization of Marx's aphorism.)
>
> *(ibid., 18, fn. 12)*

10.4324/9780429341427-32

Unlike many others who, like me, are "red diaper babies," I did not "throw the baby out with the bathwater" in a profound rejection of communist ideals faced with the realities of how dirty and full of the blood of innocents this bathwater had become. Rather, I tried to rethink the communist vision for the twenty-first century. Now I am convinced I was not deluded, as we face the great bifurcation: either a plunge in the abyss of climate hell or an ecosocialist horizon, followed by the solar communist horizon (Schwartzman 2013a).

Communism, its roots, and twentieth-century socialism

There are biblical roots to the communist vision, and its trajectory runs through the French Revolution, followed by utopian socialism critiqued by Marx and Engels (Schwartzman and Schwartzman 2019, 236–237). Socialism richly deserves a brutal critique of its successes and failures, but it is important to point out that Marxian communism has never existed – only deformed socialist societies led by communist parties. This common conflation of party leadership with the political economic system of twentieth-century state socialism should be avoided. Within the Left, we find two extremes regarding the experience of twentieth-century socialism, in particular its manifestation in the Soviet Union. Slavoj Žižek wrote off the Soviet experience as a huge disaster, while my father, Max Schwartzman, a staunch Marxist Leninist to his dying day, responded to my question about Stalin's role in the following way: "Stalin, he made some mistakes." (I am leaving out the even more extreme position of those who have deified Stalin as an infallible leader.) I think that an accurate account is somewhere between these two extremes. Here is more context:

> Communism was an inspiring vision to millions in the twentieth century, with well-known failures in its realization as "really existing socialism," or what some Marxists prefer to call "state capitalism." The construction of socialist societies in the twentieth century occurred in a very unfavorable context, under continuous attack by capitalist powers – starting with initial intervention soon after the Russian revolution, followed by the Second World War and the cold war. Socialist societies of the twentieth century had both real internal achievements and well-documented state-sanctioned mass suffering and death, in parallel with immense positive impacts on global politics, including the defeat of fascism, the post-Second World War end of colonial oppression, and the development of social-welfare programmes by capitalist states challenged by the benefits for working classes in socialist countries (e.g., West Germany and the German Democratic Republic). . . . Really existing twentieth-century "socialism" (and its survivals into the twenty-first century) combined characteristics of communism, capitalism and state capitalism. But this is no surprise, given the impurity and complexity of a real transition from capitalism, potentially into communism.
>
> *(Schwartzman 2016, 145)*

The thermodynamics of communism for the twenty-first century

I focused on the relevance of thermodynamics with its concept of entropy to a rethinking of communism in my article "Solar Communism" (SC) (Schwartzman 1996). Twelve years later, I again addressed the concept of entropy because even some Marxists did not get it right (Schwartzman 2008). In SC I discuss the lack of a full conceptualization of the technological basis of an ecosocialist transition to a future global society, arguing that an historical materialist account of this transition and a vision of this future global society should encompass its full materiality, in both the technological and social senses of that term. Socialist or Marxist political economy is necessary but not sufficient in itself to advance a vision of twenty-first century socialism. The natural, physical and informational sciences in particular, climatology, ecology, biogeochemistry, and thermodynamics must be fully engaged. These sciences will inform the technologies of renewable energy, green production, and agroecologies, whose infrastructure are to replace the present unsustainable mode. An agroecology is the mode of ecological agricultural production to ensure the health of people and thriving global ecosystems, informed both by science and the knowledges of long-standing Indigenous practices (e.g., Altieri 1995). Since the publication of SC in 1996, renewable energy technologies have advanced to a degree that they are being rapidly implemented around the world, but this growth is still coupled with the unsustainable rise of global fossil fuel consumption.

Rethinking twenty-first century communism requires consideration of the quality and quantity of the energy supply for global civilization: i.e., the thermodynamics of communism. In this context, we must confront the ever-narrowing window of opportunity to prevent the onset of catastrophic climate change (C3). The three critical material requirements for preventing C3 and achieving a communist future are global demilitarization, a transition of the current mode of agriculture to agroecologies, and solarization of the global energy infrastructure with a greater capacity than the present level. The science of thermodynamics demonstrates that a solar energy source is fundamental to truly green growth, industrial ecologies, and the phaseout of extractive mining. The ecosocialist movement should critique the misleading specter of entropy, drawn from Georgescu-Roegen's fallacious fourth law of thermodynamics, which has had strong influence on the degrowth movement and ecological economics. Likewise, the argument that "perpetual growth on a finite planet leads inexorably to environmental calamity" (Monbiot 2019) fails to deconstruct the qualitative aspects of growth, what is growing, what should degrow, and under what energy regime.

According to the second law of thermodynamics, entropy is a measure of the loss of energy available to do work. The production of energy from burning fossil fuels as well as nuclear fission generates an incremental heat flux from the earth's surface unlike, to a good first approximation, the tapping of solar radiation to do work. The latter outcome is non-incremental because the interaction of low entropy visible light with the low albedo, relatively dark earth's surface generates a corresponding flux of high entropy heat (infrared radiation) whether work is done for human civilization or not, with this heat flux escaping to space. In addition, burning fossil fuels is a prime source of anthropogenic carbon emissions in the atmosphere, driving global warming which is amplified by the melting of high albedo sea ice and surface snow at high latitudes. Hence, global solar power will then pay its "entropic debt" to space as non-incremental waste heat, without driving us towards tipping points of even more catastrophic climate change than has happened over the past few decades (Schwartzman 1996, 2008).

Coming out of this transition, which will need to be ecosocialist, a steady-state biophysical economy can emerge in a global solar communist society (Schwartzman 1996; Schwartzman and Schwartzman 2019).

In SC, I recognized the growing threat of rising greenhouse gases derived from burning fossil fuels. In the 23 years since this essay was published, this threat has grown into a global crisis of unprecedented magnitude. Humanity and existing biodiversity are now facing a huge challenge in this the first half of the twenty-first century. Shall civilization emerge in a new mode, with the end of what Marx called our prehistory, the rule of capital on our planet, or shall we plunge into a deep abyss of climate hell, climate catastrophe, for the few who survive? This is the great bifurcation ahead, and the outcome is not possible to predict. Only transnational class struggle on a scale not witnessed in human history has any chance of avoiding the abyss (Schwartzman 2013a, 2013b). There are actually two major threats to human civilization. The first is that of nuclear war, which would be deadly even if localized because of resulting climatic impacts as a result of immense amounts of soot injected into the stratosphere, with a regional conflict using less than 1 percent of the worldwide nuclear arsenal potentially having adverse consequences for global food security unmatched in modern history (Jägermeyr et al. 2020).

The second threat is C3. C3 is very likely inevitable if carbon emissions to the atmosphere are not rapidly and radically reduced and if the already unsafe atmospheric level of carbon dioxide is not reduced by sequestration technologies to a safe level. Paradoxically, however, we are also privileged to confront this challenge since the process of removing these threats raises the possibility of ending the rule of capital on our planet.

But it is increasingly clear that only with a radical shift to a global regime of peace and cooperation will it be possible to implement an effective C3 prevention programme. The threats of C3 and nuclear war pose an unprecedented opportunity to end the rule of capital because the main obstacle to elimination of these threats is the MIC, the military-industrial (fossil fuel, nuclear, state terror, and surveillance) complex, at the core of real existing capitalism (Schwartzman 2009). Thus, the challenge to dissolve the MIC puts an ecosocialist transition on the agenda for humanity – an ecosocialist transition out of prehistory and into a new global civilization, solar communism in the twenty-first century.

The global North should transition to a much less resource-consuming lifestyle – e.g., mass transit, bicycle transport in place of cars, smaller homes, and a great reduction in the consumption of obsolescent gadgets – the endless creation of manufactured desire driven by the reproduction of capital.

If the transformation of agriculture to fossil fuel–free agroecologies that are situated closer to population centres is included, this degrowth translates into lower energy consumption in the global North, resulting in a higher quality of life with a more healthful environment, cleaner air and water, and organic food, free of the chemicals and genetic contamination now inherent in industrial/GMO agriculture.

But the world needs more, not less energy consumption than now, with most of humanity living in the global South receiving a significant increase, reaching the rough minimum of 3 to 3.5 kilowatts per person. Note that reaching this minimum is necessary but insufficient for acquiring the highest life expectancy, as several petroleum-exporting countries in the Middle East as well as Russia now fall well below that goal. Life expectancy in the United States is likewise below most industrial countries of the global North. Income inequality is robustly correlated with bad health and must be reduced to achieve the world standard life expectancy and quality of life.

A rough calculation can provide some idea of energy needs 20 to 30 years from now.

This estimate must take into account the incremental energy needs of this century as well as the energy saving derived from a 100 percent renewable supply electrifying the end use, with a significantly greater fraction of primary energy consumption going to doing work rather than released as waste heat as it is now in the combustion of gasoline in automobile transport. Assuming a minimum of 3 kilowatt per person minimum for achieving the highest world standard life expectancy, nine billion people, and an efficiency gain of 30 percent for a 100 percent renewable energy transition, 3 x 9 x 0.70 = 19 terawatts (in power units), which is equal to present global primary energy consumption. But this is a minimum, to which must be added incremental needs, climate mitigation and adaptation, biosphere cleanup, and ecosystem restoration. The additional energy required will likely approach the order of 25 terawatts or more (Schwartzman and Schwartzman 2019). This is the critical challenge of a global Green New Deal (GGND).

The GND has been recently proposed both in the US and in Europe as a solution to the current crisis and as a green Keynesian reprise of the 1930s New Deal in the US. The GND has the potential of generating millions of new jobs, in both the energy conservation and clean energy sectors and for the repair of physical infrastructure, while at the same time confronting the climate crisis by decarbonizing energy supplies. The vision of a GGND first received global attention from the UNEP 2008 initiative and now is being revisited precisely because the current deep health, economic, social, and environmental crisis is global in scope (Schwartzman 2011, 2021).

Creating a renewable energy infrastructure to supply this level of energy consumption is not only possible but also imperative in order to eliminate the energy poverty now affecting the majority of humanity living in the global South and to confront the challenge of implementing a prevention programme that has any chance of avoiding C3. Given the global North's historic responsibility for the threat of C3, transfer of wind/solar capacity to the global South from the global North is imperative. With 1 to 2 percent of current annual consumption of energy (80 percent derived from fossil fuels) being used for wind/solar power creation per year, a global-scale transition can be achieved in no more than 20 to 30 years, with the complete elimination of anthropogenic carbon emissions derived from energy consumption to the atmosphere and the provision of the minimum per capita energy consumption level required for a state-of-the-science life expectancy level for all (Schwartzman and Schwartzman 2011, 2013, 2019). Now we will examine why "solar" is added to communism as an imperative description for the twenty-first century.

Why *solar* communism?

Solar is by far the most abundant source of energy, and the technologies to harness it are already growing at a near-exponential global rate. Given a robust social management process during its life cycle, solar power also has very low negative health and ecological impacts. Moreover, a global transition to a solar power infrastructure is actually achievable within the time frame necessary to avoid catastrophic climate change. Under really existing capitalism, solar power is also the energy source most compatible with decentralized, democratic management and control and relatively free of MIC dictates, compared to fossil fuels and nuclear power. Finally, a solar "clean energy" transition is a critical component of the GGND and an ecosocialist path out of capitalism, and if realized, it will be the product of bottom-up struggles, a profoundly democratic process. An ecosocialist strategic perspective of a GGND is elaborated later.

The ecosocialist path to solar communism: From the Capitalocene to the Solarcommunicene

Many have contributed to the ecosocialist perspective in recent years, and this *Handbook* provides in-depth accounts of this work. Some of the most relevant to this discussion include Kovel (2002; see also Saul 2019), Löwy (2015), Angus (2017), Magdoff and Williams (2017), Wallis (2018), and Landis (2020). Ecosocialism is advanced as the alternative to capitalism, the killer of people and nature.

Strategic thinking is imperative for the climate justice movement to have any chance to meet the 1.5 degrees Celsius warming target of the IPCC, which, if breached, will generate climate catastrophe with much worse consequences than we now witness. It is not enough to recognize that the present global system is unsustainable and that a rapid and radical transformation is necessary. And such a strategy requires a plausible plan to bring about such change – recognizing the obstacles and how to overcome them, as well as having the capacity, when implemented, to ignite the imaginations of millions around the world to form a collective transnational subject, arguably the only force capable of simultaneously preventing climate catastrophe and opening a path for global ecosocialist transition. In this regard, I have proposed a strategic perspective regarding the GGND as a venue for class struggle driving it forward, utilizing a broad front including green capital to defeat militarized fossil capital, thereby preparing the way for the next stage of a global ecosocialist transition led by the transnational working class and its allies (Schwartzman 2011, 2021). I have argued that defeating militarized fossil capital is critical to make a truly just GGND, confronting the challenge of resource extraction, especially in the global South, while ending energy poverty and having the capacity for effective climate mitigation and adaptation in a 100 percent global wind/solar energy transition. "Green" capital must be challenged all along the way, to optimize environmental, worker, and community protection while eroding its power to manage this transition until full social management is in place. We can plausibly anticipate a future of sharp qualitative changes in this erosion of the power of capital, rather than a steady and progressive process.

I am convinced that a critical component of this strategy must be bottom-up movements of all those bearing the weight of this world of extreme disparity and inequality: the exploited and the oppressed. Cultural and material living models in embryo of a radically different future are prefigurations as conceptualized by Bloch (1986) and Kovel (2002). They are being created all over the world – whether they be farming or solar power cooperatives, worker-owned industries, or community land trusts creating affordable housing outside the constraints of capital-driven speculation in urban areas. As Ernst Bloch (1986) argues, the recognition of the "Not-Yet" is an anticipation founded on the principle of hope, the title of his seminal three-volume magnum opus. Like Bloch, I am inspired by this quote from Lenin: "We ought to dream!" (Lenin 1929, 158), along with the inspiration of "Imagine" (John Lennon's well-known song).

Socialist political economy informed by a Marxist perspective is necessary but not sufficient in itself to advance a vision of twenty-first-century ecosocialism. The natural, physical, and informational sciences in particular, climatology, ecology, biogeochemistry, and thermodynamics must be fully engaged. Along with the wisdom derived from the experience of thousands of years of Indigenous peoples' agriculture and culture, these sciences will inform the technologies of renewable energy, green production, and agroecologies whose infrastructure is to replace the present unsustainable mode.

Aside from Kovel's (2002) comprehensive critique of capitalism, the main emphasis of ecosocialist thought is on the necessity for democratization in the production and consumption spheres at all spatial scales, in an ecological framework. A concise statement of this is found in the Belem Declaration of 2007, which I had the honor of signing along with many others around the world (Löwy 2015). Ecosocialists are, of course, committed to class struggle as well as creating material and social prefigurations of what is imagined for a future ecosocialist civilization. In regard to the foundational importance of the energy basis of society, Löwy makes a very welcome recognition of the importance of a solar transition, stating that the "issue of energy is decisive because fossil energy . . . is responsible for much of the planet's pollution as well as for the disastrous climate change" (Löwy 2015, 23). But the quality of the energy supply with regard to thermodynamic considerations is critical as well, to confront both the challenge of climate change and extractive industries in the global GND and an ecosocialist future. Löwy (2015, 11), without citation, points to solar communism: "The revolutionary utopia of green socialism or of solar communism does not imply that one should not act right now." If we assume green socialism is equivalent to ecosocialism, then for Löwy, ecosocialism is not a transition to communism; rather, ecosocialism is essentially equivalent to communism as a utopia.

The penetrating analyses of ecofeminists must be noted, with their critiques of patriarchy and oppression of women and sexual minorities, along with their development of social reproduction theory (Vogel 2013; Federici 2014; Salleh 2017; Bhattacharya 2017; Gimenez 2019; Brownhill and Turner 2020). These contributions of ecofeminists must inform the ecosocialist vision, as well as the creation of prefigurations of solidarity and equality, including communal childcare, cooking, and housework, within the present context of capitalism. One current example of this effort stands out in Rojava, Syria, with the Kurds inspired by Öcalan's teachings (e.g., Öcalan 2013). We should also learn from the positive if imperfect achievements of social equality in state socialist countries, notably the Soviet Union, the German Democratic Republic, and of course Cuba.

The example of Cuba – arguably now on the cutting edge of ecosocialist transition – is instructive. Cuba's world standard health and education system, along with its agroecological transformation of agriculture and development of solar energy, especially in rural areas, are remarkable achievements for a country under US imperialist embargo for the last 60 years (Funes et al. 2002; Perfecto, Vandermeer, and Wright 2009; Fernandez et al. 2018; Yaffe 2020). The people of Cuba do well with much lower energy consumption than the US. But Cuba, in spite of an outstanding standard health and education system, suffers from energy poverty, which translates into an average life expectancy virtually tied with that of the US. Cuba has a world ranking of 46 (77.8 years), and the US is ranked 40 (78.5 years), according to pre-pandemic 2019 data (WHO 2020). It is likely that life expectancy in Cuba is now even higher than in the US, given the much higher US death rate from COVID-19. Of course, in the case of Cuba, we know that the US embargo is responsible for its ranking, coupled with energy poverty. Cuba falls well below the minimum energy consumption per capita required for the highest world standard life expectancy. (Japan is close to the minimum energy consumption per capita, with the highest life expectancy in the world.) With this example in mind, an ecosocialist transition in coming decades will be a multi-year process of transformation in the best of circumstances.

In most cases, communism is barely mentioned in ecosocialist discourse, with a few exceptions, such as Kovel's (2002) discussion of the Christian communism of the Bruderhof. I take this virtual silence as motivated by the attempt to avoid confusion of ecosocialism with the common perception of communism as the same as examples of state socialism of the twentieth century, led by communist parties. For example, for many in the US, the belief that "communism doesn't work" is informed by the one-sided, distorted picture they get from mass media describing the "communist" countries of the twentieth century. Contrary to the positing of ecosocialism as the alternative to capitalism, I follow Marx's definition of socialism as the lower stage of communism, a transition from capitalism to communism. And within the context of struggles for a global GND is an opportunity to confront the climate crisis and open up a path for ecosocialist transition. This conception of ecosocialism as a transition is important to make, given the radical changes in the political and physical economies necessary to eliminate the legacy of extreme inequalities created by fossil capitalism. It is not possible to accomplish all these tasks at once. Advancing to one stage potentially opens a path to another (Schwartzman 2021). And this transition will entail real growth of renewable energy and the capacity to meet long-standing needs in the education, nutrition, and health sectors, especially in the global South. Simply saying that capitalism must be replaced by socialism is mere rhetoric, not a strategy to make this goal possible. But at the same time, the embryos, prefigurations of communism, must be seen within the womb of capitalism, the ongoing struggles to expand the commons as Hardt and Negri (2017) have long promoted in their writings, as has Federici (2014).

Conclusion

To summarize, solar communist civilization will arguably require the following to fully emerge from an ecosocialist transition terminating the rule of capital:

1 Social governance of production/consumption on every scale from the local to the global
2 Global equity, elimination of global North/South disparities in health and education
3 Disarmament (complete demilitarization of global society)
4 Expansion and social management of communally owned land.
5 The end of production of value based on labour time.

For my contribution to the James Lovelock Centenary celebration in July 2019, I coined the term Solarcommunicene, referring to the potential future global solar commons emerging out of an ecosocialist transition, terminating the rule of capital in the present epoch of the Capitalocene, the relabeling of the Anthropocene, emphasizing the hegemony of capital reproduction in political economy. Jodi Dean, in her invocation of a communist horizon (Dean 2012), reasserts the vision of radical materialist utopia that has been buried and reburied yet never extinguished. But I submit that a communist horizon in the twenty-first century should be reconceived as *solar* communist. Projecting an ecosocialist horizon is now imperative to prevent and prefigure activity embodied in multi-dimensional class struggles in our world dominated by the rule of capital; to prevent catastrophic climate change and, along the way, demilitarize, solarize, and transform agriculture with agroecologies; and to prefigure the future in the present by expanding the commons, virtual and material, reaching the ecosocialist horizon and going beyond it towards the solar communist horizon. In this context, solar communism was invoked in the Combined Strategy and Plan of Action document coming out of the founding of the 1st Ecosocialist International, 31 October to 3 November 2017, Cumbe of Veroes, Bolivarian Republic of Venezuela (Saul 2018). The Solarcommunicene awaits us, but this hopeful future is highly contingent on multi-dimensional class struggle on every scale, from the neighbourhood to the globe, and at every intersection of the oppressed and the exploited, "race," gender, sexual orientation, ethnicity, citizenship status, (a)religion, age, degree of able-bodiedness.

Designating the global solar commons as equivalent to solar communism, I said the following, in looking back from year 2121 in the epoch of the Solarcommunicene:

> What would earn this designation, which would have been the subject of ridicule by most of our ancestors? There are no remaining disparities of health, education, quality of life in any region of human civilization, with all these measures corresponding to the highest levels achieved in human history, as a result of timely applications of information and renewable energy technologies and of course the termination of class exploitation and multiple oppressions.
>
> *(Schwartzman 2021, 58)*

To repeat Lenin's advice, "One ought to dream."

References

Altieri, Miguel A. 1995. *Agroecology: The Science of Sustainable Agriculture*. Boulder, CO: Westview Press.
Angus, Ian. 2017. *A Redder Shade of Green: Intersections of Science and Socialism*. New York: Monthly Review Press.

Bhattacharya, Tithi, ed. 2017. *Social Reproduction Theory: Remapping Class, Recentering Oppression*. London: Pluto Press.

Bloch, Ernst. 1986. *The Principle of Hope*. Vol. 3. Cambridge, MA: MIT Press.

Brownhill, Leigh, and Terisa E. Turner. 2020. "Ecofeminist Ways, Ecosocialist Means: Life in the Post-Capitalist Future." *Capitalism Nature Socialism* 31 (1): 1–14.

Dean, Jodi. 2012. *The Communist Horizon*. London: Verso.

Federici, Silvia. 2014. "Marx, Feminism, and the Construction of the Commons." In *Communism in the 21st Century*, edited by Shannon K. Brincat, 171–194, Vol. 1. Santa Barbara, CA: Praeger.

Fernandez, Margarita, Justine Williams, Galia Figueroa, Garrett Graddy-Lovelace, Mario Machado, Luis Vazquez, Nilda Perez, Leidy Casimiro, Graciela Romero, and Fernando Funes-Aguilar, eds. 2018. "New Opportunities, New Challenges: Harnessing Cuba's Advances in Agroecology and Sustainable Agriculture in the Context of Changing Relations with the United States." *Elementa Science of the Anthropocene* 6: 76. https://doi.org/10.1525/elementa.337

Funes, Fernando, Luis Garcia, Martin Bourque, Nilda Perez, and Peter Rosset, eds. 2002. *Sustainable Agriculture and Resistance: Transforming Food Production in Cuba*. Oakland, CA: Food First Books.

Gimenez, Martha E. 2019. *Marx, Women and Capitalist Social Reproduction*. Chicago: Haymarket.

Hardt, Michael, and Antonio Negri. 2017. *Assembly*. New York: Oxford University Press.

Jägermeyr, Jonas, Alan Robock, Joshua Elliott, Christoph Müller, Lili Xia, Nikolay Khabarov, Christian Folberth, Erwin Schmid, Wenfeng Liu, Florian Zabel, Sam S. Rabin, Michael J. Puma, Alison Heslin, James Franke, Ian Foster, Senthold Asseng, Charles G. Bardeen, Owen B. Toon, and Cynthia Rosenzweig. 2020. "A Regional Nuclear Conflict Would Compromise Global Food Security." *Proceedings of the National Academies of Science* 117 (13): 7071–7081.

Kovel, Joel. 2002. *The Enemy of Nature: The End of Capitalism or the End of the World?* New York: Zed Books.

Landis, Tina. 2020. *Climate Solutions Beyond Capitalism*. San Francisco: Liberation Media.

Lenin, Vladimir I. 1929. *What Is to Be Done?* New York: International Publishers.

Löwy, Michael. 2015. *Ecosocialism: A Radical Alternative to Capitalist Catastrophe*. Chicago: Haymarket.

Magdoff, Fred, and Chris Williams. 2017. *Creating an Ecological Society: Toward a Revolutionary Transformation*. New York: Monthly Review Press.

Monbiot, George. 2019. "The Problem Is Capitalism." www.monbiot.com/2019/04/30/ the-problem-is-capitalism (accessed 1 April 2021).

Öcalan, Abdullah. 2013. *Liberating Life: Woman's Revolution*. Cologne: International Initiative and Mesopotamian Press.

Perfecto, Ivette, John Vandermeer, and Angus Wright. 2009. *Nature's Matrix: Linking Agriculture, Conservation and Food Sovereignty*. London: Earthscan.

Salleh, Ariel. 2017. *Ecofeminism as Politics*. London: Zed Books.

Saul, Quincy. 2018. "Reflections on the Founding of the First Ecosocialist International." http://ecosocialisthorizons.com/2018/09/reflections-on-the-founding-of-the-first-ecosocialist-international/ (accessed 29 March 2021).

Saul, Quincy, ed. 2019. *The Emergence of Ecosocialism: Collected Essays by Joel Kovel*. New York: 2Leaf Press.

Schwartzman, David. 1992. "To Each According to Her Needs." *Dialogue & Initiative* 4: 16–19.

Schwartzman, David. 1996. "Solar Communism." *Science & Society* 60 (3): 307–331.

Schwartzman, David. 2008. "The Limits to Entropy: Continuing Misuse of Thermodynamics in Environmental and Marxist Theory." *Science & Society* 72 (1): 43–62.

Schwartzman, David. 2009. "Ecosocialism or Ecocatastrophe?" *Capitalism Nature Socialism* 20 (1): 6–33.

Schwartzman, David. 2011. "Green New Deal: An Ecosocialist Perspective." *Capitalism Nature Socialism* 22 (3): 49–56.

Schwartzman, David. 2013a. "The Great Bifurcation and Prospects for Solar Communism in the Twenty-First Century." *International Critical Thought* 3 (4): 480–495.

Schwartzman, David. 2013b. "4 Scenarios for 2050." *Capitalism Nature Socialism* 23 (1): 49–53.

Schwartzman, David. 2016. "Beyond Eco-Catastrophism: The Conditions for Solar Communism." in *Socialist Register 2017*, edited by Leo Panitch and Greg Albo, 143–160. New York: Monthly Review Press.

Schwartzman, David. 2021. *The Global Solar Commons, the Future That Is Still Possible: A Guide for 21st Century Activists*. Galesburg, IL: The Solar Utopia.org Press. www.theearthisnotforsale.org/solarcommons2021.pdf (accessed 1 April 2021).

Schwartzman, Peter, and David Schwartzman. 2011. "A Solar Transition Is Possible, Institute for Policy Research and Development." http://solarUtopia.org (accessed 29 March 2021).

Schwartzman, Peter, and David Schwartzman. 2013. "A Rapid Solar Transition Is Not Only Possible, It Is Imperative!" *African Journal of Science, Technology, Innovation and Development* 5 (4): 297–302.

Schwartzman, Peter, and David Schwartzman. 2019. *The Earth Is Not for Sale: A Path Out of Fossil Capitalism to the Other World That Is Still Possible.* Singapore: World Scientific.

Vogel, Lisa. 2013. *Marxism and the Oppression of Women.* Chicago: Haymarket.

Wallis, Victor. 2018. *Red-Green Revolution: The Politics and Technology of Ecosocialism.* Toronto: Political Animal Press.

WHO. 2020. "Life Expectancy and Healthy Life Expectancy Data by Country." *World Health Organization.* https://apps.who.int/gho/data/node.main.688 (accessed 31 March 2021).

Yaffe, Helen. 2020. *We Are Cuba!* New Haven, CT: Yale University Press.

28
TECHNOLOGY AND ECOSOCIALISM

Victor Wallis

Ecosocialist technology is technology guided by universal human need and by concern for the health of the natural environment. It contrasts with capitalist technology, which is driven above all by the imperatives of cost reduction and profit maximization within a market whose contours – including incentives and subsidies – are shaped by the owners or state agents of capital. The contrast between ecosocialist and capitalist technology appears across all sectors of production and services. I explored the basics of this panorama in an earlier essay (Wallis 2018, 74–92). Here, in sketching out a theoretical overview, I will draw examples from the sectors of transport, housing, and food production and will then reflect on the newest technologies of communication, surveillance, and artificial intelligence.

Capitalist versus ecosocialist technology

Technological advancement is typically associated with reducing the amount of human labour power (and time) needed to accomplish a given task. Other things being equal, such reduction is beneficial. But other things usually are not equal. Less human labour power, except when made possible by animal labour or by such mechanical devices as levers or pulleys or screws or wheels or bicycles, has to be weighed against increased use of all the other available sources of energy, which entail – in the case of the currently dominant modes – unwanted "externalities" such as greenhouse gases, toxic emissions, and dangerous radiation. The various forms of "clean" energy likewise have side effects (or requirements for raw materials, manufacturing, storage, or transmission) that put limits on the scope of their implementation (Magdoff and Williams 2017, 261).

Beyond the labour-saving, time-saving, or cost-reduction considerations, but not unrelated to them, capitalist technology is also driven by the constant search for new marketable commodities. These respond to a whole gamut of needs or desires, often deliberately stimulated, having to do not only (or even mainly) with personal consumption but also with the requirements of administering, protecting, and expanding the power of capital, including by military means (Wallis 2018, 23–26). Thus, even apart from any possible labour-saving features, capitalist technology spawns machinery that facilitates financial speculation, ubiquitous surveillance, and mass killing[1] – as well as razing forests, decimating marine life, exhausting the soil, producing

10.4324/9780429341427-33

unhealthy food and addictive drugs (like opioids), excreting vast amounts of garbage, and sustaining an extraordinarily wasteful system of privatized daily transport.

This is not to deny that some of the technology developed under capitalism has been beneficial. Where this is the case, however, the challenge is to preserve whatever advances it embodies while reshaping its applications so as to stem the boundless plunder of resources and all the noxious externalities. In the case of vital but costly medical technologies, for example, this would mean reducing the frequency of their use by improving the general conditions affecting public health (mental as well as physical), including diet, exercise, working conditions, vacation time, air and water quality, and access to nature. With regard to mechanized transport and other services requiring energy inputs, the idea would be to switch, wherever feasible, from individual to collective usage. Underlying all such projections is the observation that historical advances in technology have not come principally from capitalists, but have more often been coopted by capital – typically on giveaway terms – from the creative input of skilled workers or from publicly funded scientific research (see, e.g., Levine 2018, 116–128). Once the new technologies are firmly in corporate hands, they are, of course, deployed for maximum profitability rather than for the efficient satisfaction of human needs.

Ecosocialist technology is perhaps most simply understood as embodying the opposite of the capitalist model.[2] Nonetheless, there is at least some area of overlap between the two, given that reduction of energy input (per unit of product) will be useful in relation not only to the ecosocialist goal of conserving resources but also, more immediately, to the capitalist objective of cutting costs. It is in this sense that ecosocialist technology embodies a dual strategy of building on capitalist achievements while also rejecting many of their applications and, more generally, being guided by entirely different priorities.

Comparing capitalism with ecosocialism in technological terms, a useful focus is their contrasting notions of efficiency. For capital, the relevant unit in which to measure efficiency is the private firm. An efficient firm is one that reduces its input costs (per unit of output) to a minimum. This requires spending as little as possible for raw materials, for infrastructure, and for labour (in terms of both sustenance and training). In other words, the most efficient capitalist firm is the one whose costs of production are most fully subsidized from without. Nature in some cases directly provides the raw materials and in every case provides the "sinks" into which effluents will be ejected. The state provides public amenities like roads, security services (police, courts, prisons), financial services (including emergency bailouts), and a military arm for securing favorable conditions in other countries and/or for guarding against systemic challenges at home. The state also pays for the education of the workforce and, in countries more enlightened than the US, for its healthcare. The workforce in turn provides its services (including the costs of its own continuous renewal) for as low a wage outlay as the firm can get away with and under working conditions as stressful as considerations of cost reduction may dictate – all of which, unless there are strong unions, signifies a heavy toll on workers' health, personal security, and general well-being. Finally, the firms' cost cutting may impose burdens on the end users of their products, ranging from health issues (in the case of processed foods) to costs of repair (in the case of machines) to time-consuming struggles with user-unfriendly websites or robotic phone operators (in the case of services).

For ecosocialism, by contrast, the unit for measuring efficiency is the totality of the social and natural environment. This whole admits of no externalities: toxins that are emitted thus have, by definition, nowhere else to go; damages done to workers are damages that register in the entire community. Most importantly, the costs of competition, which may entail international conflict, including war, cannot be dismissed as if they were unrelated to the ordinary routines of doing business. The cheap prices that capitalist firms charge for certain consumer goods are not just a tribute to smart engineering; they also reflect a whole apparatus of resource extraction and wage suppression, both internally and internationally. The benefit derived from any product must then be assessed not only with reference to the needs it fulfils, but also in terms of what its production subtracts – in the form of depleted reserves, polluted and irradiated surroundings, workplace pain and suffering, or war casualties – from the infrastructure of our existence.

Transportation technology and housing

How will the battle of the contending technologies – capitalist and ecosocialist – play itself out in practice? As we have already noted, the ecosocialist approach has no choice but to take capitalist technology as its point of departure. It will then selectively embrace, add to, or reject the various specific technologies bequeathed by capital. The technologies themselves are in turn defined only in part by particular physical devices; more important is the way the various devices, together with their spatial and physical infrastructures, fit into a larger configuration which determines how they will be deployed (in what quantities, in what proportions, and with what implications for the natural environment and for social life).

Transportation technology is a case in point. Every type of device, from bicycles to cars to trains to planes, is already in existence,[3] but system configuration based on market pressures and corporate lobbying is sharply skewed – especially in the United States, where capital is least restrained – to favor planes and cars over trains and bikes. Any move to invert this pattern would necessarily involve shifts not only in the spatial layout of human settlements but also in the cultural assumptions guiding both public policymaking and personal choices. There is a long tradition of debate on these issues, going back to early critiques of the automobile (Mumford 1963), extending forward to unmasking the ideological agenda behind suburbanization (Baxandall and Ewen 2000), and now, in recent years, surveying significant steps taken in certain locales to implement healthier, safer, and less wasteful approaches based on free public transit (Dellheim and Prince 2018).

Despite the obvious social as well as ecological benefits of doing away with massive car traffic, the politics of such a shift are daunting. Above all, it flies in the face of market pressures. The interests of capital divert the discussion – and the outcomes – along a number of axes. First of all, the automobile-petroleum complex itself, despite its clear responsibility for catastrophic climate change, continues to rush full speed ahead with its drilling (including fracking), its spill-prone pipelines, its highway projects (Weissman and Casale 2019), and its relentless promotion of the car culture. To all this, it has recently added a political campaign to equate acts of resistance (e.g., blocking pipeline construction) with "terrorism" (Cagle 2019). Second, within the more environmentally aware sectors of mainstream debate, attention to remedial measures focuses overwhelmingly on alternative energy sources and fails to question the prodigious consumption of materials and space entailed by the "one person, one car" habit.[4] Third, with specific reference to technology, preservation of the existing privatized system diverts research away from the design of collective alternatives – encompassing urban planning as well as public conveyances – and concentrates instead on devising ever-more-sophisticated gadgets for keeping car owners wedded to their "personal" vehicles.[5]

But transportation technology, despite its powerful impact and its conspicuous presence in our daily lives, is still only one component of a larger complex which, as an aspect of its irrationality, erects formidable obstacles – psychological as well as physical – to its replacement. Parallel to the assumption of privatized motor transport is the complementary phenomenon – amplified by suburbanization – of viewing the home as the principal site for social interaction, as opposed to relying more for this on spaces that are publicly accessible (cf. Oldenburg 1989). The volume of interior residential space then takes on greater importance than it would otherwise have. This, of course, leads to a need for more construction materials as well as more energy for heating and cooling. But it also means that people's social lives will be more exclusively defined by "intentional" contacts and less open to being broadened by the diversity of chance encounters (an effect that is reinforced by residential segregation and by encroachments on public space). Private homes, like private cars, frame a disproportionate share of their owners' waking hours, keeping them more insulated than they would otherwise be from the full range of possible social contacts. This configuration has evolved within a real estate market whose competitive dynamic, by making housing ever more costly, renders increasing numbers of people homeless (see Brenner, Marcuse, and Mayer 2012). The cultural basis for mobilizing around a common alternative approach is thus triply undermined – by the competitive insulation/atomization of those who are housed, the dispersion of those who are unhoused, and the chasm between the two.

Farming and food production

The technological dynamic illustrated in the sectors of transport and housing finds parallels in other dimensions of the battle between capitalism and ecosocialism. As capitalism evolves and the scope for its expansion shrinks, its practices become ever more extreme and severe. This has the dual effect of increasing the appeal (and the urgency) of ecosocialism while at the same time reinforcing all the aggressive and repressive practices that seek to cut off that option (see Giroux 2018). Although the latter response reflects the interests of capital while the pro-socialist approach is grounded in the popular majority (the working class and portions of the middle class), there is a vast overlapping sector of the population in which the antagonistic approaches play out side by side, in the form of contradictions residing within each individual. Simply put, most people recognize – even if not explicitly – the need to pursue an ecosocialist agenda, but our daily habits and practices tend to perpetuate existing arrangements.[6]

This pattern, more than any explicitly articulated strategy, is what capitalism depends on for maintaining its hold on power. The population is so deeply enmeshed in everyday routines that its opinions carry little weight: what we do has more impact than what we say. Technology, in the broad sense in which we're considering it here, has a big role in this. The fixtures of buildings, roads, and cars, whatever their inherent advantages or disadvantages, are already present, and breaking with the routines that they dictate is something that cannot be fully realized through a mass of individual decisions but can only come via a collective (i.e., political) decision to reconfigure, if not the entire complex at first, at least units within it that are large enough to link together all the main dimensions of human activity.

Such a decision must take into account the cultural as well as the material basis on which people are drawn into their acceptance of existing dominant practices. Thus, reliance on cars is conditioned not only by the distances to be traversed and the paucity of public transport, but also by the superficial impression of autonomy conferred by placing oneself in the driver's seat. As driving skills are increasingly rendered superfluous by "driverless" cars, what will be preserved are the accouterments of privacy and, for an important stratum of car owners, the symbols of personal achievement (which include the size and power of a vehicle as well as its comfort and gadgetry). A technology that burdens its users with heavy responsibility and financial pressure – all the costs of car ownership – thus at the same time offers those who can afford it (or who hope they can) an image of luxury and prestige.

A similar inertial dynamic can be found in connection with farming and food. Although quick changes may be somewhat more feasible in matters of diet than with regard to housing and transport, their scope is nonetheless limited – for the majority – both by direct economic pressure (organic foods costing more; poor neighbourhoods dependent on extortionate retailers) and by the addictive properties of salty or sweet processed food and drinks. Majority tastes in food, especially in the United States, are conditioned by the technologies of agribusiness and the capitalist meat and fishing industries, with practices that include (or generate) chemical fertilizers and pesticides, genetically modified crops,[7] over-irrigation, growth hormones, water pollution, air pollution, and the direct depletion of marine life – not to mention the poisoning of the latter by mercury and plastics (see Wallis 2018, 74–77). Not only do these practices undermine public health (Patel 2007); they are also devastating to biodiversity. On the consumer side of the equation, however, we find that unhealthy tastes are reinforced by the routines of daily life, which, under the constraints we have noted, allow most people neither the money nor the time to pursue sounder nutritional practices.

Restoring biodiversity, healthy food, and healthy community life clearly requires a transformation that encompasses technology as well as social relations. What is noteworthy, however, is that the components of the transformed technology are not all new. Many entail a retrieval of methods and practices that had been shunted aside by the priorities of (especially) capitalist agribusiness, with its reliance on crop specialization, heavy machinery, and toxic chemicals (see Magdoff 2015). The implementation of a more locally diversified approach requires a rethinking of the ways in which responsibility for various vital human activities is allocated. While the capitalist labour market assigns people to determinate "jobs," a socialist economy would permit greater flexibility in work assignments. Food need not be grown only in rural areas or only by full-time agricultural workers. Individuals' work responsibilities can be seasonally adjusted where appropriate.[8] People can be educated and encouraged to cultivate small gardens without recourse to chemical inputs. Pollinators and pest predators would once again be able to thrive, and a diversity of plant life would improve water retention, reducing the need for irrigation.

The global relevance of a return to certain older farming practices was recently shown in an unexpected way by scientists studying Cuba's coral reefs. It turns out that this coral, which had been dying prior to the 1990s, at least in part as a result of runoffs from nitrogen-based fertilizers, enjoyed a startling recovery when Cuba switched over to organic agriculture after its chemical imports from the Soviet Union were cut off (BBC Earth 2018).

Information technology, communication, and surveillance

The positive turn suggested by the Cuban case does not, unfortunately, predominate at the present time. Cuba represents the closest current approximation to a socialist society,[9] but while it retains its exemplary significance in certain spheres – public health and education as well as environmental policy and also for fostering an ethic of global solidarity – it enjoys little respite from the relentless campaign of sabotage pursued against it by the US government. The US agenda of isolating Cuba economically continues to have an impact that is in most respects debilitating.[10] It also impedes other societies from familiarizing themselves with Cuban practice. This is pertinent to our present discussion because a transition from capitalism to ecosocialism cannot be expected to take place everywhere at once; it will first need to be implemented in a succession of zones or countries that are big enough to encompass a diversity of major economic sectors. So long as such zones remain (like Cuba) threatened by hostile forces, their value as examples will be partly compromised. Without them, however, the obstacles to implementing a revolutionary alternative would appear more forbidding.

"Information age" technology tends to further reinforce existing structures. Its impact occurs along several dimensions, some of which we have already noted. Most generally, its astonishing achievements create a climate of popular awe at the presumed power of technological ingenuity to solve every imaginable problem. Capitalist interests are then emboldened, for example, to respond to the climate crisis not by reducing emissions or restoring biodiversity, but rather by advancing such risky but potentially lucrative boondoggles as shooting sun-blocking reflectors into the stratosphere (Klein 2014, 256–268). Beyond its cult of spectacular breakthroughs, the new technology bolsters existing power structures directly with its potent brew of ever-more-sophisticated surveillance practices and weapons systems. At the same time, since its basic tools – laptops and smartphones – are widely distributed, it creates an illusion of popular empowerment when, in fact, our almost obligatory utilization of those tools not only subjects us to constant surveillance (see Zuboff 2019), but also increasingly separates us – its users – into narrow, algorithmically linked threads of communication (O'Neil 2017), thereby reinforcing the compartmentalizing impact of the general shrinkage of public space.

Those who are committed to resisting these trends face an awkward dilemma, inasmuch as the fastest and most broadly accessible communications channels for any counterhegemonic project – Google, Facebook, Twitter, YouTube, etc. – are designed, owned, and administered by the emblematic embodiments of corporate capital.[11] These channels admittedly provide platforms which can be used by anyone – and which have indeed enabled an amplification of progressive networking – but the control exercised by their owners over cyberspace is such as to constitute a permanent existential threat to the outreach of any radical entity that would gain enough traction to be seen as challenging capitalist hegemony. This reflects the fact that Google et al. have the power, based on comprehensive monitoring, to establish the algorithms for Internet searches (along with other cyber links) and even to suppress certain displays or sources altogether.[12]

The control function of this apparatus works best, however, if outright suppression is kept to a minimum. More effective than overt restrictions for maintaining ideological hegemony is if the desired conformity emerges as the unplanned outcome of multiple voices, each speaking for itself (Wallis 2018, 189–190).[13] And yet within this capital-controlled framework, which shapes the selection of news for increasing numbers of people, it is unimaginably easy – thanks to there being no effective barriers – to diffuse completely fabricated "news" reports (let alone incitements to racist, misogynist, homophobic, and xenophobic outbursts). Thus, in a study of Twitter "rumor cascades," MIT researchers found that "the fake stories spread about six times faster than the real ones" (Singer and Brooking 2018, 130). What all this points to is widespread cynicism about public affairs, reflecting a sense of powerlessness that is fed by the corporate media's constant celebration (or sometimes bashing) of prominent figures.

Popular engagement with social media builds in part on a long-standing practice, prevalent in commercial culture, of twisting the concept of democracy from one of citizen participation in governance to one of preoccupation with – and disclosures about – the personalities and private lives of key politicians. Technologies that, within a socialist framework, could perhaps be used to facilitate collective governance (Dyer-Witheford, Kjøsen, and Steinhoff 2019, 154) are instead applied to the furtherance of mass distraction (e.g., complex video games), if not goals that are even more misanthropic. Under current conditions of environmental and related emergencies, however, mass distraction is already bad enough as it obstructs the development of creative political alternatives to the status quo. The difficult question for an ecosocialist transition is how social media can be turned into an extension of public space rather than an infringement on it. What arguments and practices can be deployed to cut back the negative and expand the positive uses of cyberspace?

Artificial intelligence and cellular technology

Marxism has displayed, from its beginnings, a dual response to technology. The Communist Manifesto expressed wonder at the achievements of bourgeois industry, but Marx himself, in his other writings, stressed the thoroughly dehumanizing impact of the reigning technology. Lenin, for his part, took the capitalist organization of production as the necessary model for Soviet economic practice (Wallis 2011, 320). Over the past century, however, the socially restrictive – and ultimately anti-revolutionary – impact of the technology bequeathed by capitalism has become increasingly evident, especially in the sense of augmenting centralized controls (including surveillance and the power to manipulate) and magnifying the gap, in the general population, between those who do and those who do not have access to the crucial instruments.

Beyond the high-tech applications that we have already noted, the advances in artificial intelligence (AI) represent a further step in the direction of a technology that seeks to replace human labour.[14] And yet socialist theorizing on AI has not been exclusively negative. Although the original institutional drive behind AI was to enhance the powers of capital, arguments have nonetheless been put forward for integrating high tech (which could include AI) into a socialist agenda. An earlier strand of this thinking involves viewing high tech as ushering in the potentially liberating scenario of a drastic reduction in work time. That vision has tended to lose currency, however, as we have become more aware of the energy costs that would be entailed by the mechanization of all routine work processes.[15] Another line of thought would inject the new technologies into collective decision-making processes (see Meiksins 1998), where they could expedite the sharing of necessary information, thereby facilitating democratic planning both within enterprises and in the economy as a whole.

Overall, however, it is clear that despite any speculative communist visions of AI – such as imagining "a computational 'public utility' . . . subject to democratic control" (Dyer-Witheford, Kjøsen, and Steinhoff 2019, 154) – the system-preserving thrust of AI remains decisive in the near term. It is thus vital, as Dyer-Witheford, Kjøsen, and Steinhoff affirm (2019, 7), "to contest the idea that AI can easily be detached, disentangled and re-appropriated from capitalism." A sinister example of current AI applications are those being developed by the Pentagon (DoD BHPC 2019), which consist essentially of surgical implants that would enable "direct data exchange between human neural networks and microelectronic systems," and thereby, in the words of the Defense Department report, "revolutionize tactical warfighter communications, speed the transfer of knowledge throughout the chain of command, and ultimately dispel the 'fog' of war" (quoted in Britzky 2019). The report includes numerous examples of what can only be termed a project to create (selectively) electronically enhanced human beings. What is striking is that the implants, as Britzky notes, "wouldn't just allow for transmission of human data to machines, but the other way around – machine to brain – as well as human to human interaction."

Whatever might be the applications of AI, we need furthermore to keep in mind the environmental underpinnings of its production (see Wallis 2018, 81–82). What is clear about the infrastructure of high-tech communications is that the more intricate the functions they can perform, the more prodigious the burden they place on the ecosystem and the more alarming their threats to human health. The world's data centres, as of 2015, had "about the same carbon footprint as the airline industry"; moreover, "the amount of energy used by data centres has doubled every four years and is expected to triple in the next ten years" (Bridle 2019, 63).

Ironically, these ultimate embodiments of capitalist development are themselves highly vulnerable to the climate impact that this development has brought:

> Data centres and individual computers generate vast amounts of waste heat, and require corresponding quantities of cooling, from the acres of air conditioning systems on industrial buildings to the fans that cool your laptop. . . . Increased air temperatures bring increased cooling costs – and the possibility of outright failures. . . . In the electromagnetic spectrum, the strength and efficacy of wireless transmission will be reduced as temperatures rise. . . . Wi-Fi, in short, will get worse, not better.
>
> *(Bridle 2019, 61–62)*

In terms of threats to human health, the carcinogenic effects of electronics work are well established (e.g., Hawes and Pellow 2006, 121; Ku 2006, 181). More generally, the long-standing marginalization of research on how our species is affected by electromagnetic fields is now being forcefully contested by scientists as the US government surges forward with the promotion of so-called 5G (fifth-generation) wireless technology, whose longer wavelength (compared to microwaves) requires a dramatic multiplication in the placement of transmission towers. As reported in *Scientific American*:

> The latest cellular technology, 5G, will employ millimeter waves for the first time in addition to microwaves that have been in use for older cellular technologies, 2G through 4G. Given limited reach, 5G will require cell antennas every 100 to 200 meters, exposing many people to millimeter wave radiation. . . . Short-term exposure [to millimeter waves] can have adverse physiological effects in the peripheral nervous system, the immune system and the cardiovascular system. The research suggests that long-term exposure may pose health risks to the skin (e.g., melanoma), the eyes (e.g., ocular melanoma) and the testes (e.g., sterility).
>
> Since 5G is a new technology, there is no research on health effects, so we are "flying blind" to quote a U.S. senator. . . . Meanwhile, we are seeing increases in certain types of head and neck tumors in tumor registries, which may be at least partially attributable to the proliferation of cell phone radiation. These increases are consistent with results from case-control studies of tumor risk in heavy cell phone users.
>
> *(Moskowitz 2019)*

The same report links to a petition challenging the implementation of this technology, which, as of July 2019, had been signed by 252 scientists, who, according to the author (who's the director of the Center for Family and Community Health in the School of Public Health at the University of California, Berkeley), "arguably constitute the majority of experts on the effects of non-ionizing radiation."

Such protests have not slowed down the corporate drive to turn 5G technology into common currency. Japan is already marketing a wearable air conditioner the size of a smartphone, while both Apple and Samsung are now working on a future "6G" phase, touted to enable such feats as downloading 142 Netflix movies in a single second (search "6G" on the internet for continuing updates). Defenders of the constant upgrades cite official agencies such as the World Health Organization (WHO) and, in the US, the Centers for Disease Control and Prevention (CDC) and the Food and Drug Administration (FDA), to affirm that none of this will threaten human health. But a vast body of research suggests otherwise.[16] The findings of this research underscore the importance of several core contentions: (a) the proposed innovations all entail massive increases in energy and resource use; (b) in cases in which technological advances carry *any* risk of producing negative health effects, the precautionary principle should govern; and (c) devices and infrastructures that would transform the conditions of life should in all cases be subject, *at the planning stage*, to full disclosure, informed public debate, and democratic resolution.

Conclusion

From an ecosocialist standpoint, it would seem that the urgency of stopping the technological onrush vastly outweighs any possible speculative benefits to which the innovations in question might lead. Not only has the environmental crisis already attained, for our species, existential proportions (Angus 2016); beyond this, as global warming heats up the electronic instruments, it will undermine further progress even of the technology itself. Although the risks to the technology obviously matter less than the risk to our lives, the prospective technological collapse nonetheless allows us to say that capitalism is bringing failure down upon us even in its own terms. It is self-destructing – not quite in the way that Marx projected (by directly revolutionizing the proletariat) – but nonetheless with ineluctable force.[17] The question is whether a collective awareness of this ongoing collapse will emerge soon enough to significantly accelerate all the other dimensions of the ecosocialist struggle.

What will continue to be vital in motivating people to act is the elaboration of a positive ecosocialist vision. Restoring biodiversity and transforming social relations will remain central. These tasks, as we have seen, can be viewed in their technological dimensions. What the most recent developments show, however, is that despite the acknowledged need to build on and supplement certain achievements of capitalist technology, the need to reject the core assumptions of that technology – to dismantle many of its artifacts and to keep its fossil power sources "in the ground" – is greater and more urgent than was previously thought.

Notes

1 Although capital no longer monopolizes the technology of mass killing, capitalist powers were certainly the first to develop and use it, particularly in the context of colonization. See Lindqvist 2014.

2 In principle, one could say this of simply *socialist* technology. The *eco-* prefix serves to underscore (a) the environmental emergency and (b) the rejection of a "developmentalist" understanding of socialism (see Wallis 2018, 17–23, 122–126; Kovel 2019).

3 Boats and ships constitute a somewhat distinct category, given that our immediate concern here is with the allocation of ground space. For small-scale water transport, however (especially for recreational purposes), wind and paddle power should obviously displace motor power as the latter not only wastes energy but also pollutes the waterways and disrupts wildlife.

4 I do not question the importance of obtaining abundant clean energy. But I do take issue with the way that that goal has been allowed to draw public attention away from all the other dimensions of an ecological agenda. See Foster 2017.

5 On socialist versus capitalist innovation, see Wallis 2008, 230f.

6 Roger Hallam (2019), co-founder of Extinction Rebellion, calls this hypocritical but recognizes that most people, including himself, are complicit and argues that what counts is what we do about it.

7 For a critique of the claims of GM advocates, see McGiffen 2005.

8 Urban agriculture is being practiced in cities such as Havana and Detroit. Seasonal shifting of work assignments was carried out in liberated regions of Spain in the 1930s. Space for urban gardens is already found on rooftops as well as in backyards; more will become available as car usage is cut back.

9 On the socialist basis for Cuba's switch to organic agriculture, see Levins 2005. For a recent assessment of Cuban socialism, see Campbell 2016.

10 The negative effects of the sanctions lie especially in their blocking of other countries' commerce with Cuba. On the other hand, Cuba is surely better off without (for example) US soft drinks and opioids, let alone any broader local presence of US capital.
11 Cf. note 6; this is the classic dilemma of dialectical analysis. See also Wallis 2020, Chapter 3.
12 Many examples of such suppression are given in Tveten 2019. On Facebook, see Conley 2018; on Google, Valovic 2019. Thanks to Norman Solomon for these references.
13 This has long been apparent in the sphere of mass media (Herman and Chomsky 1988).
14 For a prescient introduction to this development, see Rifkin 1995.
15 A further counter to the vision of liberation from labour is the transfer of the most arduous and hazardous manufacturing jobs to countries of the global South.
16 Dr. Martin L. Pall, Professor Emeritus of Biochemistry and Basic Medical Sciences at Washington State University, has published a 90-page annotated compendium of pertinent studies (Pall 2018).
17 A telling (though not unprecedented) example of such self-destruction is the backfiring of an attempt to reduce the mosquito population via genetic modification; the transgenic mosquitoes ended up strengthening the species' resistance (Monzon 2019).

References

Angus, I. 2016. *Facing the Anthropocene: Fossil Capitalism and the Crisis of the Earth-System*. New York: Monthly Review Press.

Baxandall, R., and E. Ewen. 2000. *Picture Windows: How the Suburbs Happened*. New York: Basic Books.

BBC Earth. 2018. *Cuba* (documentary film).

Brenner, N., P. Marcuse, and M. Mayer, eds. 2012. *Cities for People, Not for Profit: Critical Urban Theory and the Right to the City*. New York: Routledge.

Bridle, J. 2019. *New Dark Age: Technology and the End of the Future*. London: Verso.

Britzky, H. 2019. "The Army Wants to Stick Cyborg Implants into Soldiers by 2050 and It's Absolutely Insane." *Task & Purpose*, 27 November. https://taskandpurpose.com/army-cyborg-soldier-2050-study

Cagle, S. 2019. "'Protesters as Terrorists': Growing Number of States Turn Anti-Pipeline Activism into a Crime." *The Guardian*, 8 July.

Campbell, A. 2016. "Updating Cuba's Economic Model: Socialism, Human Development, Markets and Capitalism." *Socialism and Democracy* 30: 3, March.

Conley, J. 2018. "… Facebook's Plan to Act as Gatekeeper of 'Trustworthy' News." www.commondreams.org/news/2018/05/03/progressive-media-outlets-including-one-decry-facebooks-plan-act-gatekeeper

Dellheim, J., and J. Prince, eds. 2018. *Free Public Transit: And Why We Don't Pay to Ride Elevators*. Montreal: Black Rose Books.

DoD BHPC. 2019. *U.S. Department of Defense: Biotechnologies for Health and Human Performance Council, Cyborg Soldier 2050: Human/Machine Fusion and the Implications for the Future of the DoD*. Washington, DC: Department of Defense. https://community.apan.org/wg/tradoc-g2/mad-scientist/m/articles-of-interest/300458

Dyer-Witheford, N., A.M. Kjøsen, and J. Steinhoff. 2019. *Inhuman Power: Artificial Intelligence and the Future of Capitalism*. London: Pluto Press.

Foster, J.B. 2017. "The Long Ecological Revolution." *Monthly Review* 69: 6, November.

Giroux, H.A. 2018. *American Nightmare: Facing the Challenge of Fascism*. San Francisco: City Lights.

Hallam, R. 2019. "Interview." 16 November. www.youtube.com/watch?v=gjbLr_gyOgk

Hawes, A., and D.N. Pellow. 2006. "The Struggle for Occupational Health in Silicon Valley." In *Challenging the Chip: Labor Rights and Environmental Justice in the Global Electronics Industry*, edited by T. Smith, D.A. Sonnenfeld, and D.N. Pellow. Philadelphia: Temple University Press.

Herman, E.S., and N. Chomsky. 1988. *Manufacturing Consent: Political Economy of the Mass Media*. New York: Pantheon.

Klein, N. 2014. *This Changes Everything: Capitalism vs. the Climate*. New York: Simon & Schuster.

Kovel, J. 2019. *The Emergence of Ecosocialism: Collected Essays*. Edited by Q. Saul. New York: 2Leaf Press.

Ku, Yu-ling. 2006. "Human Lives Valued Less Than Dirt [RCA Workers in Taiwan]." In *Challenging the Chip: Labor Rights and Environmental Justice in the Global Electronics Industry*, edited by T. Smith, D.A. Sonnenfeld, and D.N. Pellow. Philadelphia: Temple University Press.

Levine, Y. 2018. *Surveillance Valley: The Secret Military History of the Internet*. New York: Public Affairs.

Levins, R. 2005. "How Cuba Is Going Ecological." *Capitalism Nature Socialism* 16 (3), September.

Lindqvist, S. 2014. *Exterminate All the Brutes*. New York: New Press.

Magdoff, F. 2015. "A Rational Agriculture Is Incompatible with Capitalism." *Monthly Review* 66: 10, March.

Magdoff, F., and C. Williams. 2017. *Creating an Ecological Society: Toward a Revolutionary Transformation*. New York: Monthly Review Press.

McGiffen, S.P. 2005. *Biotechnology: Corporate Power Versus the Public Interest*. London: Pluto Press.

Meiksins, P. 1998. "Work, New Technology, and Capitalism." In *Capitalism and the Information Age: The Political Economy of the Global Communication Revolution*, edited by R.W. McChesney, E.M. Wood, and J.B. Foster. New York: Monthly Review Press.

Monzon, I. 2019. "Scientists Admit Flawed Experiment May Have Created 'Super Mosquitoes'." *International Business Times*, 21 November. www.ibtimes.com/scientists-admit-flawed-experiment-may-have-created-super-mosquitoes-2871328

Moskowitz, J.M. 2019. "We Have No Reason to Believe 5G Is Safe." *Scientific American*, 17 October. https://blogs.scientificamerican.com/observations/we-have-no-reason-to-believe-5g-is-safe/

Mumford, L. 1963. *The Highway and the City: Essays*. New York: Harcourt, Brace & World.

Oldenburg, R. 1989. *The Great Good Place: Cafés, Coffee Shops, Community Centers, Beauty Parlors, General Stores, Bars, Hangouts, and How They Get You Through the Day*. New York: Paragon House.

O'Neil, C. 2017. *Weapons of Math Destruction: How Big Data Increases Inequality and Threatens Democracy*. New York: Broadway Books.

Pall, M.L. 2018. "5G . . . Compelling Evidence for Eight Distinct Types of Great Harm Caused by Electromagnetic Field (EMF) Exposures and the Mechanism That Causes Them." www.jrseco.com/wp-content/uploads/Martin_Pall_PhD_5G_Great_risk_for_EU_US_and_International_Health-Compelling_Evidence.pdf

Patel, R. 2007. *Stuffed and Starved: The Hidden Battle for the World Food System*. Brooklyn, NY: Melville House.

Rifkin, J. 1995. *The End of Work: The Decline of the Global Labor Force and the Dawn of the Post-Market Era*. New York: G.P. Putnam's Sons.

Singer, P.W., and E.T. Brooking. 2018. *LikeWar: The Weaponization of Social Media*. New York: Houghton Mifflin Harcourt.

Tveten, J. 2019. "US Media Lament Internet Censorship – in China, Not US." 16 April. https://fair.org/home/us-media-lament-internet-censorship-in-china-not-us/

Valovic, T. 2019. "Google Becomes Evil." www.commondreams.org/views/2019/09/22/google-becomes-evil

Wallis, V. 2008. "Innovation." *Historical Materialism* 16 (3).

Wallis, V. 2011. "Electrification." *Historical Materialism* 19 (1).

Wallis, V. 2018. *Red-Green Revolution: The Politics and Technology of Ecosocialism*. Toronto: Political Animal Press.

Wallis, V. 2020. *Socialist Practice: Histories and Theories*. London: Palgrave Macmillan.

Weissman, G., and M. Casale. 2019. "Highway Boondoggles 2019: Nine Projects That Should Never Happen." https://usa.streetsblog.org/2019/06/18/

Zuboff, S. 2019. *The Age of Surveillance Capitalism*. New York: Public Affairs.

29

MODEL LEARNING SPACES TOWARDS TRANSITIONING TO ECOSOCIALISM

Miguel Angel Núñez

Neoliberalism versus another possible civilization

The political events of the last months of 2019 in several Latin American countries show how the process of building another kind of civilization moves forwards and backwards. The coup in Bolivia; the fierce popular awakening Chile; the new fourth Mexican transformation; hopeful possibilities in Argentina; the betrayal in Ecuador; the faltering steps in Uruguay; the discrediting of the Colombian narco-pseudo-democracy; and government corruption scandals in Paraguay, Peru, and Brazil all show that neoliberal regimes are heading into roads with no return. Ecologically, and as an unprecedented international dispute, this road of no return is particularly obvious with the Brazilian government's intention of definitively surrendering the Amazon's riches through its destruction. More and more people are convinced that the neoliberal extractivist model has been exhausted, that it has been a failure, and that it cannot continue to be applied. This loss of credibility is due to the fact that the extractivist model produces no guarantee of security and even less of social welfare for most of the affected communities. It serves only the interests of a privileged minority who impose social inequality among our peoples.

We live at a time when more people are explicitly working towards achieving autonomy and seeking progressive justice and social equality, which allow the enjoyment of the same levels of human dignity for all. The struggle is about knowing how to continue with our survival as a species and strive for self-preservation with the few natural resources we have left by advancing other possible forms of civilization that we dream of building. From various perspectives, visions, and ideologies, several scholars and researchers have developed proposals for a new social contract or alternative civilizational model. For example, to Hathaway and Boff (2014), key characteristics for an alternative vision are sustainability; justice and economic equity; biological diversity; being rooted in place; self-reliance and openness; democracy, participation, and subsidiarity; cooperative self-organization; sharing of knowledge and wisdom; responsibility and rights; and balance. Magdoff and Williams (2017) dedicate a couple of chapters to delineating an ecological society. They add the following to the features listed: providing a full life to all, planning for people and the environment, worker- and community-based control of workplace and resources, worker rotation and responsibilities, and coordination of inter-regional relations to mutual benefit. Ecofeminists Prats, Herrero, and Torrego (2016) provide a substantive contribution to such civilizational alternatives towards making changes to socially

10.4324/9780429341427-34

constitutive paradigms, pointing out that there are complex transition processes in different areas of knowledge that need to be reconciled. They highlight the need to favor the predominance of citizens' democratic power based on fundamental principles such as the defense of life within the framework of a relationship between social evolution and nature. Other principles helpful towards a paradigmatic shift are the following: a long-term systemic vision; sufficiency; resilient biomimicry; circular closures of life cycles; eco-efficiency; integrated management of life systems; caution; an open, multilevel, and global society based on the spatial proximity of production-consumption cycles and local-regional democracy; and innovative governance and legal-political frameworks consistent with the coevolution of society and nature.

From a theological perspective, Pope Francis's Ecological Encyclical, *Laudato Si*, is oriented to a new ecological paradigm, according to which all beings are interdependent and related. That is to say, a problem and its solution cannot be approached without thinking and including others, whether human or not. In this way, it overcomes the anthropocentrism that imagines that all beings have value only to the extent that they are used by human beings. Pope Francis reaffirms the intrinsic value of each being because it reveals something of the mystery of the universe and entails a revelation of the creator. Finally, as a holistic vision of interdependence, the Action Plan of the 1st Eco-Socialist International (2017) is defined according to interrelated space-time scales. This is another alternative paradigm proposal that takes into account the characteristics of ancestral Indigenous peoples' worldviews, such as the principles of "good living" (*sumak kawsay*) and "living well" (*sumak qamaña*) among Andean peoples.

Ecosocialist Venezuela

The Venezuelan revolutionary process has been evolving from such and other perspectives. Since June 2012, the process has been taking on a new momentum and content, emerging through the Patriotic Plan and its 5th Historical Objective, which is dedicated to saving planet earth and preserving our species. In 2019, the plan was enshrined in constitutional law (Constitutional Law of the National Patriotic Plan) and is the basis for the 2019–2025 Bolivarian government programme.[1] It stands out as a new civilizational trajectory, proper to its genre and unique in the region. This new civilizational proposal, which many Venezuelans have called ecosocialism, has become a political and economic reference point in Latin America and departs definitively from neoliberal recipes, which have manifestly failed.

The context of this Bolivarian revolutionary programme must also be understood in the context of constant US belligerence and internal sabotage since the first election of Hugo Chavez. This includes, more recently, former US President Obama's 2015 decree on Venezuela as an "unusual and extraordinary threat," followed by the succeeding US administration's real military threats. Harm from interventionism and economic warfare by the US and its government allies is compounded by sabotage in all aspects of Venezuelan society by internal reactionary forces. These have been waging permanent terrorist violence, abetted by aggressive international and national media campaigns to discredit and demonize the new Bolivarian vision. The consequence has been that Venezuelans live under conditions of hardship and economic, social, and political uncertainty. Despite these attempts at undermining the social fabric of Venezuela, it has been possible to secure basic resources for different production sectors, acquire medicines, and produce and distribute food, while Chavismo, as a solid set of principles and vision of the future, has continued to play a very important role in resisting a severe, unfair, and induced economic war supported by extreme economic sanctions. This is where ecosocialism is situated as a new civilizational initiative.

On ecosocialist principles

Ecosocialist literature has grown and relies on different areas of knowledge and practices, both in the social and ecological sciences. Drawing from such work, I conceptualize ecosocialism in terms of six guiding principles for action: (a) self-containment, (b) caution; (c) inter-culturalism, (d) eco-ethics, (e) social equity, and (f) participation. I have expressed them as a complex process of social, economic, scientific, technological, and political transition, which is tied to a high motivational power being exercised in collective and daily action, taking into account the ecological (eco-regional) context, being consubstantiated with intercultural rationality to transcend the moral field and build an eco-ethic based on the complexities between society and nature. Such principles must be present in production processes and in the orientation of the interrelations and interactions represented by the implementation of different strategies and actions. The six principles are shared, evaluated, and discussed among followers of ecosocialism and can be used especially as a methodology, which I herein call the making of spaces of ecosocialist learning. This methodology has been developed, shared, and applied through different workshops, courses, and seminars where ecosocialism is taught and practiced in Venezuela. I refer to them as ecosocialist learning spaces because any social management process and its reproduction imply several aspects: a production and service experience; technical innovation; the production of knowledge, including scientific; and people's knowledge and democratic participation, which must be systemic and consolidated.

Ecosocialist learning spaces

In Figure 29.1, STE-1 (the making of learning spaces for an ecosocialist transition) represents an approximation of how the dynamics of action and reflection can be conceived in the

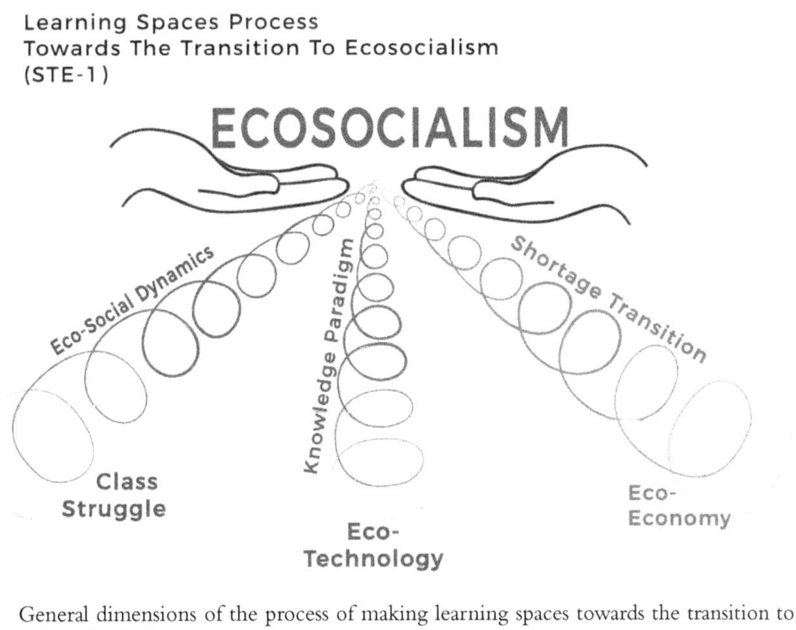

Figure 29.1 General dimensions of the process of making learning spaces towards the transition to ecosocialism (STE-1)

construction and operation of the learning spaces of ecosocialism. Social processes that emerge with their local agendas, from the bottom up, provide daily stimulations and input for social movements. We have characterized some initiatives of work, struggles, and integration as an open spiral called "spaces of ecosocialist learning," explained in detail later.

There are three dimensions that build the bases of ecosocialism. One is defined by ecosocial dynamics, identified by environmental and social tensions in the heat of class struggle, to advance and provide for vital human needs. These dynamics are linked to a communal or community reality on a human scale and energized by diverse social and alternative movements. Another dimension is a scarcity-related transition. This is made up of new eco-economic actions or dynamics that different and emerging movements are already putting into practice within and between communities or communes (*comunas*) and that have as a priority the movement towards ecosocialism. The third dimension is a knowledge paradigm. This is manifested in the different eco-technologies that have been consolidated in existing productive spaces and that respond to emerging social and technological issues that continue to arise and may need to be urgently taken care of.

Figure 29.2 shows examples of the kinds of component actions and activities, or eco-patterns, that subtend and define each type of learning-space process (STE-2). These eco-patterns are viewed as inter-related, and a community or working group identifies and incorporates eco-patterns of relevance to them. The model shown on Figure 29.2 is a way to describe the dynamics involved in the construction of or transition to ecosocialism. Some of the components (eco-patterns) are also quantifiable, as discussed next. In applying this methodology in a

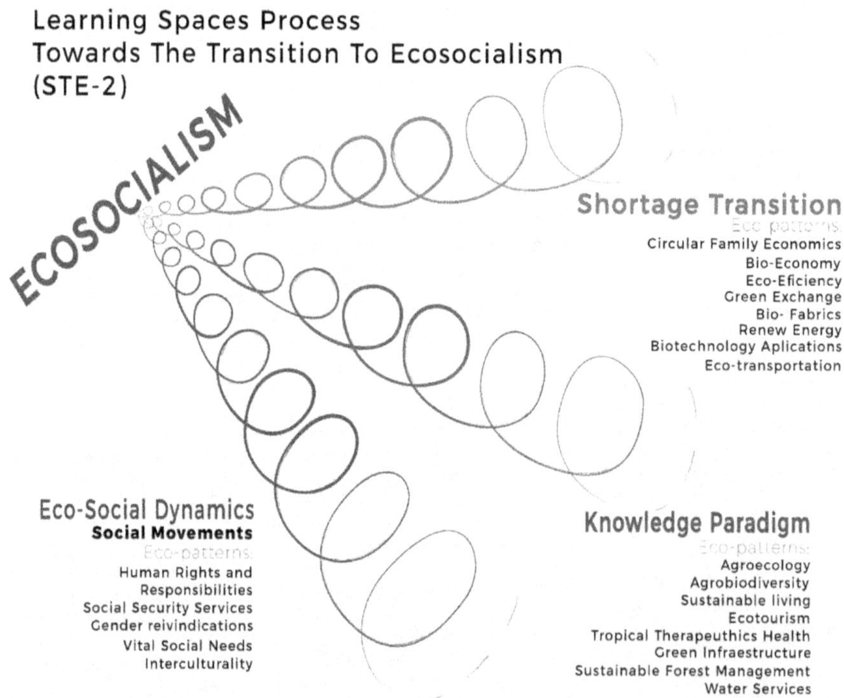

Figure 29.2 Examples of processes involved in making learning spaces towards the transition to ecosocialism (STE-2)

Table 29.1 Eco-patterns used in the methodology

Participation of communities in the practice of ecosocialism							
Eco-patttern	*Community names*	*Locations*	*What actions runs to be taken*	*What ecosocialist principles apply?*	*What methods are to be used?*	*How do they do that?*	*Achievements and advances in what places?*
Water quality							
Eco-efficiency							
Agroecology							
Agrobiodiversity							
Green exchange							
Eco-tourism							
Number of employees							

community, it is working groups who are part of that community who choose and prioritize the most important eco-patterns for themselves. As a way of being able to assess the combined effects of various eco-patterns, each eco-pattern is given a numerical value representing relative importance in the community's judgement. The formation of these learning spaces (where workgroups identify and discuss eco-patterns) starts with participants' agreeing to define and contextualize eco-patterns meaningful to them in their community. In other words, communities or workgroups within communities freely define eco-patterns according to the dynamics of their lived reality.

Each workgroup or community identifies and organizes eco-patterns progressively. For instance, an eco-pattern can be defined according to whether actions: (a) contribute to generating technologies and an internally coherent new economy helpful in overcoming the social conflicts generated by capitalism, contributing thereby to the formation of a new social fabric; (b) allow for the establishment of dynamic ecosocial models that draw from natural and cultural heritage to achieve local social and environmental benefits, where environmental education contributes to conserving and protecting biodiversity; and (c) allow the characterization and valorization of different transition processes towards ecosocialism. Table 29.1 shows a way in which the information from discussions on eco-patterns is registered. It contains some examples of eco-patterns that have been selected through deliberations by workgroups or communities and the specific characteristics to be evaluated. The eco-patterns shown in Table 29.1 turn out to be those most commonly used in meetings with communities in my experience, and it is expected that community members living in other contexts may come up with additional or entirely different eco-patterns.

Advantages of the eco-pattern methodology

We conceptualize this approach as a set of inter-related methods, where a community's condition is highlighted as a factor of leadership, permanence, and relevance, resulting from the continuous nature of learning spaces. Table 29.2 provides a summary of methods and types of information collected from Table 29.1. They include dates of completion, states, the communities selected, and exercise location. Likewise, selected topics are listed that are derived from the selection and quantification of eco-patterns chosen by the participants. Except for the

Table 29.2 Five states in Venezuela where the method has been tried out

Summary of the application of the learning spaces methodology						
State	*Dates*	*Community and institution*	*Issues*	*Number of eco-patterns discussed*	*Number of quantified eco-patterns*	*Number of Participants*
Barinas	Sep 2019	IALA in several established konucos	Food productions	5	Number of crops (agrodiversity)	34 producers
Capital District	June 2018	Canteras-Antimano	Ecological rehabilitation	3	Number of sown cultivars and number of remediated spaces	14 workers
Mérida	Sep 2017	Lomade pueblo	Agroecology	9	Number of crops (agrodiversity)	14 producers
Mérida	Sep 2017	Mano a mano	Agroecology	8	Number of crops (agrodiversity) and green exchange	36 producers
Miranda	Oct 2018	Chaguaramas Altagracia-Montaña	Agroecology	5	Number of crops (agrodiversity)	23 producers
Zulia	Sep 2017	Congo Mirador-Puerto Concha	Eco-tourism	32	Number of new jobs	300 fishing families in two schools with 4th-, 5th-, and 6th-grade students

Congo Mirador community, where more than 300 families and a good group of schoolchildren were involved, some of the 121 participants have been collaborating in the compilation of the requested data and, of course, they are community members destined to lead the spaces of ecosocialist learning.

Table 29.2 presents five states in Venezuela where the method has been tried: six production and community organizations and a government institution, the Agroecological Latin American University Paulo Freire (IALA), whose students and peasant organizations are involved in continuing their process. IALA's contributions have been systematized by defining ecosocialist learning spaces through agrodiversity measurements in some konucos. The topics most discussed were food production, agroecology, ecological remediation, and eco-tourism.

As can be seen in Table 29.2, the agrodiversity eco-pattern was the most prominent and represents the konuco as the primary food production modality for peasant households. This coincides with the Venezuelan government decree on wartime crops, which constitutes the primary food production of peasant households. According to decree No. 16 (Official Gazette No. 6450 of 17 April 2019), the crops covered are bananas, beans, corn, auyama, cassava, quinchoncho, sweet potatoes, potatoes, beans, ocumo, and aji, as well as calves of goat, poultry, rabbits, sheep, and pigs. These are the kinds of foods primarily produced by konuqueros. In these production situations, these crops and livestock should comprise the bulk of food produced. In other words, the aim is to raise crop and livestock diversity per unit of land.

Table 29.2 records 107 farmers who work between 6 and 42 crop associations (intercropping) in different production areas. We have valued the eco-pattern of agrodiversity as the most prominent in the communities studied. This konuquera production modality, in which agrodiversity is valued as a natural principle, has several implications. According to Nichols and Altieri (2013), agrodiversity is a condition that makes agroecosystems more resilient in the face of climatic change. The greater the number of crop associations, the greater the self-regulative capacity of a productive agroecosystem.

Some reflections on the methodology's application

Out the community exercises carried out, most of the eco-patterns used come from those indicated in the models represented in Figures 29.1 and 29.2. As shown in Table 29.1, the exercises generated much significant and detailed information for discussion and deeper analysis. The knowledge and wisdom shown by the participating communities shows that the methodology is flexible and participatory and tends to give coherence to the groups' work process. Likewise, the methodology expressed a teaching dynamic sensitized to the environment and to each specific situation. A lot of information was exchanged, which is a positive, but it should be summarized to enable a more accurate discernment of eco-patterns. We thus plan to make further adjustments to the methodology in future. It is evident that while eco-pattern identification makes for a more precise description of a situation, its interpretation is more appropriate if done according to the evaluated context and integrated with other eco-patterns. For example, in the case of food production, peasant konukos clearly demonstrate to us the eco-pattern of agrodiversity. This can be related, as shown on Figure 29.2, to vital social needs, healthy food, and social security and its relationship with the eco-technologies expressed in agrodiversity-agroecology. On the other side of the STE-2 spiral, there is also a connection brought to light regarding the family economy, eco-efficiency, green exchange, and many others that could be selected and as appropriate to community dynamics. These relationships may arise as the exercise continues, in a synergistic interactive process. Due to the richness of the debates and the immense amount of information that is collected in these work meetings, it is clear to us that the notion or concept of eco-patterns has several notional, political, and organizational connotations, which must be considered and incorporated into contributions to theoretical reflections and practices in the processes of transition towards ecosocialism. Eco-patterns are perceived as readings and effective interpretations of different behaviours within a community and specific eco-pattern components as played out among different social and environmental tensions. Simultaneously and sequentially, these permanent tensions lead to a whole series of relationships and interactions between them until specific goals and objectives are achieved. Several participants consider that, even if the concept of eco-patterns is true, it is useful to know that the behaviour of the many relationships that can be brought to light. It is also true that, as the approach is systematized, it raises the level of knowledge and evolves so that other types of eco-patterns may come to light. These are at the level of the group, the individual, and the collective, and they can become manifest in linear or cyclic ways. To reiterate, on the long road to ecosocialism, eco-patterns can give us an estimate of the evolution and transition states of a community, locality, or workgroup. Some participants have suggested that in the representations, descriptions, analyses, and eco-pattern identification offered through our model (Figures 29.1 and 29.2), some of the components for the creation of a scientific model are combined, where the reality created in the spaces is being determined by means of ecosocialist learning. This suggestion allows us to continue to

Table 29.3 Ecosocialist policies

Criterion	*Policies*
Wisdom	Application of a plurality of wisdom
Strategies	Multi-dimensional
Organization	Horizontal
Freedom-liberty	Built-in through participation
Coexistence	Ideological tolerance
Quality of life	Adapted to natural conditions
Interculturality	Revalorization of cultural diversity
Ideology	Combatting consumerism
Legal	Human rights compliance
Eco-economy	Circular, social, and solidarity
Technology	Energy efficiency

deepen the theoretical basis of our version of ecosocialism. Throughout our meetings so far, the importance of the ecosocialist principles described earlier has been discussed, and this is helpful in elaborating and supporting proposals for ecosocialist policies (Table 29.3).

The approach and methodology we have been proposing and the adjustments that may well arise through implementation help us perceive how ecosocialism can be and is being built. The exercise aids in defining spaces of ecosocialist learning by specifying the most immediate demands from our communities for the advancement of ecosocialism.

Final thoughts

The 2019 revolts in several Latin American countries may have shared an oppositional stance to extractivist neoliberalism, even if they were also very different kinds of revolts. Yet what is happening is actually not exceptional. It is a global phenomenon that not only responds to climate change but that also rejects "normal" or "traditional" institutions and mainstream political parties and the ways in which they have been confronting the structural problems that define our collective human future. In this, conventional political parties have lost legitimacy, and people do not feel represented by them. Political parties are reduced to politicians and state reformism. People are rejecting such parties and do not believe in them, seeing in them a political class who have locked themselves in. They time and again come up with few alternatives amounting to little more than an occasional electoral market and offering examples of corrupt ways to get rich. Citizens have lost their trust in their governments, parliamentarians, ministers, and presidents. Governments have lost political legitimacy and have generated a fragmented and chaotic transformation of politics. Therefore, people can't stand it anymore; they get tired and rebel, often expressing their feelings through various forms of violence. Spaces emerge where provocateurs, infiltrators, looters, and criminals are found alongside snipers and militias involved in vandalism and organized looting. These expressions of violence do not correspond to the essence, responsibilities, and functions of the social movements that have led such rebellions.

In contrast to political parties, social movements are indicating a search for new cultural values and principles, articulated with technical-political proposals. They have already been presenting and even practicing solutions to many of the demands and problems that the state has not yet been able to resolve. These solutions range from the replacement of fossil energies with renewables to agroecology and bio-economy, which are the basis for the eco-patterns model introduced and discussed in this chapter. We believe that, in addition to continuing to push for these eco-alternatives, there must also be other measures aimed at curbing the obscene benefits to powerful interests and their political influence.

Ecosocialism in the face of the global ecosocial-climatological crisis requires placing as another central element the redistribution of wealth and the structural problems that occur to us. We must guard against the risk that our efforts just become another type of reformism, generating benefits for a few and not for the majority that are claiming it.

Ecosocialism and its advancement indicate possibilities for building a new kind of civilization, governed by eco-ethical principles founded on honesty, simplicity, and sufficiency. These ethics imply that one must speak to people and especially the youth in a transparent way, encouraging them to create and consolidate learning spaces where ways of living are developed that are based on lowering consumption rates, preventing pollution, raising everyone's well-being, and increasing available time for relationships and participatory democracy. As expressed in the different chapters of this *Handbook*, ecosocialism is a proposal for a new model of civilization in permanent construction, seeking to replace traditional notion of human relationships as economic, social, cultural, and political entities isolated from the natural environment with the integration of humans within and as part of the natural world. This is fundamental to building mutual understanding and weighing the current impacts of our actions relative to the future. This new model implies practicing true independence and sovereignty, aiming to promote collective structural transformation processes to effect positive changes for both society and nature.

Note

1 See www.mppp.gob.ve/wp-content/uploads/2019/04/Plan-Patria-2019-2025.pdf (accessed 14 April 2021).

References

Hathaway, Mark, and Leonardo Boff. 2014. *El Tao de la Liberación* [*The Tao of Liberation*]. Madrid: Trotta.

Magdoff, Fred, and Chris Williams. 2017. *Creating an Ecological Society: Toward a Revolutionary Transformation*. New York: Monthly Review Press.

Nichols, Clara I., and Miguel A. Altieri. 2013. *Agroecologia y Cambios Climáticos* [*Agroecology and Cliimate Change*]. Lima: RedAgres, Cyted, and Socla.

Prats, Fernando, Yayo Herrero, and Alicia Torrego. 2016. *La Gran Encrucijada: Sobre la Crisis Ecosocial y el Cambio de Ciclo Histórico* [*The Great Crossroads: On Ecosocial Crisis and Historical Cyclical Change*]. Madrid: Libros en Acción.

30

ECOSOCIALISM AND THE GREEN NEW DEAL

Natasha Heenan

Introduction

This chapter will consider the opportunities presented by the emergent politics of the Green New Deal (henceforth GND) for the wider dissemination and discussion of ecosocialist ideas and praxis. The first section introduces the politics of the GND and sketches an ecosocialist orientation to it. The second section provides an overview of existing social and political movements for an ecosocialist GND and considers debates over political strategy. The final section evaluates key issues and critiques of the GND from an ecosocialist perspective. Ecosocialists face a choice between critically engaging with and promoting a popular, socialist politics for the GND or missing an unprecedented opportunity to bring our ideas out of academic obscurity and into the mainstream.

The present struggle

We find ourselves at a conjuncture where the contradictions of capitalism can no longer be sustained without making it clear that our lives are being sacrificed at the altar of capital accumulation. That the cracks in the system are beginning to reveal themselves to more people does not make a revolution or transition to socialism inevitable. If anything, situated as we are at the end of a long period of global economic stagnation and accelerating ecological breakdown, a future of barbarism in the form of a nascent ecofascism seems equally likely. The exploitation of labour and the rest of nature has caused irreparable damage and loss of human and non-human life. We have been fighting on too many fronts for far too long: for living wages, for decent housing, for clean air and water, for land rights, for an end to incarceration, for freedom of movement, for just cities and food sovereignty. Mainstream economic debates remain focused on how to tweak the system to avoid sub-optimal outcomes like inequality, while liberal environmentalists uncritically support market mechanisms for climate change and dream of a technocratic, capitalist-led energy transition. Even today, both orthodox economists and environmentalists fail (or are unwilling) to see the class struggle for the trees.

In the global labour movement, the demand for a *just transition* has existed for decades while successful examples of it remain few and far between, and climate change has been used as a wedge by employers to pit workers against each other (Russell 2018). Radical parts

10.4324/9780429341427-35

of the climate movement have long called for *climate justice*, while big ENGOs are unable to move beyond unspecified and vague calls for climate action. Notably, this "climate action" is sought from a range of actors not limited to states but including factions of capital for whom climate change is opening up a new frontier of accumulation. Over the past few decades, these social movements have surged and waned, in particular the upsurge and subsequent fading of the climate movement prior to 2009 (Rosewarne, Goodman, and Pearse 2014). In the vast majority of countries, labour unions have been steadily weakening since the 1980s (OECD 2020).

However, recent strike waves concentrated in the US education sector provide some cause for hope of a revival of social movement unionism, as seen in teachers' demands for wages alongside critiques of privatization (Tarlau 2019). Just as impressive is the new global movement of strikes for the climate spurred by young women and the indignation of school students who face a future of catastrophic climate breakdown. The year 2019 also witnessed an upsurge in struggles over social reproduction in various places around the world, including protests against transport and fuel fee increases in Brazil and Ecuador and cuts to pensions in France. These insurgencies are occurring against a backdrop of widespread and deepening economic precarity and social fragmentation. While spurring revolutionary action, this has also laid the groundwork for the rise of Right-wing populism from the US to the Philippines, India, and Brazil. In response to climate change, various states seek to balance legitimacy and accumulation by committing to inadequate emissions targets and meaningless climate emergency declarations while others rely on increasingly authoritarian moves to silence dissent (Bruff 2014).

Enter the Green New Deal

The recent re-emergence and popularity of the concept of a GND, beginning with the Sunrise Movement in 2017 and culminating in a resolution for a ten-year climate mobilization introduced to the United States Congress by Alexandria Ocasio-Cortez in early 2019, signals that the time is ripe for a popular climate politics that can connect struggles over labour, nature, and social reproduction. Ecosocialists, who have long stressed the importance of the relations between the personal, communal, and ecological contradictions in capitalism, are well placed to embrace such a moment. The resolution outlined a plan to create millions of green jobs in service of decarbonizing the US economy by investing heavily in an expansion of renewable energy, public housing, transport, infrastructure, health, and education. The text also invoked both the labour and climate movements' demands for a *just transition* and *climate justice*, placing them together in a concrete plan to transition to a low-carbon future. This is significant in that it signals a rejection of the jobs-versus-environment impasse and shifts political terrain by making the case for a just transition on an economy-wide scale, not just for one industry or worksite.

The 14-page resolution covers issues spanning agricultural land use, universal healthcare, and access to nature. It also begins with an acknowledgement that climate change is one of several related crises and that "frontline communities," including "indigenous peoples, communities of color, migrant communities, [and] deindustrialized communities" are already and will continue suffering the impacts of climate change disproportionately (Ocasio-Cortez 2019a, 4). The rights of labour are also in focus, and the GND calls for the creation of good union jobs, reforms to increase the collective bargaining power of workers, and "wage and benefit parity for workers affected by the transition" (Ocasio-Cortez 2019a, 12). Notably absent are any mechanisms by which the fossil fuel industry would be held accountable for the climate crisis. The resolution also does not rule out the use of nuclear power; however, a fact sheet released by Ocasio-Cortez's office includes stronger language against nuclear (Ocasio-Cortez 2019b, 2).

If the purpose of the New Deal was to restart the engine of capitalist growth following the Great Depression, the GND might be seen as a response to the lingering economic impacts of the global financial crisis (GFC) and looming climate breakdown. It is also a pre-emptive move, one that anticipates the Right's exhaustion of climate denial rhetoric and the potential for highly conservative responses to climate change once the tactics of climate denial have been abandoned. As Klein (2017) has argued, states and capital routinely seize on moments of crisis to restructure economies in ways that allow for new rounds of accumulation. In the early weeks of 2020, the Right-wing Australian Prime Minister Scott Morrison responded to the bushfire crisis by arguing that building dams and clearing forests was "climate action" (Morrison 2020). Vague calls for "climate action" are easily deployed in reactionary ways, divorced from climate justice and any critical analysis of the root causes of the crisis. Action on climate including appeals to science, far from being politically neutral, can be shaped by technocratic, capitalist, and ecofascist ideas. The importance and centrality of the notion of climate justice to the GND therefore cannot be overstated. Even more important is the need to ground discussion of the GND in ecosocialist ideas, beginning with an analysis of the interconnected crises we face as fundamental *crises of value* in capitalist social relations of (re)production.

An ecosocialist orientation to the Green New Deal

A decade ago, David Schwartzman (2011, 56) concluded that "the prospect of a Green New Deal gives us a powerful wedge if we choose to use it." The argument I want to make here echoes that call for a strategic approach to the politics of the GND. The GND provides a new opening to further advance and popularize a structural critique of "climate-changing capitalism" (Bryant 2019). Contrary to the orthodox economic understanding of climate change as a market failure (i.e., a bug in an otherwise functioning system), the production of nature in capitalism requires widespread socio-ecological destruction (Stern 2007; Smith 1984). Both ecosocialists and ecofeminists have argued that the extraction of surplus value in the labour process (including social reproduction) and production for exchange relies on the exploitation of human beings and the rest of nature, effectively alienating people from nature and one another (Mies 1986; Salleh 1997; O'Connor 1998; Foster 2002; Giacomini et al. 2018). The strength of the ecosocialist critique is in the way it links this exploitation of labour and nature, revealing that they are two sides of the same coin.

Appraising the Green New Deal from an ecosocialist perspective requires us to analyze the political and material conditions we are faced with today. Climate change threatens to make large parts of the earth uninhabitable by the end of the twenty-first century (IPCC 2018). Additionally, the dangers of a warming world have been known for decades (Garnaut 2008). During that time, the strategies employed by environmentalists, climate activists, various UN bodies, and the scientific community have not resulted in a response anywhere near approaching the scale that is necessary to confront such a challenge. In addition, Saed (2019) observes that even if we succeed at arriving at our ecosocialist destination in the near future, the conditions of ecological breakdown, including climate change, will not suddenly cease to exist. The consequences of capitalism are not bound to the past and present but stretch out like a relentless fog, tendrils creeping into our socialist future. This makes the task of winning an ecosocialist world all the more urgent and demands a serious engagement with the politics of the GND.

Discussion of the GND should not be limited to the proposals laid out in any one document, and the global politics of the GND remains an open and contested field. We should be clear that the GND we seek looks very different from technocratic plans that embrace market mechanisms and nuclear power and ignore the global unevenness of capitalist development and climate change. The Ecosocialist Working Group of the Democratic Socialists of America has begun this work already, devising a set of ecosocialist guiding principles for the GND (EWG-DSA 2019). They are as follows:

1 Decarbonize the economy fully by 2030
2 Democratize control over major energy systems and resources
3 Centre the working class in a just transition
4 Decommodify survival
5 Reinvent our communities to serve people and planet, not profit
6 Demilitarize, decolonize, and strive for a future of international solidarity and cooperation
7 Redistribute resources from the worst polluters

They are also clear about the political strategy behind ecosocialist support for a GND, stating that:

> Creating a fully ecological society will require a revolutionary transformation to replace the capitalist social order based on exploitation and oppression with a new society based on cooperation, equity, and justice. A Green New Deal must serve as a bridge toward this future.
>
> *(EWGDSA 2019)*

These principles and the strategy of a transitional demand or non-reformist reform deserve serious consideration if we are committed to the spread of ecosocialist ideas. James O'Connor (1998) argued that in undermining the conditions of production, capitalism would create social, ecological, and economic crises: ruptures in the fabric of social relations. These ruptures can be opportunities for capitalist "fixes" and state restructuring to lay the path for further accumulation, but equally, they can provide openings for new social movements to contest the stranglehold capital holds over life. Today, children striking for the climate, prison abolitionists, workers picketing for fair pay and conditions, tenant organizers, and drought-affected communities organizing mutual aid all struggle in response to a series of interconnected crises, to reproduce life against capital. An ecosocialist GND is a chance to unite these struggles against a common enemy, for a common cause.

A key area for ecosocialist researchers and activists to develop is the reimagining of work and social reproduction. For some, this looks like a job guarantee and collectivizing unpaid reproductive work; for others it requires a demand for a universal basic income (Weeks 2011). Both proposals have merits and problems, and ecosocialists should consider what ground there is between workerism and post-work to demand secure incomes for all. Given that dreams of full automation are just that, how do we organize and distribute socially necessary labour? Is there a path to worker ownership of the means of production via full employment? Old debates about the sources of value in capitalism, and therefore the revolutionary subjects capable of overthrowing it, must be refurbished through materialist analysis of the present conditions with which we are faced.

Instead of work that is meaningless and destructive, the future of work in a warming world can attend to the care and regeneration of people and planet, repairing our frayed social fabric and the scarred ecologies in which we are embedded. How we utilize technology in order to aid large reductions in working time is something that should be determined democratically, not by the market. The realm of individual freedom represented in that hard-won eight hours each day for "what we will" should be expanded and extended to the people, many of them women, who have never really had it. Beyond work, what would a GND based on the plurality of ecosocialist concepts, including *buen vivir*, commoning, and public luxury, look like? Scholars and writers like Alyssa Battistoni and Thea Riofrancos (see Guy 2019), Nick Estes (2019), Tithi Bhattacharya (2019), Leah Temper and Sam Bliss (2019) have begun the work of imagining what an ecosocialist, feminist, and anti-colonial approach to the GND might look like, in detail. Central to these writings are the principles of decommodification, decolonization, decarbonization, and democratization.

Cooperation on a global scale through international solidarity across the working class would require the dismantling of violent borders and recognition of climate refugees. Wealthy countries like the US, the UK, Canada, and Australia have a responsibility to provide shelter to those seeking safety. The Blueprint for Europe's Just Transition, a detailed GND-type draft plan released for public consultation in 2019, promises "mechanisms of accountability for Europe's historic role in resource extraction in the Global South" and includes a detailed section on international justice (Adler, Wargan, and Prakash 2019, 6). Another important aspect of the global politics of a GND is a transformation in the way we provide nutritious food for everyone on this planet. The complexities of this task cannot be ignored, and the decades of advocacy by groups such as La Via Campesina should be the starting point for a GND politics based on Indigenous sovereignty, land rights, and agrarian justice (Ajl 2018).

Lastly, how can we decarbonize our societies according to the twin principles of climate justice and just transition? Some proposals preference nationalization and centralized ownership of energy, but an ecosocialist GND could draw on the pioneering work of Trade Unions for Energy Democracy to argue more broadly for social ownership and decentralized control of renewable energy (Sweeney, Benton-Connell, and Skinner 2015). Energy democracy puts workers and communities at the centre of decision-making processes about how to produce and distribute energy. This precludes a technocratic, capitalist-led transition and requires a global justice approach, due to the supply chains implicated in an expansion of renewable energy and storage. As Sica (2019) argues, an ecosocialist approach to energy and the GND that focuses on good union jobs and reducing the cost of power for families would be enormously popular. The idea of "taking back the grid" and providing power to the people is perhaps the most inspiring part of the GND.

Political strategies to win a Green New deal

In the past two years, there has been a rapid emergence of social movements using explicitly ecosocialist language and politics to argue for a GND. There is not enough space to detail them all here, and many of these groups remain small and on the fringes of the climate movement. However, in the US and the UK, ecosocialists have been successful in promoting their version of a GND within large political organizations. In the US, the aforementioned DSA Ecosocialist Caucus successfully pushed for the adoption of ecosocialist guiding principles for a Green New Deal at the 2019 national conference of the DSA, an organization representing nearly 60,000 members (EWGDSA 2019). The ecosocialist caucus group Labour for Green New Deal was similarly instrumental in pushing the UK Labour party, which has a membership of over 500,000, to adopt a GND prior to the 2019 general election (Jacobs 2019). In light of Labour's defeat in the 2019 election, the latter case demonstrates why such organizing must occur both within and outside electoral politics. The key question will be what emphasis is placed on electoral strategies to win a GND versus a focus on building democratic working-class organizations on a mass scale capable of challenging the capitalist state.

The GND signals a shift in climate politics after a long period in which the Left has failed to produce credible and popular alternatives to market-based climate solutions. For at least two decades, the mainstream climate movement has focused on international agreements to curb emissions and individual consumer action while welcoming empty gestures from corporations engaged in greenwashing. During that time, alternatives to liberal environmentalism have been put forward by Indigenous groups, peasants, feminists, and coalitions such as the World Social Forum. These efforts have focused on a critique of capitalism and colonialism originating "from the margins" and frontlines of the climate crisis. Matt Huber convincingly makes the case that the former, "lifestyle environmentalism," is counterproductive, and the latter, "livelihoods environmentalism," is insufficient as a basis for building a mass movement (Huber 2019). Instead, he argues that we need an "ecological politics for the working class" based on materialist politics with broader appeal (Huber 2019).

While I agree with Huber's critique, I would add two small caveats. First, the majority of people who are engaged in livelihoods environmentalism outside academia are indeed part of the global, multi-racial working class, and their demands centred around *climate justice* should therefore form part of the basis for an ecological politics for the working class (Salleh 2010). Indigenous people's demand for sovereignty, land rights, and self-determination will be a crucial determinant of plans for the transformation of land use, access to nature, and virtually every other aspect of the GND. To this end the Red Nation, an Indigenous organization in Albuquerque, has proposed a "Red Deal" as both a critique of and an alternative to the GND (TRN 2019).

Second, a movement for the working class can only be built by acknowledging the contradictory role that trade unions play in capitalism and a commitment to working within those limitations while struggling to transcend their tendency towards class compromise (Humphrys 2019). Just as our theoretical understanding of work must be updated to account for current material and political conditions, ecosocialists must engage Marxist debates over the state in the context of the emergent politics of the GND. Debates among Leninists, anarchists, and democratic socialists looking to thinkers like Miliband and Poulantzas for strategies "within and against" the state rage on, largely absent a historical materialist analysis of the present conditions of ecological breakdown. How does an ecosocialist understanding of climate change inform and require a rethinking of Marxist analyses of the state and, therefore, our formulation of political strategy?

Ecosocialist critiques of the Green New Deal

There are two main critiques of the GND from an ecosocialist perspective. The first argues that the GND is tantamount to green capitalism – specifically, green Keynesianism – and cannot be reconciled with the ecosocialist principle of anti-capitalism. The second, associated with degrowth, argues that the GND would involve an expansion of economic growth, production, and consumption that is ecologically unsustainable. These critiques should be used to strengthen rather than abandon an ecosocialist politics for the GND. This is a strategic necessity in the absence of credible and popular alternatives which would satisfy the requirements of the GND's critics.

Ecosocialists such as John Bellamy Foster and ecofeminists like Ariel Salleh have cautioned against the embrace of a GND, arguing that it could amount to a fix for capitalism, a plan of class compromise to smooth out the contradictions that would have a negative impact on the global South (see Triantafyllou 2019; Salleh 2010). Instead, Foster calls for a "long ecological revolution" for the establishment of "a really efficient society by a needs-based criterion [which] would provide people with food, shelter, health care, education, meaningful work and all the other things that people desperately need but don't have in large numbers" (cited in Triantafyllou 2019, 1). However, the outcome of this revolution is closely aligned with the stated goals of ecoscocialist proponents of the GND. The disagreement seems to stem from the political strategy we use to achieve our end goal (ecosocialism) and what is possible, given our political and material conditions today.

Salleh's argument that existing GND proposals do nothing to address the "systematic exploitations of race and gender that underpin the global economy" (2010, 15) is the strongest critique of nationally focused plans. The European Blueprint for a Just Transition provides an example of how to include global justice in the GND, and their provision for climate reparations is a step in the right direction. In addition, Salleh highlights the exclusion of people engaged in unpaid reproductive work and highly exploitative work in "geographic and domestic peripheries" from GND programmes overly focused on a just transition for workers in fossil fuel industries in the global North. An ecosocialist GND therefore requires a fundamental transformation in human-nature relations, beginning with an understanding of where "hidden" value is currently produced and extracted in the global economy (Federici 2012).

Ecosocialist advocates of degrowth contend that the GND threatens to replicate the endless growth paradigm of capital accumulation and aspires to increases in production and consumption that are ecologically unsustainable based on "technological hubris" (Dale 2019, 1). One of the key problems they point to is the expansion of renewable energy required to power the GND and the global mining supply chains and attendant emissions (not to mention human rights abuses) implicated in this (Bernes 2019). The concept of degrowth championed by scholars such as Giorgos Kallis takes as axiomatic the ecological concept of limits, or carrying capacity. From this given limit, degrowth theorists seek to determine how we can live without causing adverse ecological impacts and insist that we can "construct a society that lives better with less" (Kallis 2011, 873).

Gareth Dale provides an excellent map of where GND proponents and degrowthers overlap and diverge from one another, demonstrating that there is ample space for the cross-pollination of ideas and political strategies between the two (Dale 2019). However, I agree with Huber (2019) that an approach to climate politics focused on reduced consumption, rather than on meeting people's needs, is difficult to rescue from claims of eco-austerity, especially during times of economic stagnation. Even though degrowth proponents *are* concerned with meeting people's needs, their focus on consumption translates too easily into an ineffective lifestyle environmentalism. In addition, the application of the ecological concept of carrying capacity to human-nature relations and politics is a fraught endeavour, which degrowthers themselves acknowledge (Sayre 2008; Kallis 2019). Our focus should be on qualitative changes in how we can live together, rather than on quantitative limits to the production of nature.

The GND would entail a huge redistribution of wealth, resources, and energy from a minority of people who consume in excess of what most people can ever imagine. Public luxury, democratic planning, and the ordinary consumption of people who are able to repair things and spend much less time at work will be more than compensated for by a complete reduction in private, excessive luxury. More importantly, if the exploitation of people and planet in capitalism is the root cause of our socio-ecological crises, our focus should be on establishing social control over the production of nature (Mészáros 2015). In the words of Neil Smith:

> Socialism is neither a utopia nor a guarantee. It is however the place and the time where and when the unity of nature becomes a real possibility. It is the arena of struggle to develop real social control over the production of nature.
>
> *(Smith 1984, 89)*

Conclusion

This moment presents an unrivalled opportunity to contest the domination of capital on a grand and coordinated scale. Ecosocialists can choose to embrace and promote a popular, socialist politics for the GND. As Thea Riofrancos has argued, the GND represents a struggle on a "new terrain of class politics" (Riofrancos 2019). Our other option is to remain in relative obscurity and allow an opportunity to step closer to the world we want to be coopted by liberal environmentalists and capitalists. An ecosocialist GND is a chance to democratically determine what our lives are for and how we want to live together. There are a million ways it could fail, especially if it is allowed to become a pet project for ENGOs, academics, and policy wonks. If we succeed, though, I can imagine a world where we have real freedom, with meaningful jobs and good housing, nutritious and sustainable food, robust public education, beautiful public spaces, and justice for the communities who have been left behind in the dying days of the current system.

Acknowledgement

None of us work alone, and I would like to acknowledge my colleague Anna Sturman for her contribution to many of the ideas in this chapter.

References

Adler, David, Pawel Wargan, and Sona Prakash, eds. 2019. "Blueprint for Europe's Just Transition." *DiEM25*, November. https://report.gndforeurope.com/

Ajl, Max. 2018. "Beyond the Green New Deal." *Brooklyn Rail*. https://brooklynrail.org/2018/11/field-notes/Beyond-the-Green-New-Deal

Bernes, Jasper. 2019. "Between the Devil and the Green New Deal." *Commune*, 25 April. https://communemag.com/between-the-devil-and-the-green-new-deal/

Bhattacharya, Tithi. 2019. "Three Ways a Green New Deal Can Promote Life Over Capital." *Jacobin*. https://jacobinmag.com/2019/06/green-new-deal-social-care-work

Bruff, Ian. 2014. "The Rise of Authoritarian Neoliberalism." *Rethinking Marxism* 26 (1): 113–129.

Bryant, Gareth. 2019. *Carbon Markets in a Climate-Changing Capitalism*. Cambridge: Cambridge University Press.

Dale, Gareth. 2019. "Degrowth and the Green New Deal." *The Ecologist*, 28 October. https://theecologist.org/2019/oct/28/degrowth-and-green-new-deal (accessed 28 August 2021).

Ecosocialist Working Group Democratic Socialists of America (EWGDSA). 2019. "An Ecosocialist Green New Deal: Guiding Principles." *EWGDSA*. https://ecosocialists.dsausa.org/2019/02/28/gnd-principles/

Estes, Nick. 2019. "A Red Deal." *Jacobin*. www.jacobinmag.com/2019/08/red-deal-green-new-deal-ecosocialism-decolonization-indigenous-resistance-environment

Federici, Silvia. 2012. *Revolution at Point Zero: Housework, Reproduction, and Feminist Struggle*. Oakland, CA: PM Press.

Foster, John. 2002. *Ecology Against Capitalism.* New York: Monthly Review Press.
Garnaut, Ross. 2008. *The Garnaut Climate Change Review: Final Report.* Cambridge: Cambridge University Press.
Giacomini, Terran, Terisa E. Turner, Ana Isla, and Leigh Brownhill. 2018. "Ecofeminism Against Capitalism and for the Commons." *Capitalism Nature Socialism* 29 (1): 1–6.
Guy, Jonathan. 2019. "Winning Friends, Crushing Enemies: An Interview with Alyssa Battisoni and Thea Riofrancos." *The Trouble*, 12 November. www.the-trouble.com/content/2019/11/12/winning-friends-crushing-enemies-an-interview-with-alyssa-battisoni-and-thea-riofrancos
Huber, Matt. 2019. "Ecological Politics for the Working Class." *Catalyst* 3 (1): 7–45.
Humphrys, Elizabeth. 2019. *How Labour Built Neoliberalism: Australia's Accord, the Labour Movement and the Neoliberal Project.* Leiden: Brill.
IPCC. 2018. *Global Warming of 1.5°C: An IPCC Special Report on the Impacts of Global Warming of 1.5°C Above Pre-Industrial Levels and Related Global Greenhouse Gas Emission Pathways, in the Context of Strengthening the Global Response to the Threat of Climate Change, Sustainable Development, and Efforts to Eradicate Poverty.* Edited by V. Masson-Delmotte, P. Zhai, H.O. Pörtner, D. Roberts, J. Skea, P.R. Shukla, A. Pirani, W. Moufouma-Okia, C. Péan, R. Pidcock, S. Connors, J.B.R. Matthews, Y. Chen, X. Zhou, M.I. Gomis, E. Lonnoy, T. Maycock, M. Tignor, T. Waterfield. https://archive.ipcc.ch/report/sr15/pdf/sr15_citation.pdf
Jacobs, Michael. 2019. "Labour's Green New Deal Is Among the Most Radical in the World – But Can It Be Done by 2030?" *The Conversation*, 25 September. https://theconversation.com/labours-green-new-deal-is-among-the-most-radical-in-the-world-but-can-it-be-done-by-2030-123982
Kallis, Giorgos. 2011. "In Defence of Degrowth." *Ecological Economics* 70 (5): 873–880.
Kallis, Giorgos. 2019. "Socialism Without Growth." *Capitalism Nature Socialism* 30 (2): 189–206.
Klein, Naomi. 2017. *No Is Not enough: Defeating the New Shock Politics.* London: Penguin.
Mészáros, Istvan. 2015. *The Necessity of Social Control.* New York: Monthly Review Press.
Mies, Maria. 1986. *Patriarchy and Accumulation on a World Scale: Women in the International Division of Labour.* London: Zed Books.
Morrison, Scott. 2020. "Scott Morrison's Plan to Make Australia 'Resilient' to Climate Change." Interviewed by Ray Hadley. *2GB*, 20 January. www.4bc.com.au/scott-morrisons-plan-to-make-australia-resilient-to-climate-change/
Ocasio-Cortez, Alexandria. 2019a. "H.Res.109–116th Congress (2019–2020): Recognizing the duty of the Federal Government to create a Green New Deal." 12 February. https://ocasio-cortez.house.gov/sites/ocasio-cortez.house.gov/files/Resolution%20on%20a%20Green%20New%20Deal.pdf
Ocasio-Cortez, Alexandria. 2019b. "Untitled Fact Sheet." 7 February. https://assets.documentcloud.org/documents/5729035/Green-New-Deal-FAQ.pdf
O'Connor, James R., ed. 1998. *Natural Causes: Essays in Ecological Marxism.* New York: Guilford Press.
OECD. 2020. "Trade Union Density [Data File]." https://stats.oecd.org/Index.aspx?DataSetCode=TUD
The Red Nation (TRN). 2019. "The Red Deal: Indigenous Action to Save Our Earth." *The Red Nation.* https://therednationdotorg.files.wordpress.com/2019/11/red-deal-final.pdf
Riofrancos, Thea. 2019. "Plan, Mood, Battlefield: Reflections on the Green New Deal." *Viewpoint Magazine*, 16 May. www.viewpointmag.com/2019/05/16/plan-mood-battlefield-reflections-on-the-green-new-deal/
Rosewarne, Stuart, James Goodman, and Rebecca Pearse. 2014. *Climate Action Upsurge: The Ethnography of Climate Movement Politics.* New York: Routledge.
Russell, Ellen D. 2018. "Resisting Divide and Conquer: Worker/Environmental Alliances and the Problem of Economic Growth." *Capitalism Nature Socialism* 29 (4): 109–128.
Saed. 2019. "James Richard O'Connor's Ecological Marxism." *Capitalism Nature Socialism* 30 (4): 1–12.
Salleh, Ariel. 1997. *Ecofeminism as Politics: Nature, Marx, and the Postmodern.* London: Zed Books.
Salleh, Ariel. 2010. "Green New Deal – Or Globalisation Light." *Arena* 105: 15–19.
Sayre, Nathan F. 2008. "The Genesis, History, and Limits of Carrying Capacity." *Annals of the Association of American Geographers* 98 (1): 120–134.
Schwartzman, David. 2011. 'Green New Deal: An Ecosocialist Perspective." *Capitalism Nature Socialism* 22 (3): 49–56.
Sica, Carlo E. 2019. "For a Radical Green New Deal: Energy, the Means of Production, and the Capitalist State." *Capitalism Nature Socialism* 31 (4): 34–51.
Smith, Neil. 1984. *Uneven Development: Nature, Capital, and the Production of Space.* New York: Blackwell.

Stern, Nicholas. 2007. *The Economics of Climate Change: The Stern Review*. Cambridge: Cambridge University Press.

Sweeney, Sean, Kylie Benton-Connell, and Lara Skinner. 2015. "Power to the People: Toward Democratic Control of Electricity Generation." Trade Unions for Energy Democracy (TUED), Working Paper No. 4. http://unionsforenergydemocracy.org/resources/tued-publications/

Tarlau, Rebecca. 2019. "What's Behind the Teacher Strikes: Unions Focus on Social Justice, Not Just Salaries." *The Conversation*, 22 February. https://theconversation.com/whats-behind-the-teacher-strikes-unions-focus-on-social-justice-not-just-salaries-111490

Temper, Leah, and Sam Bliss. 2019. "A Green New Deal for an Ecological Economy." *Uneven Earth*. http://unevenearth.org/2019/10/a-green-new-deal-for-an-ecological-economy/

Triantafyllou, Vaios. 2019. "A Green New Deal Is the First Step Toward an Eco-Revolution." Interview with John Bellamy Foster. *Truthout*, 9 February. https://truthout.org/articles/a-green-new-deal-is-the-first-step-toward-an-eco-revolution/

Weeks, Kathi. 2011. *The Problem with Work: Feminism, Marxism, Antiwork Politics, and Postwork Imaginaries*. Durham, NC: Duke University Press.

31

BIOENERGY AND ECOSOCIALISM

Nadia Singh

Introduction

Bioenergy occupies an important place in the sustainable development discourse and is touted as a "silver bullet in the energy market" (Dauvergne and Neville 2009, 1087). Since early 2000, there has been massive euphoria about these projects from both by corporates and national governments, especially in the context of the developing countries of the world. It was argued that commercial production of bioenergy would enable these countries to create a green energy market, meet their climate change commitments, and reduce their widespread energy poverty (Neville 2015). Bioenergy projects also formed a part of the United Nations Sustainable Development Agenda. The UN declared 2014–2024 as the "decade of sustainable energy for all" (UN 2012, 2).

Bioenergy projects have been promoted across many developing countries of the world in the last two decades, including in China, India, Brazil, Thailand, and Malaysia (High Level Panel of Experts on Nutrition and Food Security [HLPE] 2013). Bioenergy policies across different countries were fostered by production and input subsidies, capital subsidies, research grants, tax concessions, and legislation (Singh and Singh 2019). The political backing for bioenergy prompted a number of large oil companies like British Petroleum, Shell, and Exxon to make investments in a range of bioenergy projects (Kolk and Levy 2001). Many of these companies began to support bioenergy projects with a view to garnering positive media attention as environmentally friendly companies (Kolk and Levy 2001, 503). For instance, British Petroleum launched the slogan "Beyond Petroleum" in early 2000 and spent $200 million to rebrand themselves (Cherry and Sneirson 2010, 999). Large agricultural and food corporations also began to promote biofuels as a new market for their product and to establish control over agricultural prices and technology (Neville 2015). In 2008, four companies, Archer Daniel Midland, Du Pont, Deere, and Monsanto, representing agrochemical, farm equipment, and seed supplies sectors, established the Alliance for Abundant Food and Energy to lobby the US government to promote biofuels (Dauvergne and Neville 2009).

Despite the excessive government support and widespread euphoria among corporates, renewable energy projects in developing countries have not taken off as expected. On the contrary, they have been associated with many adverse impacts, including threats to food security across many developing nations (Hammod and Bo 2016) and land grabs by bioenergy

10.4324/9780429341427-36

producers in Asia and Africa, as well as excessive deforestation and land use change (Baka 2014). Large agroindustry players in the global energy markets have captured the gains from bioenergy developments (Neville 2015). Little benefit has accrued to the energy poor population in developing countries (Hunsberger and Ponte 2014). On the contrary, these vulnerable groups have emerged as net losers from bioenergy projects. They have been subject to rising food price inflation, exploitative working conditions in bioenergy plantations, and growing incidents of land appropriation by biofuel producers. Some studies have revealed that the working conditions in these bioenergy plantations are very poor. According to the First Information and Action Network (FIAN) International (2008), bioethanol plantations in Brazil infringed on basic human rights of workers. Workers were subjected to serious occupational hazards and poor housing and dietary conditions, which compromised their physical integrity. The study also found that the bioethanol industry was the source of the largest number of cases of child labour and slavery in Brazil.

It is in this context that the present chapter analyzes the weaknesses of bioenergy reforms situated within a standard neo-classical policy framework and how an ecosocialist paradigm can offer an alternative framework to pursue technocratic projects such as bioenergy. While ecosocialists welcome the development of renewable resources currently taking place in modern capitalist economies, they believe that these reforms by themselves are an inadequate solution to the ecological crisis. In contrast, ecosocialists advocate a "strong sustainability" approach, rooted in the dialectical link between humans and nature. They contend that this idea of sustainability can be achieved by moving beyond the micro definition of environmental concerns as an isolated project to the macro understanding of environmental issues to incorporate people's concerns such as unemployment, livelihood security, and poverty. Morrison (1995, 101) has said that in the long run, ecological problems can only be solved through "the rise of a new political economy, rooted in respect for the interdependence of social, political and economic realms and their connection to the encompassing social and natural ecologies."

The promise of bioenergy developments

Bioenergy reforms form a part of the larger discourse of "green" capitalist reforms in the world economy. Green capitalism is based on modifications of the traditional theory of economic growth to account for the role of natural capital in the production process in society. These reforms aim to provide a pathway for "transition to bio-based economies" while keeping the existing production and consumption structures of the global economy unchanged (Ponte 2014, 262). A major plank of these reforms comprises mainstream green technologies and renewable energy prerogatives. Renewable energy is being promoted as the next "green revolution," which would replace fossil fuels with "clean environmentally friendly sources of energy" (Friedman 2008, 24).

Since early 2000, the global energy sector has been faced with multiple challenges. These include declining fossil fuel reserves and the peak oil[1] crisis, increasing threat of climate change, and the widespread energy poverty plaguing many developing nations. Statistics reveal that close to 1.2 billion people across developing countries of the world live "off the grid" and do not have access to modern energy sources. Eighty percent of these people live in developing countries (IEA 2016).

In response to this crisis, bioenergy projects began to be promoted across many developed and developing countries of the world. Bioenergy had four main properties which made it a credible alternative to fossil fuel energy (HLPE 2013). Firstly, bioenergy could be used as a fuel for cooking and heating purposes and could also be employed as a transport fuel.

Secondly, direct carbon emissions from biofuels were believed to be insignificant compared to those from fossil fuels. Thirdly, bioenergy was perceived as a means to reduce the energy dependence of oil-importing countries and strengthen their energy security since it could be produced locally with home-grown feedstock. Fourthly, it was seen as a means to boost rural electrification and promote a new source of livelihood for farmers in developing agricultural economies.

Bioenergy represented a potential solution to the energy crisis which could be conveniently instituted within the existing structures of the industrial economies. It did not require any fundamental changes in the patterns of fuel consumption and production (Dauvergne and Neville 2009, 1090). Owing to their reputation as a "clean" fuel, bioenergy projects also garnered the support of many international NGOs between 2000 and 2006 (Neville 2015). NGOs such as Greenpeace UK saw biofuels as a local alternative energy initiative which would bring down carbon emissions and at the same time help counter the hegemony of large corporations. Fueled by government and industry support, there has been massive rise in bioenergy production across the globe in recent years. (See Table 31.1.)

Various types of bioenergy sources are being promoted by national governments and large business corporates in recent years. These include modernization of traditional biomass sources, as well as development of technologically enhanced sources of bioenergy (HLPE 2013). Some scholars contend that with adequate technical progress, the drawbacks of traditional biomass energy can be overcome, and it may emerge as a "key source of locally produced energy in developing countries" (Gomerio 2015, 8520). A number of initiatives promote technically enhanced biomass energy for household use and electricity generation in rural areas of developing countries. These initiatives include installation of improved biogas stoves in rural areas of developing countries, promotion of the sustainable harvesting of wood, improvement in fuel processing technologies, and setting up biomass power-generation projects (Singh 2017). For instance, these initiatives are being extensively promoted by the Indian government. Statistics revealed that between 2008–2009 and 2017–2018, the growth rate of power generation based on bioenergy was estimated to be 22.5 percent per annum (Energy Statistics of India 2020). Liquid biofuels are also being promoted. The main sources of liquid biofuels include ethanol and biodiesel. Biofuels based on agricultural residues, organic components of municipal wastes, and algal matter, referred to as second-generation biofuels, are being researched as well (Andree, Diego, and Koomen 2017). However, these are still in the nascent stage of production and currently account for only 0.25 percent of the total bioenergy production across the world (BP 2020).

Table 31.1 Biofuel production across the world (1,000 barrels of oil equivalent)

Region	*2010*	*2015*	*2019*
North America	542	656	726
Latin America	205	265	296
Asia Pacific	69	161	294
Africa	6	7	9
Europe	205	265	296
World	1,025	1,560	1842

Source: BP (2020, 56)

Contestations surrounding bioenergy developments

The tall promises surrounding bioenergy developments are belied by a number of adverse outcomes. By early 2007, a number of negative repercussions of bioenergy developments came to the forefront. An analysis of the bioenergy developments across the world reveals that there are serious concerns with respect to bioenergy projects, in terms of their social, economic, and ecological sustainability.

From a purely economic point of view, with the current level of technological development, the energy returns from bioenergy are quite low while their cost of production is substantial. This makes it highly unlikely that bioenergy will offer a credible energy alternative in the near future. This has been shown by a number of studies. Smil (2008) estimated that the power density of biofuels was two times less than hydro and solar energy and three times less than that of fossil fuel energy. Other studies have shown that net energy returns on investment (EROI) for biofuels are quite low (Gomerio 2015). Carriquiry, Du, and Timilsina (2011) conducted a detailed study on the cost of production and relative economic merits of fossil fuels and biofuels. Their findings showed that at the time the cost of producing biofuels was five times higher than fossil fuels. More recent studies have shown that the cost of production of biofuels continues to be two to two and a half times higher than fossil fuels (IEA 2020). This makes it unlikely that biofuels will be a credible energy alternative in the near future.

The effect of biofuels on the mitigation of pollution and the reduction in GHG emissions also remains questionable. Some studies based on life cycle analysis (LCA) have demonstrated that the GHG emissions from the use of fossil fuels are only marginally lower than those of biofuels (Farrell et al. 2006; Searchinger et al. 2008; de Castro et al. 2014). Fargione et al. (2008, 1238) propounded that biofuel production will create a "biofuel carbon debt" by releasing a stock of soil and biomass carbon that is 17 to 420 times higher than the annual GHG reductions that bioenergy developments would bring about through the replacement of fossil fuels. Another major concern with the production of biofuels is that chemical plants involved in the production of ethanol produce large amounts of polluted water and chemical toxins, which are harmful to the natural environment. Pimentel and Patzek (2005) showed that the production of 1 liter of ethanol generates 13 liters of polluted water. Their research also revealed that there are currently no proper procedures in place for the disposal of waste generated through biofuel production.

There are also serious concerns about bioenergy expansion from a social sustainability prerogative. Biofuel production competes for land use with other agricultural activities like production of food crops and also with other environmental objectives like protection of rainforests and tropical lands for maintaining biodiversity. Fischer et al. (2002) and Field, Elliott Campbell, and Lobell (2008) conducted two comprehensive studies based on satellite imagery to assess the global land availability to sustain the current levels of production of biofuels. Both these studies found that almost all the available land for bioenergy production across Asia, Africa, and America is either under cultivation or under forests. Thus, expansion of biofuel can only come through substitution with other crops or bringing forests, grasslands, and pastures into production of crops. This may lead to two types of changes in land use patterns: namely, direct land use change (DLUC)[2] and indirect land use change (ILUC).[3]

Another serious concern with the commercialization of bioenergy is the impact on food security and food price inflation, especially in developing countries. The massive rise in global biofuel production between 2007 and 2008 was accompanied by the steepest increase in global food prices since 1980s (HLPE 2013). In 2008–2009, cereal prices rose by two to two and a half times while sugar prices rose by eight to ten times in the global food markets (Abbott 2011). Prices continued to remain volatile in the subsequent years, leading to severe food shortages and sparking food riots across many developing nations. The rapid expansion in production of corn and vegetable oil for use as a fuel was a key factor in triggering the crisis. Biofuel production was described as an "amplifying" factor in the global food price surge (Abbott 2011, 54). The inter-linkages between biofuel production and global food price rises have been confirmed by many international agencies. According to Action Aid (2013, 27), "The current EU biofuel production could produce enough food to produce 185 million hungry people across developing nations."

An ecosocialist perspective on bioenergy reforms

The key question that emerges from this discussion is what led to bioenergy being reduced from the "fuel of the future" to a highly contested energy alternative? Ecosocialist scholars contend that the source of these contestations stems from the larger framework of "green capitalism," within which these bioenergy reforms are located (Venkatachalam 2007). This framework locates the economic explanation of the degradation of nature in terms of the standard "market failure" argument, implying that due to the nonexistence of markets for environmental goods, people failed to recognize the otherwise positive prices of environmental goods and services (Baumol and Oates 1988). Therefore, solutions to the environmental crisis comprise assigning a price to environmental goods and services through a range of instruments like carbon taxes, subsidies, direct state regulation, and the creation of "green" markets for products such as bioenergy.

Ecosocialists critique this as a reductionist approach to sustainability as it does not take into account of "the socioeconomic as well as ecological inequalities of commodity production under capitalism" (Luke 2006, 101), which have manifested in crises like the situation in the present times. In contrast, ecosocialist scholars argue that ecological concerns are not divorced from larger societal concerns. Therefore, the power structures and inequities across class, race, and gender divisions in capitalist societies are the central parameters in explaining the process of ecological degradation in society.

Ecosocialists have developed a new perspective on the environment by analyzing the problems of ecological degradation through a Marxian political economy lens so that ecological challenges can be located within the existing class and power structures in society. They envision an ecologically sustainable economy, founded on socialist principles of egalitarianism, elimination of poverty, equitable resource distribution, and democratic control over lives and communities (Pepper 1998, 234). This model endeavours to recognize basic societal needs as well as the prerogatives of environmental protection. The basic foundations of ecosocialism are (a) ecologically rational production processes, (b) democratic planning of investment and production decisions along with preservation of ecological equilibrium, and (c) new technological structure of productive forces which recognize the limits to the availability of natural resources (O'Connor 1998, 278).

Ecosocialists recognize that the transition to an ecosocialist society is a long-term process and "a multi-faceted struggle for a new social order" (Foster 2015). As a first step in this process, green reforms such as bioenergy projects in capitalist economies should be made more democratic in nature. This approach to technocratic reforms is cognizant of the social, economic, institutional, and ethical aspects of sustainability. According to Löwy (2005, 21):

> [T]he struggle for ecological reforms under capitalism can be a vehicle for dynamic change, a transition between minimal demands and maximal program, provided that

one rejects the pressures of the ruling class for competitiveness and modernization in the interests of the rules of the market.

The ecosocialist paradigm does not treat technocratic projects like bioenergy reforms as a mere ecological fix but attempts to embed these green energy imperatives within the larger political economy of a region. This can be accomplished through a "bottom up" policy design, rooted in democratic and decentralized institutions, community participation in such initiatives involving grassroot environmental movements, labour movements, and socially responsible citizens and corporations. One example of such grassroot green energy imperatives is illustrated by Singh and Singh (2019) through a case study of bioenergy imperatives in the Punjab region of India. This work demonstrated how community-based organizations and environmentally conscious citizens were creating small-scale models of green energy production, rooted in ingenious technology and locally available resources. The research provided empirical evidence of how the involvement of ground-level stakeholders can create alternative trajectories of participatory and inclusive models of sustainable energy deployment.

Such an approach to green energy developments will facilitate the formulation of people-centric, inclusive policies that are responsive to the energy needs and aspirations of local populations at the ground level. This will help garner popular support for these imperatives and overcome many of the socio-ecological challenges with which bioenergy projects are currently plagued.

Conclusion

This chapter traced the development trajectory of bioenergy sources in the global economy in the last two decades. Bioenergy went on from being the "fuel of future" to a highly controversial energy alternative, with a number of sustainability challenges. The analysis demonstrated how these challenges essentially spring from the agenda of green capitalism, which implements these reforms as a technical fix. We then went on to see how ecosocialism could provide a more inclusive framework to implement such technical reforms so that issues of social and ecological justice are embedded within the policy framework. However, a major challenge with respect to the application of ecosocialist perspective is that a fully developed blueprint for transformation to an ecologically sustainable society is still in the process of being constructed. Ecosocialists believe that the fully developed blueprint evolves in praxis. Its best practices that address the nuts and bolts of the policy are slowly beginning to emerge, such as the mechanisms to involve local stakeholders in renewable energy deployment and garnering popular public support from such campaigns, the mechanisms of addressing conflicts between the different sections of the community, and the prerogatives of balancing needs. As the institutions of participatory democracy are strengthened and respond to the challenges of environment in their contexts, policy practices will be established.

Notes

1 Peak oil refers to the destruction of oil reserves after reaching a point of maximum extraction of fossil fuel energy. Charles King Hubbert first described this phenomenon in 1956. He studied the pattern of domestic oil production in USA and concluded that US oil production would "peak" in 1970s, while the world oil production would "peak" between 1995 and 2000 (Hubbert 1956). Although Hubbert's initial calculations proved wrong, since then a number of scholars have refined Hubbert's calculations and predicted that peak oil would be reached around 2050.

2 DLUC occurs when feed stocks for biofuel production are grown on forests, rainforests and tropical lands which were previously not under crop production (HLPE 2013).
3 ILUC occurs when biofuel production leads to changes in land use, not on site but elsewhere, due to the need to compensate foregone production (HLPE 2013).

References

Abbott, Phillip. 2011. *Stabilization Policies in Developing Countries After the 2007–2008 Crisis.* Paris: OECD.

Action Aid. 2013. *Adding Fuel to the Flame: The Real Impact of EU Biofuel Policies on Developing Nations.* Johannesburg: Action Aid. www.actionaid.org/sites/files/actionaid/adding_fuel_to_the_flame_actionaid_2013_final.pdf

Andree, Bo P.J., Vasco Diego, and Eric Koomen. 2017. "Efficiency of Second-Generation Biofuel Crop Subsidy Schemes: Spatial Heterogeneity and Policy Design." *Renewable and Sustainable Energy Reviews* 67: 848–862.

Baka, Jennifer. 2014. "What Wastelands? A Critique of Biofuel Policy Discourse in South India." *Geoforum* 54: 315–323.

Baumol, William J., and Wallace E. Oates. 1988. *The Theory of Environmental Policy*, 2nd ed. Cambridge: Cambridge University Press.

British Petroleum. 2020. "BP Statistical Review of World Energy, 2016." www.bp.com/content/dam/bp/business-sites/en/global/corporate/pdfs/energy-economics/statistical-review/bp-stats-review-2020-full-report.pdf (accessed 14 March 2021).

Carriquiry, Miguel A., Xiaodong Du, and Govinda R. Timilsina. 2011. "Second Generation Biofuels: Economics and Policies." *Energy Policy* 39 (7): 4222–4234.

Cherry, Miriam A., and Judd F. Sneirson. 2010. "Beyond Profit: Rethinking Corporate Social Responsibility and Greenwashing after the BP Oil Disaster." *Tulane Law Review* 85: 983–1038.

Dauvergne, Peter, and Kate J. Neville. 2009. "The Changing North-South and South- South Political Economy of Biofuels." *Third World Quarterly* 30 (6): 1087–1102.

de Castro, Carlos, Oscar Carpintero, Fernando Frechoso, Margarita A. Mediavilla, and Luis J. de Miguel. 2014. "A Top-Down Approach to Assess Physical and Ecological Limits of Biofuels." *Energy* 64 (1): 506–512.

Energy Statistics of India. 2020. *Energy Statistics of India 2019–20.* New Delhi: Ministry of Statistics and Programme Implementation.

Farrell, Alexander E., Richard J. Plevin, Brian T. Turner, Andrew D. Jones, Michael O'Hare, and Daniel M. Kammen. 2006. "Ethanol Can Contribute to Energy and Environmental Goals." *Science* 311 (5760): 506–508.

Fargione, Joseph, Jason Hill, David Tilman, Stephen Polasky, and Peter Hawthorne. 2008. "Land Clearing and the Biofuel Carbon Debt." *Science* 319 (5867): 1235–1238.

FIAN International. 2008. *Agrofuels in Brazil: Report on the Fact Finding Mission of the Impact of Public Policies Encouraging Production of Agrofuels on the Enjoyment of Human Rights to Food, Work and Environment Among Peasants, Indigenous Communities and Local Workers in Brazil.* Heidelberg, Germany: FIAN International. www.fian.at/assets/AgrofuelsInBrazil.pdf (accessed 13 December 2015).

Field, Christopher B., J. Elliott Campbell, and David B. Lobell. 2008. "Biomass Energy: The Scale of Potential Resource." *Trends in Ecology and Evolution* 23 (2): 65–72.

Fischer, Gunther, Harrij van Velthuzien, Mahendra Shah, and Freddy Nachtergaele. 2002. *Global Agro-ecological Assessment for Agriculture in the 21st century.* Laxenburg and Rome: International Institute for Applied System Analysis and Food and Agriculture Organisation of the United Nations. http://pure.iiasa.ac.at/id/eprint/6667/1/RR-02-002.pdf

Foster, John Bellamy. 2015. "Marxism and Ecology: Common Fonts of a Great Transition." *Monthly Review* 67 (7). http://monthlyreview.org/2015/12/01/marxism-and-ecology/#en20 (accessed 12 March 2016).

Friedman, Thomas L. 2008. *Hot, Flat and Crowded: Why We Need a Green Revolution and How it Can Renew America.* New York: Strauss and Giroux.

Gomerio, Tiziano. 2015. "Are Biofuels an Effective and Viable Energy Strategy for Industrialized Societies? A Reasoned Overview of Potentials and Limits." *Sustainability* 7 (7): 8491–8521.

Hammod, Geoffrey F., and Bo Li. 2016. "Environmental and Resource Burdens Associated with World Biofuel Production Out to 2050: Footprint Components from Carbon Emissions and Land Use to Waste Arising from Water Consumption." *GCB Bioenergy* 8 (5): 894–908.

HLPE. 2013. *Biofuels and Food Security: A Report by the High-Level Panel of Experts on Food Security on Food Security and Nutrition of the Committee of World Food Security*. Rome: Committee of the World Food Security.

Hubbert, M. King. 1956. "Nuclear Energy and the Fossil Fuels." Meeting of the Southern District Division of Production, American Petroleum Institute San Antonio, Texas, Publication No. 95. Shell Development Company, Exploration and Production Research Division, Houston.

Hunsberger, Carol, and Stefano Ponte. 2014. "Sustainable Biofuels in the Global South." *Geoforum* 54: 243–247.

IEA. 2016. *World Energy Outlook*. Paris: IEA.

IEA. 2020. *World Energy Investment, 2020*. Paris: IEA.

Kolk, Ans, and David Levy. 2001. "Winds of Change: Corporate Strategy, Climate Change and Oil Multinationals." *European Management Journal* 19 (5): 501–509.

Löwy, Michael. 2005. "What Is Eco-Socialism?" *Capitalism Nature Socialism* 16 (2): 15–24.

Luke, Timothy W. 2006. "The System of Sustainable Degradation." *Capitalism Nature Socialism* 17 (1): 99–112.

Morrison, Roy. 1995. *Ecological Democracy*. Boston: South End Press.

Neville, Kathryn J. 2015. "The Contentious Political Economy of Biofuels." *Global Environmental Politics* 15 (1): 21–40.

O'Connor, James. 1998. *Natural Causes: Essays in Ecological Marxism*. New York: The Guildford Press.

Pepper, David. 1998. "Sustainable Development and Ecological Modernization: A Radical Homocentric Perspective." *Sustainable Development* 6: 1–7.

Pimentel, David, and Tad W. Patzek. 2005. "Ethanol Production Using Corn, Switchgrass and Wood: Biodiesel Production Using Soybean and Sunflower." *Natural Resources Research* 14 (1): 65–76.

Ponte, Stefano. 2014. "Roundtabling Sustainability: Lessons from the Biofuel Industry." *Geoforum* 54: 261–271.

Searchinger, Timothy, Ralph Heimlich, R.A. Houghton, Fengxia Dong, Amani Elobeid, Jacinto Fabiosa, Simla Tokgoz, Dermit Hayes, and Tun-Hsiang Yu. 2008. "Use of US Croplands for Biofuels Increases Greenhouse Gases from Emissions through Land Use Change." *Science* 39 (5867): 1157–1268.

Singh, Nadia. 2017. "Sustainability Crisis: A Critical Evaluation of Green Energy Policies." *Economic and Political Weekly* 52 (3): 66–69.

Singh, Nadia, and Pritam Singh. 2019. "Political Economy of Bioenergy Transitions in Developing Countries: A Case Study of Punjab, India." *World Development* 124 (C). https://doi.org/10.1016/j.worlddev.2019.104630

Smil, Vaclav. 2008. *Energy in Nature and Society: General Energetics of a Complex System*. Cambridge, MA: The MIT University Press.

U.N. 2012. "Sustainable Energy for All." UN Press Release, 21 December. www.un.org/press/en/2012/ga11333.doc.htm (accessed 1 May 2017).

Venkatachalam, L. 2007. "Environmental Economics and Ecological Economics: Where They Can Converge?" *Ecological Economics* 61 (2–3): 550–558.

32

THE ENVIRONMENTAL BENEFITS OF SOCIALIST STATES

Salvatore Engel-Di Mauro

Challenging prevailing accounts of socialist states

Traversing multiple currents of ecosocialism is an attempt to break from aspects of previous forms of socialism. This is particularly the case with respect to state socialism, including on account of social horrors in a minority of cases. As the social aspects have been amply studied and debated, only the environmental record is discussed here. There are legitimate reasons to distance oneself from the sort of socialism that has left a far-from-stellar imprint on the various landscapes and ecosystems within its direct reach. Many socialists and now ecosocialists, especially in the core capitalist countries, refuse to recognize even the USSR as having been socialist or consider the negative repercussions of such socialist formations as outgrowths of a deformed, productivist form of socialism.

There are multiple ways of understanding the matter, depending on how one defines socialism. Here, state socialism is regarded as a social system whose trajectory depends on varying outcomes of class struggles, always underlain by multiple forms of oppression. It is defined by a relatively centralized state ownership of the main means of production, justified through Marxist Leninist principles by a self-described and exclusively ruling communist or like party and legitimated in terms of acting on behalf of working classes to build communism. Social policies are characterized by commitments to socio-economic equality, full employment, job and housing security guarantees, universal health and education provisions, and collective forms of consumption. Historically (so far), socialist states have arisen at the capitalist world's margins and under horrific social conditions, resulting from decades of colonial plunder and imperialist intervention by free-market democracies and/or from conflagrations induced by capitalist states (e.g., Chase-Dunn 1982; Lane 2014; O'Connor 1998; Szalai 2005).

In prevailing narratives, it is socialist states that continue to be disproportionately charged with all sorts of gruesome misdeeds, including on the environmental front. While some of the charges are certainly legitimate, they apply to a minority of socialist states and largely over delimited historical periods. Moreover, such perspectives should be sharply tempered by the recurring ecocides and genocides committed by liberal democracies for some centuries. The instances of awful and, on occasion, disastrous environmental impacts through state socialism were neither pervasive nor intrinsic to that form of socialism. This is very much unlike capitalist democracy, through which the worst ecological disasters and associated losses and debilitation

10.4324/9780429341427-37

of human life have happened, like the continuing horrors at Bhopal (Dhara and Dhara 2002), Hanford (Brown 2013), the Niger Delta (Sojinu et al. 2010), Lago Agrio (Coronel, Au, and Izzotti 2020), and Fukushima (Kasar et al. 2020). Of course, the causes of the now decades-long planetary disaster euphemistically called climate change long preceded socialist states. Moreover, as Baer observes, more than twice the amount of CO_2 (1,518,826 compared to 624,081 million tonnes) has been emitted globally since the disappearance of almost all socialist states (Baer 2018, 32; sums calculated from Ritchie and Roser 2017).

The net environmentally positive effects of socialist states have been for decades demonstrated empirically in plain contradiction with stated claims (e.g., Barr 1988; Dominick 1998; Goldman 1972; Pryde 1991), but data have been presented in scandalously biased form and are scattered across publications in different fields of the biophysical sciences, at times requiring a technical background to catch the wider implications of the analyses presented (see Engel-Di Mauro 2021). The more recently expanded availability of environmental records has even prompted at least some scholars, writing almost solely on the USSR, to recognize the falsity of the notion that socialism per se is intrinsically damaging to the biophysical environment (Bruno 2016; Weiner 2017). One can claim much more than that, though. Most indicators point to socialist states having had, on balance, a superior environmental track record to even liberal democracies. And this is with all the resources liberal democracies have stolen over hundreds of years through colonialism, the pollution such democracies offload to much of the rest of the world, and the fancy high-technology mitigation apparatuses developed in liberal democracies thanks to the capital accumulation from ransacking much of the planet. Given the efforts and, at times, technical background needed even to detect contrary evidence and tendentious arguments, it is understandable that many have fallen for the self-serving exaggerations promoted in the early 1990s by capitalist institutions like the World Bank and the International Monetary Fund (Peterson 1993).

Exposing and challenging decades of distortions on socialist states is no mere exercise in improving scientific accuracy. It helps us identify, appreciate, and learn from state socialist accomplishments that can serve as models for present ecosocialist struggles. It is hardly by chance that the only country where improvements in people's well-being have been met in an ecologically sustainable way is Cuba (Moran et al. 2008), whose public health strides are world renowned (Fitz 2020). This kind of achievement is common to state socialism. By the 1980s, state-socialist countries' health indicators largely surpassed those of capitalist countries at similar levels of economic development (Cereseto and Waitaki 1986; Navarro 1992).

The aim of a reassessment of state socialism as offered here is to encourage more careful reflection on as well as more context-sensitive studies of socialist states to provide a sturdier grounding to develop ecosocialist strategies. Two aspects are examined for this purpose. One is an abbreviated rundown of the biophysically and socially constructive legacies of socialist states (see Engel-Di Mauro 2021 for a fuller analysis and discussion). Reestablishing them, even if very differently, would support a transition towards ecosocialism much more than any form of liberal democracy. The current example set by Cuba demonstrates the potential to move in that direction. The other issue is the need for a much wider and relational view attentive to the different experiences of state socialism and the constraints resulting from capitalist imperialism and a global capitalist economy. State socialism has been much wider geographically than the suite of usual suspects studied. Rarely is the same degree of Leftist attention showered on the environmental records of Cuba, Ethiopia, Madagascar, Mongolia, or Vietnam, to name a few. The wealth of lessons from those histories should be drawn from and built on, rather than jettisoned, if one is truly interested in developing and actualizing ecosocialist strategies.

Revisiting state-socialist environmental records

It should be relatively easy to point out the ecocidal nature and results of free-market democracies and capitalism generally, but we are instead usually presented with an inversion of reality. This includes the omission, if not outright denial, of any ecologically constructive outcomes from state-socialist systems, exacerbated by the erasure of the diverse practices and effects among the more than 20 state-socialist countries that have existed and that, by the 1970s, included about a third of humanity.

The first socialist state was at the forefront of ecologically sustainable development. Environmental issues were priorities even during the Russian Civil War. Under Bolshevik rule, ecology as a science thrived and became the most advanced in the scientific world. Environmentalist currents were intrinsic to state socialism, at times shaping national policies if not putting up successful pressures from below (Elie and Coumel 2013; Gare 2002; Kelley 1976; Suing and Dedaj 2018; Weiner 1999). Over a roughly 70-year period, the net effects of such currents were constructive not only for the USSR, but also for most of the socialist states that arose since.

Many species were saved from the brink of extinction or protected by means of large preserves, especially in the USSR and Cuba. Preserves were or have been expanded in number and areal extent over time, with some exceptions, and this enables the protection of entire ecosystems along with their diverse soils and surface waters. To make this possible, millions of people have been mobilized to environmental causes and educated formally and informally on the importance of environmental protection, leaving lasting legacies of environmental literacy and sensibility (Colwell et al. 1997; Elie and Coumel 2013; Goldman 1972; Ostergren and Shvarts 2000; Roman 2018; Rosset and Benjamin 1994; Weiner 1999).

Soil conservation measures were applied as much as feasible and were, on the whole, successful, given severe economic constraints and inherited low levels of productivity (Betancourt 2020; Chendev et al. 2015; Liu and Tian 2010; Rosset et al. 2011; Wuepper et al. 2020). A late 1980s Food and Agriculture Organization study on soil degradation ranked the USSR 28th in the world, compared to the US's 52nd place (Bot, Nachtergaele, and Young 2000). The southeast Asian state-socialist countries also ranked close to both Koreas and higher than neighbouring Thailand. Within Europe, results generally indicated less soil degradation in the industrialized state-socialist countries than in their liberal democratic counterparts (for a critical appraisal of that FAO study, see Engel-Di Mauro 2014). In Cuba, the struggle for biophysically sustainable farming within the socialist state bore major successes with the development and spread of food production systems based on agroecological principles and traditional peasant knowledge systems. Thanks to these strides, Cuba is the epicentre of agroecological development and application and, with that, of soil conservation. What is more, advances in urban farming are showing the rest of the world how to integrate the urbanization process with more environmentally sustainable living and direct grassroots participation (Fernandez et al. 2018; Koont 2011; Sosa et al. 2019; Rosset and Benjamin 1994; Rosset et al. 2011).

Logging was mainly kept within the limits of forest regeneration. There were many instances of major afforestation efforts as well that reduced soil erosion and protected waterways, as well as the safeguarding of habitats for many species (Biró et al. 2013; Brain 2011; Liu and Tian 2010; Potapov et al. 2015; Rosset et al. 2011). As soon as it was established, the socialist state of Burkina Faso (1983–1987) featured environmental protection as a priority, alongside gender equality, public health, literacy, and national self-reliance. Until Sankara was assassinated in 1987, mass mobilizations led to planned wood cutting and livestock movement as well as reforestation to contain desertification (Biney 2018). With the restoration of the French neocolonial yoke, forest area contracted by 23 percent (World Bank 2021). This is but one salient example of the potential for lasting ecological benefits afforded by the establishment of socialist states.

Prioritizing the development of collective consumption infrastructure is another important aspect of socialist states. Public transportation was privileged in urban centres over individual motorized vehicle use, consequently always highly attenuated. City planning included provisions and actualizations of ample green spaces. Internal travel and internal population migration tended and tends to be restricted while housing was and is often centrally planned and largely guaranteed, all of which contributes to reducing if not preventing urban expansion and to keeping air pollution in check in areas between cities (Goldman 1972; Josephson et al. 2013, 91–93; Koont 2011, 175–176; Pryde 1972). Per-capita resource consumption was low, and materials recycling was well developed and highly encouraged, at least until the 1980s, in state-socialist Central and Eastern Europe (Birman 1989; Dudenkov 1985; Gille 2004; Goldman 1972; Krausmann et al. 2016; Peterson 1993, 130).

Air and water pollution from industries, though mainly confined to relatively small areas, became problematic, especially where state-socialist systems undertook rapid industrialization, mainly in the 1970s, but measures were adopted that led to substantial improvements in reducing pollutant output and remediating polluted sites (Iakovleva, Guerra, and Shklyarskiy 2020; Izmerov 1973; Kan 2009; Li et al. 2019). There were also little-known yet crucial early advances, like the banning of leaded petrol and paint, in the USSR by the 1940s (Thomas and Orlova 2001). To put the matter in wider perspective, such pollution problems did not arise in most state-socialist countries in Africa and over the first few decades in Southeast Asia, where pollution was largely the result of years of US military assaults (Zierler 2011). In the industrializing and industrialized state-socialist countries, measures to tackle pollution were developed. They included switching to natural gas where feasible and retrofitting industrial plants with less-polluting equipment. In countries initially lacking basic sewerage collection and treatment infrastructure, it was socialist states' introduction of water purification plants that improved public hygiene (Dominick 1998; Goldman 1972; Muldavin 2000; Placeres, Melián, and Toste 2011; Pryde 1972).

Thanks to productivity gains, cutting-edge environmental monitoring programmes and stations were developed in state-socialist countries that industrialized. Monitoring was distributed as densely as possible even over vast territories (Laity 2008; Lioubimtseva 2014; Permitin and Tikunov 1992; Pryde 1991; Whittle and Santos 2006). Spearheading several international environmental treaties, socialist states promoted biodiversity protection, the reduction of transboundary air pollution and greenhouse gas emissions, desertification containment, and soil conservation worldwide (Elie 2015; Josephson et al. 2013, 196–197; Oldfield 2018; Sokolovsky 2004).

The ecologically constructive impacts of socialist states, within their borders as well as globally, are in part traceable to the environmental sensibility pervading sections of what became ruling parties. In the USSR, that sensibility endured even at the height of civil war and internal, mass-murderous repression (Foster 2015; Weiner 1999). Socialist states that, like the PRC, were turned into one-party governing institutions presiding over capitalist economies (Long, Herrera, and Andréani 2018) have demonstrated an environmentally disastrous tendency instead. In the PRC, the struggle within the party and society at large over the meaning and practice of ecological civilization, which means returning to a socialist path, seems decisive in stemming and reversing the environmentally destructive tide (Gare 2012; Li 2016). To some extent, this tendency towards greater environmental harm with more capitalist reform is evident when pro-capitalist technocrats within a ruling party have a greater say in running an economy within state-socialist countries. However, the tendency was or had been held in check by combinations of environmental movements, concerned or sympathetic high-ranking party officials, and regional to local bureaucracies standing to lose from environmental problems (see also McIntyre and Thornton 1978). In Cuba, current environmental protection accomplishments are part of long-standing concerns with undoing the damaging and deadly environmental and social legacies of the capitalist past, including the struggle to overcome centuries of degrading practices grounded in sugar exports (Rosset and Benjamin 1994; Monzote 2008; Whittle and Rey Santos 2006). Cuba has thereby become a beacon of ecological sustainability under adversity and, arguably, the only remaining model of state socialism (Cabello et al. 2012; Levins 2005).

The capitalist context of socialist states

These variable but mainly positive environmental impacts of socialist states are remarkable when wider contemporary contexts and capitalist historical legacies are considered. Internal political struggles and external military and economic assaults have made ecologically constructive relations extremely tough to achieve or maintain. At the same time, socialist states inherited situations of much environmental devastation and widespread social deprivation. State-socialist countries have also been very socially and ecologically diverse. A few were already industrialized, but most were largely agrarian, and many of those had also been looted by colonizing liberal democratic empires.

There were certainly major environmentally destructive aspects to state socialism, mainly through the introduction and expansion of increasingly export-oriented mining, fossil fuel production and use, and agrochemical-based and mechanized farming, as well as manufacturing and processing industries. These, however, are typical negatives encountered in any capitalist country, more so in countries reliant on raw material and/or intermediate manufacturing products like state-socialist countries.

The matter must be contextualized. Unlike the wealthiest capitalist countries (mainly liberal democracies), state-socialist countries' sometimes inadequate capacity for pollution prevention or remediation is traceable to an inability to accumulate or direct enough resources to those ends. This is especially understandable when it is viewed relative to the high risk associated with reducing the level of military defense to parry liberal democracies' belligerence, never mind the historical lack of colonies to pillage and to offload problems.

Too often it is underappreciated or even forgotten that the new revolutionary formation that became the USSR was invaded militarily soon after its establishment and that it was constantly besieged economically and threatened militarily by liberal democracies until its dissolution. At the same time, the Bolsheviks in power did not have the allegiance of most of the peasant majority. It is in this context of internal political weakness and enormous external pressures that the USSR's degeneration into an ever more centralized, paranoid, partially self-destructive, and, for a time, terrorizing set of institutions came about (Getty and Naumov 1999; Weiner 2017). In China, the historical lesson was incorporated to develop strategies to ensure the backing of most peasants, but the revolutionaries had to contend with a long tradition of non-capitalistic commercial orientation as well (D'Mello 2009; Marks 2017). Similarly precarious situations permeated all other socialist states.

These aspects alone should remind ecosocialists of the high stakes and the immediate, mass-murderous capitalist reprisals that accompany any successful or even potentially successful revolutionary activity. Out of the experience of the Russian Revolution, one could say that achieving socialism (and eventually communism) hinges at the very least on obtaining mass support and securing international coordination pre-empting devastating capitalist reaction. This requirement of at least neutralizing the most militarily and economically powerful capitalist formations is something that Marx and Engels had pointed out by the 1880s (Rodney 2011). Subsequent revolutions nevertheless benefited greatly from the support of what survived of the USSR, even if eventually splits emerged, as with the PRC, Yugoslavia, and Albania. From these splits, there is much that still needs to be learned about the advantages and dangers of state power and political fragmentation in an international scene characterized by increasing centralization and concentration of capitalist power embodied in gigantic corporations.

The development of technology and expansion of productive capacity were primary objectives in socialist states not only in raising the material well-being of the majorities whose lives were mainly awful in pre-revolutionary times, but also in terms of sheer survival under siege from "the most advanced countries" (Weiner 1988, 233) and from inimical forces within state-socialist countries. A contradiction ensued from this general set of conditions that should be much more appreciated. The connection between developing the forces of production and living standards improvements and self-defense was made amply clear not too long after the first successful socialist revolution in what came and went as the USSR.

The legitimate claim for self-defense and survival concealed internal strife and increasing centralization of political decision-making processes, which were also precipitated by capitalist encirclement. The matter was especially urgent in the early decades, but the threat and massive pressures lasted even while chances of invasion were markedly reduced by means of the insanity of nuclear warhead détente. One can point to the 1930s purges or even earlier, with the violent suppression of the Kronstadt Soviet and the Tambov peasant revolt, as the beginning of the end or of the development of a degenerating system, but the state of siege was no mere Bolshevik hallucination or ideological expedient. All countries that underwent similar or similarly inspired revolutions faced similar challenges and attacks directly from liberal democracies or through their proxies. The Cuban state, as argued here the only socialist state left, is a living example of what will happen to a successful revolution elsewhere, if any comes about of that calibre again.

These comprise major social contradictions that are still rarely confronted by ecosocialists. They are manifested in three main ways. One is the unfavourable contexts wherein socialist movements were able to bring about revolutions and the socialist states (instead of, for example, confederations of workers councils) that emerged from those revolutions. The present neoliberal capitalist predominance throughout most of the world and thereby exacerbated or newly caused global environmental degradation constitute a major set of conditions for current struggles. Another process is the socialism-undermining interconnectivity between state-socialist and capitalist countries, which are former or present colonial powers. This was a challenge in the past and will be so for any future successful revolution, including between socialist states or formations. Another major context is composed of pre-existing and newly formed internal social conflicts, including decolonization and class struggles, which are always constituted through multiple forms of oppression. These are not going away and were sets of relations that posed great challenges to socialist movements, whether they succeeded in bringing about revolutions or not.

These three overarching processes combined in the past to undermine or redirect in unexpected negative directions whatever promising paths socialist revolutions or socialist states were undergoing. But they were also met with the resolve of many pushing against those destructive aspects and succeeding in eventually arriving at better living conditions and net positive environmental outcomes. Those processes underlie two major contradictions that should be addressed in any ecosocialist project.

One is having to build defenses to fend off capitalist powers while bringing sufficient material well-being to all and less, not more, militarism. Part of this is also ensuring that basic infrastructure is functioning or even exists at all. The Norilsk smelters, the coal mines of the German

Democratic Republic, the development of processing plants in Vietnam, and the PRC's Great Leap Forward are among the numerous examples that followed from these pressures and attempts to improve living standards and the state of a country with available technologies.

The second main contradiction is in building the foundations of a classless society, inimical to endless capital accumulation, while having to exert the utmost effort to survive in a capitalist world economy. To start building such foundations, most socialist revolutions had to face and try to overcome the oppressive and at times ecologically unsustainable social structures in which many revolutionaries were raised. Many revolutions occurred in mainly agrarian or peasant-majority societies with semi-feudal arrangements heavily influenced by capitalist relations. The undermining and transformation of peasant systems had been well underway prior to those revolutions and, in most cases, under the brutal and stultifying conditions of colonialism. Just to enable people to raise their material well-being required technological and economic resources unavailable in the countries where the revolutions succeeded. Having to import manufactures or technologically crucial parts often necessitated raw material export under declining terms of trade. The raw material export aspect is one example of pressures to introduce or expand environmental damage. Something similar occurred by way of loan repayment and joint company schemes in the industrialized state-socialist countries, where environmentally unsound production was, to a major degree, to satisfy capitalist countries' market demands. Simply delinking from the capitalist world economy can lead to a drastic reduction of lifespans or worsening living standards, the stuff that creates the mass discontent useful in helping reactionaries restore capitalism.

To summarise, three broad, inter-related factors have shaped the relationship between state socialism and the environment. One is the effects of the shifting context of socialist states in a capitalist world economy. Another is the changing interconnections between socialist states and capitalist (and often former colonial) powers. A third is the pre-existing and newly formed internal social conflicts faced in and between state-socialist countries following successful revolutions or systemic changes, including class struggles. These three processes inform overarching quandaries that need to be faced by any current socialist formation: building the foundations for a future classless and state-free society and the defenses against capitalist powers and survive within a capitalist world economy while bringing sufficient material well-being to all and, in the process, avoiding environmental harm. All this is necessary while at the same time ensuring close coordination and mutual support among socialist states and socialist movements. Addressing these quandaries involves two forms of inter-related but different struggles: a social and an ecological one. Socialist states exemplified this combined struggle. They still offer much, not only in terms of signposts about what to prevent, but also of potential to overcome capitalist relations in ecologically sustainable ways.

Conclusions

Overall, socialist states, when not regressing to capitalist ways, were able to expand or keep natural preserves, protect most species, keep consumption levels low, produce more ecologically sustainable cities than the most industrialized capitalist countries, expand environmental monitoring capacity, and much else, all under immense pressures from within and without throughout their existence. One should draw from and build on such historical experiences and achievements in ways that can help develop an ecological socialism, rather than jettisoning state socialism entirely.

There is much that can be built or rebuilt from the decades of experiences and successes in state-socialist countries. It is neither necessary nor advisable to start from scratch if the desire is to move as quickly as possible beyond capitalism. Rejecting state socialism is not much of an option, considering how state-socialist systems helped reduce or mitigate the destructive tendencies of capitalism and given the deterioration of social and ecological conditions worldwide, especially since the 1990s. What should be done instead is to revisit and learn from socialist states so as to build on their strengths and overcome their awful aspects. That is, each state-socialist case can be studied in context, as was to some extent done here, to learn what worked well towards biophysical sustainability and determine the reasons for such successes. The same should be carried out regarding negative outcomes.

There are certainly other facets that should be investigated for a critical reappraisal of the relationship between state socialism and the environment, such as failures to recruit scientists in the biophysical and engineering fields, the variable and shifting biophysical processes with which socialist states have dialectically related, studying the effects of biophysical processes independent of human action, and accounting for the irreversibility of some destructive environmental impacts. Such challenges cannot be examined here but should be kept in mind in the quest for an ecosocialist future, if it is to be truly ecologically oriented as well as socialist.

As the Burkinabé revolutionaries understood, ecological sustainability is a political struggle and socialism its linchpin. Inheriting and elaborating on a century of socialist experiences and centuries of decolonization struggles, they had embarked on a promising adventure cut short by violent French neocolonial intervention. Nevertheless, they showed, risking their lives and being martyred, that there is little prospect for an ecologically sustainable society without socialism, without decolonization. Like some of the early Bolsheviks, they showed how socialism is imbued with ecological thinking and how constructive state institutions can be in the quest for social equality and ecological sustainability. State socialism, as an intermediate phase, can have and has had ecologically and socially beneficial effects. This can be claimed without downplaying state-socialist problems and occasional horrors. The issue is recognizing what has gone well and facing up to what has gone wrong, with an aim to inform current and future efforts, strategies, and struggles for an ecologically sustainable, egalitarian, classless, state-free society.

References

Baer, Hans. 2018. *Democratic Eco-Socialism as a Real Utopia: Transitioning to an Alternative World System*. New York: Berghahn.

Barr, Brenton M. 1988. "Perspectives on Deforestation in the USSR." In *World Deforestation in the Twentieth Century*, edited by John F. Richards and Richard P. Tucker, 230–261. Durham, NC: Duke University Press.

Betancourt, Mauricio. 2020. "The Effect of Cuban Agroecology in Mitigating the Metabolic Rift: A Quantitative Approach to Latin American Food Production." *Global Environmental Change* 63: 102075.

Biney, Ama. 2018. "Madmen, Thomas Sankara and Decoloniality in Africa." In *A Certain Amount of Madness: The Life, Politics and Legacies of Thomas Sankara*, edited by Amber Murray, 127–146. London: Pluto Press.

Birman, Igor. 1989. *Personal Consumption in the USSR and the USA*. New York: Palgrave Macmillan.

Biró, Marianna, Bálint Czúcz, Ferenc Horváth, András Révész, Bálint Csatári, and Zsolt Molnár. 2013. "Drivers of Grassland Loss in Hungary During the Post-Socialist Transformation (1987–1999)." *Landscape Ecology* 28: 789–803.

Bot, Alexandra J., Freddy O. Nachtergaele, and Anthony Young. 2000. *Land Resource Potential and Constraints at Regional and Country Levels*. World Soil Resources Report 90. Rome: Food and Agriculture Organization of the United Nations, www.fao.org/3/a-x7126e.pdf (accessed 11 November 2020).

Brain, Stephen. 2011. *Song of the Forest: Russian Forestry and Stalinist Environmentalism, 1905–1953*. Pittsburgh: University of Pittsburgh Press.

Brown, Kate. 2013. *Plutopia: Nuclear Families, Atomic Cities, and the Great Soviet and American Plutonium Disasters*. Oxford: Oxford University Press.

Bruno, Andy. 2016. *The Nature of Soviet Power: An Arctic Environmental History*. Cambridge: Cambridge University Press.

Cabello, Juan José, Dunia Garcia, Alexis Sagastume, Rosarlo Priego, Luc Hens, and Carlo Vandecasteele. 2012. "An Approach to Sustainable Development: The Case of Cuba." *Environment Development and Sustainability* 14: 573–591.

Corset, Shirley, and Howard Waitzkin. 1986. "Capitalism, Socialism, and the Physical Quality of Life." *International Journal of Health Services* 16 (4): 643–658.

Chase-Dunn, Christopher. 1982. "The Transition to World Socialism." In *Socialist States in the World-System*, edited by Christopher Chase-Dunn, 271–296. Beverly Hills, CA: Sage.

Chendev, Yuri G., Thomas. J. Sauer, Guillermo H. Ramirez, and Charles L. Burras. 2015. "History of East European Chernozem Soil Degradation; Protection and Restoration by Tree Windbreaks in the Russian Steppe." *Sustainability* 7: 705–724.

Colwell, Mark A., Alexander V. Dubynin, Andrei Yu. Koroliuk, and Nikolai A. Sobolev. 1997. "Russian Nature Reserves and Conservation of Biological Diversity." *Natural Areas Journal* 17 (1): 56–68.

Coronel Vargas, Gabriela, William W. Au, and Alberto Izzotti. 2020. "Public Health Issues from Crude-Oil Production in the Ecuadorian Amazon Territories." *Science of the Total Environment* 719: 134647. https://doi.org/10.1016/j.scitotenv.2019.134647 (accessed 4 November 2020).

D'Mello, Bernard. 2009. "What Is Maoism?" *Economic and Political Weekly* 44 (47): 37–48.

Dhara, V. Ramana, and Rosaline Dhara. 2002. "The Union Carbide Disaster in Bhopal: A Review of the Health Effects." *Archives of Environmental Health* 57 (5): 391–404.

Dominick, Raymond. 1998. "Capitalism, Communism, and Environmental Protection: Lessons from the German Experience." *Environmental History* 3 (3): 311–332.

Dudenkov, Staly V. 1985. "The Recycling of the Wastes of Production and Consumption as an Aspect of the Environmental Protection in the USSR." In *Appropriate Waste Management for Developing Countries*, edited by Kriton Curi, 95–100. Boston: Springer.

Elie, Marc. 2015. "Formulating the Global Environment: Soviet Soil Scientists and the International Desertification Discussion, 1968–91." *The Slavonic and East European Review* 93 (1): 181–2014.

Elie, Marc, and Laurent Coumel. 2013. "A Belated and Tragic Ecological Revolution: Nature, Disasters, and Green Activists in the Soviet Union and the Post-Soviet States, 1960s–2010s." *The Soviet and Post-Soviet Review* 40: 157–165.

Engel-Di Mauro, Salvatore. 2014. *Ecology, Soils, and the Left: An Eco-Social Approach*. New York: Palgrave Macmillan.

Engel-Di Mauro, Salvatore. 2021. *State Socialism and the Environment*. London: Pluto Press.

Fernandez, Margarita, Justine Williams, Galia Figueroa, Garrett Graddy-Lovelace, Mario Machado, Luis Vazquez, Nilda Perez, Leidy Casimiro, Graciela Romero, and Fernando Funes-Aguilar. 2018. "New Opportunities, New Challenges: Harnessing Cuba's Advances in Agroecology and Sustainable Agriculture in the Context of Changing Relations with the United States." *Elementa Science of the Anthropocene* 6: 76.

Fitz, Don. 2020. *Cuban Healthcare: The Ongoing Revolution*. New York: Monthly Review Press.

Foster, John Bellamy. 2015. "Late Soviet Ecology and the Planetary Crisis." *Monthly Review* 67 (2): 1–20.

Gare, Arran. 2002. "The Environmental Record of the Soviet Union." *Capitalism Nature Socialism* 13 (3): 52–72.

Gare, Arran. 2012. "China and the Struggle for Ecological Civilization." *Capitalism Nature Socialism* 23 (4): 10–26.

Getty, J. Arch, and Oleg Naumov. 1999. *The Road to Terror: Stalin and the Self-destruction of the Bolsheviks, 1932–1939*. New Haven, CT: Yale University Press.

Gille, Zsuzsa. 2004. "Europeanising Hungarian Waste Policies: Progress or Regression?" *Environmental Politics* 13 (1): 114–134.

Goldman, Marshall I. 1972. *The Spoils of Progress: Environmental Pollution in the Soviet Union*. Boston: The MIT Press.

Iakovleva, Emiliia V., Daniel D. Guerra, and Andrey Y. Shklyarskiy. 2020. "Alternative Measures to Reduce Carbon Dioxide Emissions in the Republic of Cuba." *Journal of Ecological Engineering* 21 (4): 55–60.

Izmerov, Nikolaj Fedotovic. 1973. *Control of Air Pollution in the USSR*. Geneva: World Health Organisation.

Josephson, Paul, Nicolai Dronin, Aleh Cherp, Ruben Mnatsakanian, Dmitry Efremenko, and Vladislav Larin. 2013. *An Environmental History of Russia*. Cambridge: Cambridge University Press.

Kan, Haidong. 2009. "Environment and Health in China: Challenges and Opportunities." *Environmental Health Perspectives* 117 (12): A530–A531.

Kasar, Sharayu, Suchismita Mishra, Yasutaka Omori, Sarata Kumar Sahoo, Norbert Kavasi, Hideki Arae, Atsuyuki Sorimachi, and Tatsuo Aono. 2020. "Sorption and Desorption Studies of Cs and Sr in Contaminated Soil Samples around Fukushima Daiichi Nuclear Power Plant." *Journal of Soils Sediments* 20: 392–403.

Kelley, Donald R. 1976. "Environmental Policy-Making in the USSR: The Role of Industrial and Environmental Interest Groups." *Soviet Studies* 28 (4): 570–589.

Koont, S. 2011. *Sustainable Urban Agriculture in Cuba*. Gainesville: University Press of Florida.

Krausmann, Fridolin, Birgit Gaugl, James West, and Heinz Schandl. 2016. "The Metabolic Transition of a Planned Economy: Material Flows in the USSR and the Russian Federation 1900 to 2010." *Ecological Economics* 124: 76–85.

Laity, Julie. 2008. *Deserts and Desert Environments*. Chichester: Wiley.

Lane, David. 2014. *The Capitalist Transformation of State Socialism: The Making and Breaking of State Socialist Society, and What Followed*. London: Routledge.

Levins, Richard. 2005. "How Cuba Is Going Ecological." *Capitalism Nature Socialism* 16 (3): 7–25.

Li, Minqi. 2016. *China and the 21st Century Crisis*. London: Pluto Press.

Li, Tiankui, Yi Liu, Sijie Lin, Yangze Liu, and Yunfeng Xie. 2019. "Soil Pollution Management in China: A Brief Introduction." *Sustainability* 11: 556.

Lioubimtseva, Elena. 2014. "Impact of Climate Change on the Aral Sea and Its Basin." In *The Aral Sea*, edited by Philip Macklin, 405–427. Berlin: Springer.

Liu, Mingliang, and Hanqin Tian. 2010. "China's Land Cover and Land Use Change from 1700 to 2005: Estimations from High-Resolution Satellite Data and Historical Archives." *Global Biogeochemical Cycles* 24: GB3003.

Long, Zhiming, Rémy Herrera, and Tony Andréani. 2018. "On the Nature of the Chinese Economic System." *Monthly Review* 70: 32–43.

Machín Sosa, Braulio, Adilén María Roque Jaime, Dana Rocío Ávila Lozano, and Peter Michael Rosset. 2013. *Agroecological Revolution: The Farmer-to-Farmer Movement of the ANAP in Cuba*. Havana: ANAP and Via Campesina.

Marks, Robert B. 2017. *China: An Environmental History*. Lanham: Rowman & Littlefield.

McIntyre, Robert J., and James R. Thornton. 1978. "On the Environmental Efficiency of Economic Systems." *Soviet Studies* 30 (2): 173–192.

Monzote, Reinaldo Funes. 2008. *From Rainforest to Cane Field in Cuba: An Environmental History since 1492*. Translated by A. Martin. Raleigh: The University of North Carolina Press.

Moran, Daniel D., Mathis Wackernagel, Justin A. Kitzes, Steven H. Goldfinger, and Aurélien Boutaud. 2008. "Measuring Sustainable Development – Nation by Nation." *Ecological Economics* 64 (3): 470–474.

Muldavin, Joshua. 2000. "The Paradoxes of Environmental Policy and Resource Management in Reform-Era China." *Economic Geography* 76 (3): 244–271.

Navarro, Vicente. 1992. "Has Socialism Failed? An Analysis of Health Indicators under Socialism." *International Journal of Health Services* 22 (4): 583–601.

O'Connor, James. 1970 [1998]. *Natural Causes. Essays in Ecological Marxism*. New York: Guilford.

Oldfield, Jonathan. 2018. "Imagining Climates Past, Present and Future: Soviet Contributions to the Science of Anthropogenic Climate Change, 1953–1991." *Journal of Historical Geography* 60: 41–51.

Ostergren, David, and Evgeny Shvarts. 2000. "Russian Zapovedniki in 1998: Recent Progress and New Challenges for Russia's Strict Nature Preserves." *USDA Forest Service Proceedings* RMRS-P-14: 209–213.

Permitin, Viktor E., and Vladimir S. Tikunov. 1992. "Environmental Monitoring in the USSR: Present State and New Tasks." *International Journal of Environmental Studies* 40 (1): 67–77.

Peterson, D.J. 1993. *Troubled Lands. The Legacy of Soviet Environmental Destruction*. A RAND Research Study. Boulder, CO: Westview Press.

Placeres, Manuel R., Maricel G. Melián, and Mireya A. Toste. 2011. "Principales Características de la Salud Ambiental de la Provincia La Habana [Main Characteristics of Environmental Health in Habana Province]." *Revista Cubana de Higiene y Epidemiología* 49 (3): 384–398.

Potapov, P.V., S.A. Turubanova, A. Tyukavina, A.M. Krylov, J.L. McCarty, V.C. Radeloff, and M.C. Hansen. 2015. "Eastern Europe's Forest Cover Dynamics from 1985 to 2012 Quantified from the Full Landsat Archive." *Remote Sensing of Environment* 159: 28–43.

Pryde, Philip R. 1972. *Conservation in the Soviet Union*. Cambridge: Cambridge University Press.

Pryde, Philip R. 1991. *Environmental Management in the Soviet Union*. Cambridge: Cambridge University Press.

Ritchie, Hannah, and Max Roser. 2017. "CO_2 and Greenhouse Gas Emissions." https://ourworldindata.org/co2-and-other-greenhouse-gas-emissions (accessed 21 October 2020).

Rodney, Walter. 2011 [1972]. *How Europe Underdeveloped Africa*. Baltimore, MD: Black Classic Press.

Roman, Joe. 2018. "The Ecology and Conservation of Cuba's Coastal and Marine Ecosystems." *Bulletin of Marine Science* 94 (2): 149–169.

Rosset, Peter, and Medea Benjamin. 1994. "Cuba's Nationwide Conversion to Organic Agriculture." *Capitalism Nature Socialism* 5 (3): 79–97.

Rosset, Peter Michael, Braulio Machín Sosa, Adilén María Roque Jaime, and Dana Rocío Ávila Lozano. 2011. "The Campesino-to-Campesino Agroecology Movement of ANAP in Cuba: Social Process Methodology in the Construction of Sustainable Peasant Agriculture and Food Sovereignty." *The Journal of Peasant Studies* 38 (1): 161–191.

Sojinu, Samuel O., Jizhong Wang, Oluwadayo Sonibare, and Eddy Zeng. 2010. "Polycyclic Aromatic Hydrocarbons in Sediments and Soils from Oil Exploration Areas of the Niger Delta, Nigeria." *Journal of Hazardous Materials* 174 (1–3): 641–647.

Sokolovsky, Valentin. 2004. "Fruits of a Cold War." In *Clearing the Air: 25 Years of the Convention on Long-Range Transboundary Air Pollution*, edited by Johan Sliggers and Willem Kakebeeke, 7–15. Geneva: United Nations.

Sosa, Dayana, Isabel Hilber, Roberto Faure, Nora Bartolomé, Osvaldo Fonseca, Armin Keller, Thomas D. Bucheli, and Arturo Escobar. 2019. "Polycyclic Aromatic Hydrocarbons and Polychlorinated Biphenyls in Urban and Semi-Urban Soils of Havana, Cuba." *Journal of Soils & Sediments: Protection, Risk Assessment, & Remediation* 19 (3): 1328–1341.

Suing, Guillaume, and Viktor Dedaj. 2018. *L'Écologie Réelle: Une Histoire Soviétique et Cubaine* [*Real Ecology: A Soviet and Cuban History*]. Paris: Delga.

Szalai, Erzsébet. 2005. *Socialism. An Analysis of Its Past and Future*. Budapest: Central European University Press.

Thomas, Valerie M., and Anna O. Orlova. 2001. "Soviet and Post-Soviet Environmental Management: Lessons from a Case Study on Lead Pollution." *Ambio* 30 (2): 104–111.

Weiner, Douglas R. 1988. *Models of Nature. Ecology, Conservation, and Cultural Revolution in Soviet Russia*. Bloomington and Indianapolis: Indiana University Press.

Weiner, Douglas R. 1999. *A Little Corner of Freedom: Russian Nature Protection from Stalin to Gorbachëv*. Berkeley: University of California Press.

Weiner, Douglas R. 2017. "Communism and Environment." In *The Cambridge History of Communism*, edited by Juliane Fürst, Silvio Pons, and Mark Selden, 502–528. Cambridge: Cambridge University Press.

Whittle, Daniel, and Orlando Rey Santos. 2006. "Protecting Cuba's Environment: Efforts to Design and Implement Effective Environmental Laws and Policies in Cuba." *Cuban Studies* 37: 73–103.

World Bank. 2021. "Forest Area (Sq. Km) – Burkina Faso." https://data.worldbank.org/indicator/AG.LND.FRST.K2?locations=BF (accessed 13 January 2021).

Wuepper, David, Pasquale Borrelli, Daniel Mueller, and Robert Finger. 2020. "Quantifying the Soil Erosion Legacy of the Soviet Union." *Agricultural Systems* 185: 102940.

Zierler, David. 2011. *The Invention of Ecocide: Agent Orange, Vietnam, and the Scientists Who Changed the Way We Think About the Environment*. Athens: University of Georgia Press.

33

ECOSOCIALISM AND SOCIALISM WITH CHINESE CHARACTERISTICS

Qingzhi Huan

An overview of ecosocialism research in China

The study of ecosocialism or eco-Marxism in a broad sense as an academic subject in China began in the mid-1980s. One of the landmark works is a 1986 article by Jin Wang, published in the journal *Teaching and Research*. In that article, he described "eco-Marxism" and "ecosocialism" as two slightly different theoretical trends of thought (Wang 1986). According to him, the former is more from the eco-Marxist school of North America, especially the writings of William Leiss and Ben Agger, while the latter is based more on the political ideology of green parties in Europe and the German Green Party in particular. Since then, ecosocialism or eco-Marxism as a burgeoning school of foreign socialism or Marxism has been attracting more and more attention from Chinese academia, and in most cases, these two terms are used interchangeably (Chen 2003; Liu 2007; Xu 2007; Wang 2009; Cai 2014).

Generally speaking, there are three major themes or foci in Chinese ecosocialism research, or the Chinese version of ecosocialism (Huan 2010 [2017]). They are Marx and Engels's thoughts on the relationship between human and nature, Western ecosocialists' or eco-Marxists' theories and writings, and ecosocialist analyses of and reflections on the theory and practice of Socialism with Chinese Characteristics.

Due to the distinct contexts of theory and reality, quite a few Chinese scholars are devoted to digging and combing through the Marxist sources of ecological thinking, or the harmonious relationship between humans and nature, insisting that scientific socialism or communism is the best form of ecology. For instance, in my own work (Huan 1994), I conceptualized Marx's *theory of humanized nature* from his basic concept of "objective activity" as four elements, and Engels's *dialectic view of nature* from his core concept of "dialectics of nature" as four other elements. The former set of elements is as follows: humanization of nature is a gradual process of history; objective necessity of natural things is still there but with a new meaning; human objective activity is the way of humanizing nature; and human society is the realistic basis of the humanization of nature. The latter set is as follows: nature is inter-related and interacting; dialectical laws are the basic laws in the development and change of nature; natural science is the realistic tool to describe the dialectical picture of nature; and the development of nature and human thinking follows the same dialectical rules. My major arguments are that there is an internal consistency between Marx's and Engels's ecological thoughts or the Marxist view of

10.4324/9780429341427-38

nature and modern ecology. Furthermore, Marx's view of humanized nature and naturalized humans and Engels's view of dialectical nature and dialectical humans are complementary rather than antagonistic from a broader perspective of ecology, especially from an eco-Marxist perspective. By and large, even while using different interpretive approaches and sources, Chinese ecosocialists share an adherence to and development of Marx and Engels's ecological thoughts or Marxist view of nature (Xie 2003; Liu 2010; Chen and Cai 2014).

Another major research area is the translation and theorization of Western ecosocialists' writings. Most well-known and representative works by ecosocialists were already starting to be translated into Chinese in the early 1990s. Among them are the works of Ben Agger (1991), William Leiss (1993 [2016]), James O'Connor (1998), David Pepper (2005), John B. Foster (2006), Saral Sarkar (2008), Jonathan Hughes (2011), Ted Benton (1991), Joel Kovel (2015) and André Gorz (2018). Meanwhile, hundreds of monographs (dissertations), journal articles, and book chapters have been published over the past decades that focus on the textual interpretation and critical analysis of Western ecosocialists' writings, especially from a classical Marxist perspective. Arguably, for most researchers in this field as well as for "outsiders," textual interpretations and critical analyses of Western ecosocialists' writings constitute the main work on and approach to the study of ecosocialism or eco-Marxism (Chen 2012; Wen 2015; Wu 2017). Admittedly, despite the considerable output, the overall caliber of such publications should not be overestimated. For a breakthrough to be achieved, in this author's estimation, current research on the Chinese context of socialist modernization needs to be integrated with these works on Western ecosocialism so as to construct a real space of critical exchange and dialogue between Western ecosocialists and Chinese scholars (Huan 2019a).

The third research area is exploring, from an ecosocialist perspective, the thematic periodization and strategies of China's socialist modernization, such as environmental protection, sustainable development, scientific development, and eco-civilization construction. Taking 2012 as a boundary, for instance, Li Zang, Yanxi Zhang and Xia Shi, Leisheng Zhang and Yu Zhang, and Yan Wang have respectively summarized the expositions of Marx and Engels on the harmonious relationship between humans and nature, environmental protection, sustainable development, and scientific development (Zang 2003; Zhang and Shi 2004; Zhang and Zhang 2006; Wang 2012). Xigang Liu, Shinan Fang, Yunfei Zhang, and Yuchen Wang have focused their efforts on the close relationship between eco-civilization and historical materialism and/or eco-Marxism, or Marx and Engels's thoughts on eco-civilization (Liu 2014; Zhang 2014; Wang 2015; Fang 2017). According to them, all efforts to keep pace with the times in the theoretical generalizations and strategic choices of the governing Communist Party of China (CPC) can find their theoretical origins in Marx and Engels's thoughts on the relationship between humans and nature, or eco-Marxism in a broad sense.

The integration of socialist eco-civilization into Socialism with Chinese Characteristics

Marked by the 17th National Congress of CPC held in 2007, the integration of the socialist eco-civilization concept and strategy into the theoretical system of Socialism with Chinese Characteristics has become an important issue for the study of ecosocialism in China. To put it differently, the theoretical and practical exploration of the socialist road with Chinese characteristics led by the CPC since the foundation of People's Republic of China (PRC) in 1949, particularly since the reform and opening up in 1978, has been entering into a new era, realizing that better ecological environment protection and governance must be prioritized, and this requires the initiation and implementation of a comprehensive or societal transformation.

In his working report to the CPC's 17th National Congress, Secretary General Jintao Hu expounded the main points of the new political ideology and general plan of governing the country through *eco-civilization construction* (2007, 20). According to him, on the one hand, China, led by CPC, will from now on be devoted to the construction of eco-civilization by gradually forming an energy- and resource-saving and environmentally friendly industrial infrastructure, mode of growth, and consumption pattern, scaling up the circular economy and the portion of renewable energy sources to the national level, getting major pollutants under control, and thereby improving environmental quality. On the other hand, the CPC and the administrative organs at different levels will work hard to instill the idea of eco-civilization in the whole society. In what is already a breakthrough for the CPC, the governments at different levels and the common people have incorporated the terms *construction of eco-civilization* and *constructing an eco-civilization* into the CPC's major policy documents, referring to both a distinctive political ecology (a "red-green" ideology) and a new public policy of environmental governance. Noteworthy, however, is the absence of terms such as *socialism* and *socialist* in such official documents.

Compared to his working report to the 17th National Congress, Secretary General Jintao Hu (2012) devoted an entire chapter rather than just a paragraph to elaborating on the necessity of and strategic measures to take in promoting the construction of eco-civilization. He systematically analyzed the general ideas for vigorously promoting the construction of eco-civilization, which aims at resource-saving and environmentally friendly spatial use patterns, industrial infrastructure, modes of production and lifestyles by implementing the basic national policy of environmental protection and the strategies of *Three Developments* (green, circular, and low carbon) and expounding on the three "overall objectives": reversing environmental deterioration trends at the source, creating better production systems and living environments for the people, and contributing to global ecological security. He described four key policy aspects to heighten the promotion of building an eco-civilization: optimizing land-use patterns and spatial arrangements, promoting resource conservation comprehensively, strengthening ecosystem protection, and strengthening the construction of the eco-civilization system. Taken as a whole, these constitute a set of guidelines for the CPC and for governments at different levels.

From an ecosocialist perspective, it is very significant that in the last paragraph, Hu's report calls for the "striving for a new era of socialist eco-civilization" (Hu 2012, 41). What is more, in its section on the general principles of the revised *Charter of the CPC*, the first sentence in clause no. 18 states: "The CPC will lead the people in building a socialist eco-civilization" (CPC 2012, 6). Arguably, by saying *socialist eco-civilization* instead of *eco-civilization*, the working report and the revised *Charter of the CPC* conveyed a clear message to the world that the construction of eco-civilization in the Chinese context is inseparable from the theory and practice of Socialism with Chinese Characteristics.

A major difference between 2012 and 2017 working reports is the signaling of a structural change. That is to say, Socialism with Chinese Characteristics is understood to have entered a new era, and this gives the interpretive context for all the detailed discussion on strategic targets and general task requirements in the chapter "Accelerating the reform of eco-civilization system, building a beautiful China" (Xi 2017a, 50–52). First, Secretary General Jinping Xi expounded on the reasons for promoting the reform of the eco-civilization system or the construction of eco-civilization at the levels of political philosophy, as well as of goal setting and strategic measures. The reform of the eco-civilization system is a very important supporting dimension, indicating that some comprehensive, fundamental, and historic changes have taken place in China over the past five years. "Adhering to the harmonious coexistence of human and nature" is one of 14 general plans in the framework of carrying Socialism with Chinese

Characteristics through a new era. The phase-based objectives of eco-civilization construction are among the key indicators to evaluate and measure China's progress towards the primary stage of socialism. Following this, in his report, Xi described the "Four strategic deployments" for promoting China's construction of eco-civilization in a new era: "promoting green development," "focusing on solving outstanding environmental problems," "strengthening the protection of ecosystems," and "reforming the environmental monitoring system." Together with the 2012 working report's descriptions of the construction of eco-civilization, the "four strategic deployments" are an authoritative expression of the CPC's political, ecology, and policy guidelines as the governing political party.

Of significance is the very first inclusion of a chapter on the outlook of socialist eco-civilization. In the last paragraph of the chapter, Xi calls for "Firmly establishing the concept of socialist eco-civilization in the society to promote the formation of a new pattern of modernization construction characterized by the harmonious coexistence of humanity and nature" (Xi 2017a, 52). The raising of these new terms and their connotations should not be overinterpreted, especially from an ecosocialist or eco-Marxist perspective. Instead, it is more reasonable to say that such new terms echo what was said in the last paragraph of the working report in 2012. Socialist eco-civilization is just another political conceptualization or summary of what was stated in the previous working report.

Nevertheless, at least for the core members of the China Research Group on Socialist Eco-Civilization (CRGSE), which was formed jointly in 2015 by Peking University and the Rosa Luxemburg Foundation, the term *socialist eco-civilization* should be defined and understood from a more radical "green-Left" perspective, and from the perspective of ecosocialism or eco-Marxism in particular (Huan 2016). For the CRGSE, there are some substantial differences between socialist eco-civilization and eco-civilization in general. The latter can be based on a political philosophy and actual politics that are at odds with socialist eco-civilization in dealing with environmental protection and governance. It is a matter of distinguishing ecosocialism from eco-capitalism (Huan 2019b).

Xi Jinping's Thought on Eco-civilization and *Socialism with Chinese Characteristics for a new era*

At the Eighth National Ecological and Environmental Protection Conference held in May 2018, the expression of *Xin Jinping's Thought on Eco-civilization* was officially recognized to conceptualize the main theoretical achievements, general plans for governing the country, and key policy measures concerning promoting the construction of socialist eco-civilization in the new era. Now *Xin Jinping's Thought on Eco-civilization* has been widely regarded as one of the key aspects or elements of *Xin Jinping's Thought on Socialism with Chinese Characteristics for the New Era*, the authoritative expression of contemporary China's political ideology established at the 19th National Congress of CPC in 2017.

In today's Chinese academic circles, there are basically two approaches to generalize or interpret the main contents of *Xin Jinping's Thought on Eco-civilization*. The first approach uses this term in a narrow sense, focusing on his internal speeches, written instructions, and writings or publications with his own signature, especially those after the CPC's 18th National Congress in 2012. Of those writings, the most representative are *Zhijiang New Language* (2007) and *The Compilation of Xi Jinping's Expositions on Socialist Eco-civilization* (2017b). In the former, Xi first proposed and elaborated on the green development concept of "Lucid waters and lush mountains are invaluable assets," which can be to a large extent considered an abbreviation of *Xi Jinping's Thought on Eco-civilization*. In the latter, one can find the most important addresses, speeches, talks, allusions, written instructions, and congratulatory letters by him on the construction of eco-civilization between CPC's 18th and 19th National Congresses.

Another important document about *Xin Jinping's Thought on Eco-civilization* is his address at the 8th National Ecological and Environmental Protection Conference, which was later published in the political magazine *Seeking Truth* (Xi 2019). In this address, besides enumerating the reform and policy measures to be implemented, Xi systematically expounded on the six key perspectives or principles which must be clearly adhered to in promoting the construction of eco-civilization. These were immediately expanded to the so-called "eight perspectives" or "eight principles which must be clearly adhered to" (Li 2018a, 2018b). Specifically, these perspectives or principles are composed of a deep view of history, in which "[e]cological prosperity leads to civilization prosperity while ecological decline leads to civilization decline"; a scientific view of nature as "[t]he harmonious coexistence between humanity and nature"; a green development view of "[l]ucid waters and lush mountains are invaluable assets"; a view of people's basic livelihood: "Good ecological environment is the most inclusive welfare of people's livelihood"; a holistic system view of "[l]andscape, forest, field, lake and grass are the community of life"; a strict view of the rule of law in "[p]rotecting the environment with the strictest system"; a view of national action in "[b]uilding a beautiful China together"; and a win-win global outlook in "[c]o-planning in the road for global eco-civilization construction."

By contrast, some scholars emphasize that *Xi Jinping's Thought on Eco-civilization* should be understood or interpreted within a larger context. Their major arguments can be summed up as the following two points. Firstly, the construction of socialist eco-civilization has experienced a process as long as the history of PRC; thus, all the theoretical achievements and practical experiences in the field of environmental protection and governance over the past seven decades should be considered and incorporated into the system of *Xi Jinping's Thought on Eco-civilization* (Zhang 2014; Chen and Qian 2016). Secondly, it is more scientific and reasonable that *Xi Jinping's Thought on Eco-civilization* be regarded as a collective theoretical achievement, just like the important expositions of Jinping Xi on other political or policy issues, as well as other CPC top leaders' theories such as Mao Zedong Thought and Deng Xiaoping Theory (Jiang and Cui 2013).

From the perspective of ecosocialism or eco-Marxism, *Xi Jinping's Thought on Eco-civilization* and its practice raises or highlights two theoretical questions that need our further attention and discussion. The first issue is the relationship between *Xi Jinping's Thought on Eco-civilization* and ecosocialism or eco-Marxism in a broad sense. On the one hand, it is explicitly acknowledged and repeatedly stressed that *Xi Jinping's Thought on Eco-civilization* is an inheritance and development of Marx and Engels's theory of the relationship between humanity and nature (Xi 2018, 21–22). In other words, the former is a contemporary Chinese version of the latter. On the other hand, however, the theoretical criticisms of capitalism and the political imaginations of socialism in the future by Western ecosocialists or eco-Marxists are still not clearly recognized as a theoretical source for *Xi Jinping's Thought on Eco-civilization* (Wang 2015, 2017). A crucial issue here is not only an appropriate attitude towards Western Marxism but also a correct position on the socialist future of green transformation in the name of socialist eco-civilization construction. For this reason, it is still a very meaningful and necessary work to enrich *Xi Jinping's Thought on Eco-civilization* by going beyond the classic works of Marx and Engels and absorbing the theoretical achievements of Western ecosocialists more actively (Huan 2019a).

Another issue is the relationship between *Xi Jinping's Thought on Eco-civilization* and Socialism with Chinese Characteristics for a new era. On the one hand, as discussed earlier, *Xi Jinping's Thought on Eco-civilization* is a very important constituent or supportive part of *Xi Jinping's Thought on Socialism with Chinese Characteristics for a New Era*, a new official expression of Socialism with Chinese Characteristics after the 19th National Congress of CPC in 2017. Thinking about *Xi Jinping's Thought on Eco-civilization* and its practice in such a grand context, a prominent issue, among others, is that the major ideas and targets of socialist eco-civilization can only be realized under a condition of comprehensive social transformation (Huan 2019b). In other words, as many ecosocialists or eco-Marxists have often pointed out, the aims of building a socialist eco-civilization will not be achieved easily without corresponding reforms or the reconstruction of economic, political, and cultural systems.

References

Agger, Ben. 1991. *An Introduction to Western Marxism.* [in Chinese] Translated by Zhi Shen et al. Beijing: Renmin University Press.

Benton, Ted, ed. 1991. *The Greening of Marxism.* [in Chinese] Translated by Rongxiang Cao and Jilong Li. Beijing: Social Science Academic Press.

Cai, Huajie. 2014. *Is Another World Possible? A Study on Ecosocialism.* [in Chinese] Beijing: Social Science Academic Press.

Chen, Xicheng, and Hutang Cai. 2014. *Marx's and Engels' Thought on Ecological Philosophy and Its Contemporary Value.* [in Chinese] Beijing: China Social Science Press.

Chen, Xueming. 2003. *On Ecosocialism.* [in Chinese] Xinbei: Yang Chih Book.

Chen, Xueming. 2012. *Who Is Culprit? Seeking the Root of Ecological Crisis.* [in Chinese] Beijing: People's Publishing House.

Chen, Zhilin, and Hangyuan Qian. 2016. "An Overview on the Research of Jiang Zemin's Ecological Thought." [in Chinese] *Theory Research* 4: 10–14.

Communist Party of China (CPC). 2012 [2017]. *The Charter of the CPC.* [in Chinese] Beijing: People's Publishing House.

Fang, Shinan. 2017. *Marx's and Engels' Thought on Eco-Civilization: A Study Based on the Collection of Marx and Engels.* [in Chinese] Beijing: People's Publishing House.

Foster, John B. 2006. *Marx's Ecology: Materialism and Nature.* [in Chinese] Translated by Rensheng Liu and Feng Xiao. Beijing: China Higher Education Press.

Gorz, André. 2018. *Capitalism, Socialism, Ecology.* [in Chinese] Translated by Shuyi Peng. Beijing: Commercial Press.

Hu, Jintao. 2007. *The Working Report to the 17th National Congress.* Beijing: People's Publishing House.

Hu, Jintao. 2012. *The Working Report to the 18th National Congress.* Beijing: People's Publishing House.

Huan, Qingzhi. 1994. *The Discovery of the Values of Natural Environment: A Study on Marx's and Engels' View of Nature from the Perspective of Contemporary Society.* [in Chinese] Nanning: Guangxi People's Publishing House.

Huan, Qingzhi, ed. 2010. *Ecosocialism as Politics: Rebuilding the Basis of Our Modern Civilisation.* Dordrecht: Springer.

Huan, Qingzhi. 2016. "Socialist Eco-Civilization and Social-Ecological Transformation." *Capitalism Nature Socialism* 27 (2): 51–66.

Huan, Qingzhi. 2017. "Eco-Marxism as a Political Philosophy." [in Chinese] *Journal of Beijing Administrative Institute* 4: 12–19.

Huan, Qingzhi. 2019a. "The Sinization of Eco-Marxism: Implication, Approach and Limit." [in Chinese] *Journal of China University of Geosciences* 4: 84–99.

Huan, Qingzhi. 2019b. "Socialist Eco-Civilization as a Transformative Politics." [in Chinese] *Marxism and Reality* 2: 21–29.

Hughes, Jonathan. 2011. *Ecology and Historical Materialism.* [in Chinese] Translated by Xiaoqiong Zhang and Xiaobin Hou. Nanjing: Jiangsu People's Publishing House.

Jiang, Li, and Minghao Cui. 2013. "An Analysis of Hu Jintao's Thought on Eco-Civilization." *The Journal of Liaoning's Institute of Socialism* 1: 55–58.

Kovel, Joel. 2015. *The Enemy of Nature: The End of Capitalism or the End of World?* [in Chinese] Translated by Yanfei Yang and Chunyong Feng. Beijing: Renmin University Press.

Less, William. 2007 [1993]. *The Gift of Nature.* [in Chinese] Translated by Changling Yue and Jianhua Li. Chongqing: Chongqing Publishing House.

Less, William. 2016. *The Limits to Satisfaction: An Essay on the Problem of Needs and Commodities.* [in Chinese] Translated by Yongxue Li. Chongqing: Chongqing Publishing House.
Li, Ganjie. 2018a. "Mobilizing the Whole Society to Build a Beautiful China with the Guidance of Xi Jinping's Thought on Eco-Civilization." [in Chinese] *China's NPC* 15: 45–50.
Li, Ganjie. 2018b. "Firmly Fighting against Pollution with the Guidance of Xi Jinping's Thought on Eco-Civilization." [in Chinese] *Report on Current Events* 14: 39–53.
Liu, Rensheng. 2007. *An Introduction to Eco-Marxism.* [in Chinese] Beijing: Central Composition and Translation Press.
Liu, Xigang. 2014. *Marx and Engels' Thought on Eco-civilization and Its Practice in China.* [in Chinese] Beijing: China Social Science Press.
Liu, Zenghui. 2010. *Marxist Ecological Thought and Its Practice.* [in Chinese] Beijing: Beijing Normal University Press.
O'Connor, James. 1998. *Natural Causes: Essays in Ecological Marxism.* [in Chinese] Translated by Zhengdong Tang and Peihong Zang. Nanjing: Nanjing University Press.
Pepper, David. 2005. *Ecosocialism: From Deep Ecology to Social Justice.* [in Chinese] Translated by Ying Liu. Jinan: Shandong University Press.
Sarkar, Saral. 2008. *Ecosocialism or Eco-capitalism? A Critical Analysis of Humanity's Fundamental Choices.* [in Chinese] Translated by Shulan Zhang. Jinan: Shandong University Press.
Xi, Jinping. 2007. *Zhijiang New Language.* [in Chinese] Hangzhou: Zhejiang People's Publishing House.
Xi, Jinping. 2017a. *The Working Report to the 19th National Congress.* Beijing: People's Publishing House.
Xi, Jinping. 2017b. *The Compilation of Xi Jiping's Expositions on Socialist Eco-Civilization.* [in Chinese] Beijing: Central Party Literature Press.
Xi, Jinping. 2018. *Address at the Conference Commemorating the Bicentenary of Marx.* Beijing: People's Publishing House.
Xi, Jinping. 2019. "Promoting the Construction of Eco-Civilization in China to a New Level." [in Chinese] *Seeking Truth* 2: 4–19.
Xie, Baojun. 2003. *The Connotation of Ecological Philosophy of Marx's View of Nature: A Theoretical Harbinger of the Combination of "Red" and "Green".* [in Chinese] Harbin: Heilongjiang People' Publishing House.
Xu, Yanmei. 2007. *A Study on Ecological Marxism.* [in Chinese] Beijing: Social Science Academic Press.
Wang, Jin. 1986. "'Ecological Marxism' and 'Ecosocialism': Comments on Two Trends of Thought Triggered by Green Movement." [in Chinese] *Teaching and Research* 6: 39–44.
Wang, Yan. 2012. *Marxism's Thought on Sustainable Development and Its Contemporary Value.* [in Chinese] Beijing: Guangming Daily Press.
Wang, Yuchen. 2009. *Ecological Criticism and Green Utopia: A Study on the Theories of Ecological Marxism.* Beijing: People's Publishing House.
Wang, Yuchen. 2015. *Ecological Marxism and Eco-civilization.* Beijing: People's Publishing House.
Wang, Yuchen. 2017. *Ecological Marxism and the Theories of Eco-civilization in the Catch-Up Countries.* Beijing: People's Publishing House.
Wen, Xiaochun. 2015. *On André Gorz's Thought of Eco-Marxism in His Later Years.* [in Chinese] Shanghai: Shanghai People's Publishing House.
Wu, Ning. 2017. *André Gorz's Ecological Marxism.* [in Chinese] Beijing: Jiuzhou Press.
Zang, Li. 2003. *Marx and Engels on Environment.* [in Chinese] Beijing: China Environment Press.
Zhang, Leisheng, and Yu Zhang. 2006. *Marx's Development Theory and the Scientific Outlook on Development.* [in Chinese] Beijing: Economic Science Press.
Zhang, Yanxi, and Xia Shi. 2004. "Marx and Engels' Thought on Sustainable Development and Its Inspiration." [in Chinese] *Contemporary Economic Research* 10: 14–17.
Zhang, Yunfei. 2014. *Eco-civilization from a Perspective of Historical Materialism.* [in Chinese] Beijing: Renmin University Press.

34

THE FOURTH INTERNATIONAL'S CONTRIBUTION TO ECOSOCIALISM

Michael Löwy and Daniel Tanuro

Leon Trotsky: A great thinker, but no ecosocialist . . .

Leon Trotsky, founder of the Fourth International in 1938, is one of the great revolutionary figures of the twentieth century and an outstanding Marxist theorist. Lenin and Trotsky were the main leaders of the October Revolution. Founder of the Red Army and implacable adversary of the Stalinist counter-revolution, Trotsky stands out for his Marxist analysis of fascism and bureaucracy, as well as his theories of combined and uneven development and permanent revolution.

By founding the Fourth International when it was "midnight in the century," he handed down the heritage of revolutionary Marxism to subsequent generations. His work covers key events such as the Russian Revolution, the rise and decline of the Communist International, the revolutionary wave of 1917 through 1923 and its ebb, fascism and Stalinism, the Popular Front and the Spanish revolution, the decline of the British Empire, and the rise of the United States.

The lack of any ecological consciousness casts a shadow over this assessment. The legacy that Trotsky handed down to his successors makes no reference to the concerns, tools, and precursory concepts of ecosocialism developed by Marx and Engels. This deficiency is especially striking in Trotsky's views on progress and the relationship between humankind and nature, in his conception of science and technology, and in his approach to the peasant question.

While their dialectical vision of progress had led the founders of Marxism to be wary of "human victories over nature" ("for each such victory nature takes its revenge on us") (Engels 1934), Trotsky extoled them without nuance:

> Through the machine, man in Socialist society will command nature in its entirety. . . . He will point out places for mountains and for passes. He will change the course of the rivers, and he will lay down rules for the oceans.
>
> *(Trotsky 1941, 5)*

The idea that the forward march of science was unraveling all the mysteries of the universe one after another was prevalent in the nineteenth and early twentieth centuries. The founders

10.4324/9780429341427-39

of Marxism did not always steer clear of this approach, but they believed in neither "absolute knowledge" nor "sovereign thought" (Engels 1947, Chapter IX). Less cautious, Trotsky loudly proclaimed his "faith in the unlimited possibilities for knowledge, prediction and mastery of matter" (Trotsky 1940, 24–31).

There are indications that Marx did not consider technology to be neutral. Trotsky's assessment is different. Asking the question, "What is the concept of class society?" he answers, "[F]undamentally this is the organization of production. . . . [T]echnology is a basic conquest of mankind" (Trostky 1962, Chapter I). In his works, one cannot find any foretaste of the precautionary principle.

True, it is important to be mindful of context. From 1923 to 1924, faced with the ebb of the world revolution and the demobilization of the Russian masses, two alternative orientations were emerging in the Communist Party. On the one hand, Stalin and Bukharin argued for building "socialism in one country at a snail's pace." Giving up on world revolution, they wagered that enriching the countryside would give the regime the means to build a new society. On the other hand, Trotsky argued for the planned development of nationalized heavy industry. In his view, this development was vital to enable the Soviet regime to resist without degenerating as it awaited a new international wave of revolutionary struggles.

Trotsky was right to promote industrial growth as a way of creating the means for development in the countryside but wrong in thinking that this meant state-led "industrialization of agricultural production" (Trostky 1962, Chapter III). Unlike Marx, he did not factor in the need to "govern the human metabolism with nature in a rational way" (Marx 1981, 959). This probably explains why Trotsky's works mention neither Lenin's policy on the protection of the "ecological preserves" (*zapovedniki*) nor the attacks on these ecological achievements initiated by Stalin when he tried to hide the failure of forced collectivization by opening these nature reserves to farming.

In Trotsky's defense, it should be noted that a technicist and modernist culture was hegemonic in the international labour movement of the time, including among Russian revolutionaries. But this context does not explain everything. On ecology, the founder of the Red Army lagged behind Marx, Engels, and even Lenin. As a result, his heirs have had to overcome these gaps in order to reappropriate and build upon the unfinished work of "Marx's ecology" (Foster 2000).

Ernest Mandel's (belated) discovery of ecology

From the 1960s until his death, Ernest Mandel (1926–1995) was the main leader and theoretician of the Fourth International. A brilliant Marxist economist, his writings were studied and admired well beyond the ranks of his movement. He also wrote the most important resolutions of the FI during these years and contributed to its adoption of a non-dogmatic and consistently democratic version of communism.

In the early 1970s, and well before other Marxists, Mandel grew alarmed at the destruction of the environment and sustained this concern until the end of his life. This comes through in his 1972 polemical response to the Meadows report on the limits to growth – as do the limitations of his thinking on these matters. His resistance to admitting the finite character of natural resources – and therefore of the growth of material productive forces – is evident when he argues that the "anarchy" of production, rather than its "unlimited" character, is what should be condemned. Moreover, he put forward typically productivist and unsustainable proposals – for instance, about agriculture, arguing for clearing of forests and irrigating huge tracts of arid lands (Mandel 1972).

The 1972 piece is shot through with the fear that talk of the limits of growth would serve as cover for an austerity offensive against workers and a planetary neo-Malthusian offensive against the poor. This fear was – and remains – quite relevant, and Mandel's answer to it was absolutely legitimate. But, in his polemic, insufficient distinction is made between the objective fact of the natural limits to production and the way this fact is commandeered for reactionary purposes. As a consequence, there is no clear distinction between capitalist pressure on workers to sacrifice their well-being and the absolute need to reduce society's consumption of, for example, energy.

Curiously, at the very moment he provides this rather disappointing answer to the Meadows report, he deals with the same document in a more cautious way in his 1972 book on *Late Capitalism*. The report argues that an extension of the American productive structure to the entire planet would deplete global raw-material reserves by the end of the century and endanger the planet's atmospheric oxygen. While suggesting that the report exaggerates its case, Mandel acknowledges that "it may be right" (Mandel 1973, 513). In the same book, he quotes from Barry Commoner's classic ecological (and socialist) work *The Closing Circle* (Commoner 1971): "The earth is polluted neither because man is some kind of especially dirty animal, nor because there are too many of us. The fault lies with human society" (Mandel 1975, 504). He also pays tribute to another pioneering socialist/ecological by Harry Rothmann (1972), which he calls "the best Marxist work, which deals with the whole problem of the dangers for the environment and the possible measures to prevent them" (Mandel 1975, 576).

To illustrate the dramatic consequences of chemical pollution on the environment and human health, Mandel refers to several examples from the USA and Germany in an extensive footnote. However, unlike Commoner and Rothmann, Mandel only affords ecology a marginal place in his economic and ideological analysis of late capitalism.

Several decades later, in something of a critical self-assessment, he wrote the following in *Power and Money*:

> Today we have become aware, with much delay, that dangers to the earth's non-renewable resources, and to the natural environment of human civilization and human life, also entail that the consumption of material goods and services cannot grow in an unlimited way.
>
> *(Mandel 1992, 207)*

Interestingly enough, in the footnote following this passage, Mandel mentions Peter Hennicke and Michael Müller's *Die Klima-Katastrophe* (Hennicke and Müller 1989) but does not tackle the issue of climate change in his own writing.

Around the same time, in an article from 1990, he makes what can perhaps be seen as his most serious assessment of the ecological crisis:

> From the start of the 20th century onwards, and beginning with Rosa Luxembourg, Marxists were right to sum up humanity's future in the formula "socialism or barbarism." . . . Today, the choice has become "socialism or death." . . . This stems from the implicit danger of extermination from the risk of intercontinental or global conventional or nuclear war in the midst of nuclear power stations . . . but there are other mortal dangers to humankind, above all the threat of environmental destruction.
>
> *(Mandel 1990, 25–30)*

In other words, while Ernest Mandel grew increasingly aware of the danger of ecological catastrophe, this did not lead him to reformulate his Marxism along ecosocialist lines. This would be a task left to the subsequent generation of Fourth Internationalists.

The Fourth International's road to ecosocialism (2001)

After May 1968, the Fourth International began to awaken to the importance of ecological issues. Its activists in France, Germany, Britain and beyond took part in environmental movements, mainly in the struggle against nuclear energy and pollution. Little by little, ecological questions began to be discussed at its congresses, but no specific resolution on ecology was adopted during those years. At the 12th World Congress in 1985, the main resolution called on the FI and its sections to take up the environmental question increasingly in propaganda and general activities and to undertake joint actions with the environmental movement.

In the late 1980s, a group of comrades from the FI wrote a document on socialism and ecology, but it was decided that the issue needed more discussion before being brought to the 13th World Congress which took place in 1991. The document, "Socialist Revolution and Ecology," which was published after the congress, is a very substantial attempt to discuss the causes and consequences of the ecological crisis, the policies of the bourgeoisie, the limits of the ecological movement, the situation in the labour movement, and a programme of action. Urgent demands requiring internationally coordinated action included a radical break with the exploitative system of agricultural export production in the dependent countries, which produces famine and poverty; an immediate ban on the entire cycle of nuclear power production; and an end to the destruction of the tropical rainforests and the deadly contamination of forests in the industrialized countries.

This 1991 document, adopted provisionally by the FI at that moment, was a great achievement but had some obvious limitations. For instance, describing the ecological crisis, it refers to air and water pollution, the destruction of the forests, and industrial disasters (meaning chemical and nuclear ones) but makes no mention of climate change. While the idea of a socialist ecology is very much present, the concept of *ecosocialism* – as a new way of understanding the socialist programme – is still absent.

Several years later, around 2001, a group of French FI members (Laurent Garrouste, Vincent Gay, and Michael Löwy, among others) decided to write a new draft, entitled "Ecology and Socialism," based on the previous document but with substantial changes. It was presented at the 15th World Congress (2003) with a report by Michael Löwy and was adopted by a large majority.

In a critical self-assessment, the resolution's introduction acknowledges that it had not only been the reformist leaderships of the labour movement who had ignored or dismissed ecological problems: "We must recognise that revolutionary currents in general – and the Fourth International in particular – were very late in integrating the ecological question." In a section on the FI and the ecological crisis, the resolution says that many comrades in our organization "have continued to look upon ecological problems as one contradiction of capitalism among many others" and had taken on ecological issues only when other social forces brought them to the fore. While other currents and individuals had been discussing ecology and socialism for decades, the FI had practically remained silent.

The resolution begins with a discussion of the ecological crisis in the capitalist world as well as in the "former bureaucratic societies" (USSR, etc.). Climate change is mentioned here, but just as one among many other problems, without any special emphasis. In a section on the "workers movement and ecology," the contributions of Marx and Engels and their limitations are discussed, and social democracy and Stalinism are taken to task for their blind productivism. Reformist ecology is also criticized, and the true alternative is presented as *ecosocialism*, an ecological current that incorporates the fundamental achievements of Marxism while casting off its productivist fetters.

Here is how the resolution summarizes its understanding of "ecosocialism":

> Eco-socialism is the current in the workers' and ecology movements most sensitive to the interests of workers and peoples of the South. It breaks with the productivist ideology of progress – in its capitalist and/or bureaucratic form (so-called "actually existing socialism") – and opposes the infinite expansion of an environmentally destructive mode of production and consumption. It understands that "sustainable development" is impossible within the framework of the capitalist market economy.
>
> *(Fourth International 2003, 12–13)*

The influence of Walter Benjamin on some of the document's authors can be seen in its criticism of linear progress. The resolution also defines ecosocialism as a *revolutionary project*:

> As revolutionaries, our objective is to join forces with this current and convince workers that partial reforms are totally inadequate. Micro-rationality must be replaced with socialist, ecologist macro-rationality, calling for a genuine change in civilisation. . . . This means the first issue at hand is the question of control over the means of production, and above all over decisions relating to investments and technological change.
>
> *(Fourth International 2003, 13)*

Explaining the meaning of ecosocialism, the 2003 resolution insists that it requires a civilizational change of paradigm:

> An overall reorganization of the mode of production and consumption is needed, based on criteria foreign to the capitalist market: people's real needs and environmental safeguards. This means an economy in transition to socialism, based on the people's own democratic choices of priorities and investments – and not the "laws of the market" or an all-seeing politburo. This would be a planned economy, able to find lasting ways of overcoming the tensions between satisfying social needs and ecological imperatives. It would be a transition leading to an alternative way of life, a new civilization, beyond the reign of money, consumer habits artificially fuelled by advertising, and the endless production of environmentally harmful goods (the private car!).
>
> *(Fourth International 2003, 13)*

The 2003 document distances itself from "workerism" far more than the draft from 1991, but it does not ignore the importance of winning the labour movement to ecosocialism:

> While criticising the ideology put forth by the leading currents of the workers' movement, they understand that workers and their organizations are an essential force for transforming the system.
>
> *(Fourth International 2003, 12)*

The "Ecology and Socialism" resolution adopted by the FI's 15th World Congress (2003) and translated into several languages is a turning point in the history of the Fourth International. For the first time, it adopted a document dealing exclusively with the ecological crisis, and for the first time, it defined itself as an ecosocialist organization. In fact, despite its belated conversion to ecology, the FI became the first international organization to adopt ecosocialism. From that moment on, ecosocialism has increasingly become a key aspect of the FI's revolutionary programme, strategic perspectives, and practical work.

While the document considers the ecological crisis "one of the main threats to humanity" (Fourth International 2003, 20) in our times, climate change is treated as only one among several other issues, and its threat to human life is hardly mentioned. This is perhaps the greatest shortcoming of the resolution. It would be repaired in subsequent years.

Alongside its own activities, the FI contributed to various international ecosocialist initiatives. A first international ecosocialist meeting took place in Paris in 2007, with the help of the French section of the FI. The conference decided to create an Ecosocialist International Network (EIN), open to anyone who agreed with the main ideas of the *First Ecosocialist Manifesto* (published by Joel Kovel and Michael Löwy in 2001).

The EIN asked Ian Angus, Joel Kovel, Michael Löwy, and Danielle Follet to draft a new manifesto, dealing mainly with the issue of climate change. In anticipation of the World Social Forum to be held in Belem do Para, Brazil, in January 2009, this document was called the *Belem Ecosocialist Manifesto*. This second manifesto was signed by hundreds of people from dozens of countries and was printed by the Brazilian Ecosocialist Network (which included several FI members and sympathizers) in English and Portuguese and widely distributed at the Belem World Social Forum.

The next initiative of the EIN took place in Copenhagen in December 2009, at the UN COP-15 Summit on Climate Change. During the protest demonstration against the Summit, which brought 100,000 people from across Europe into the streets, the EIN handed out an ecosocialist "comic-strip" called "Copenhagen, April 12, 2049." The handout showed what Copenhagen would look like submerged by the sea. The leaflet was printed and distributed by the Danish Socialist Workers Party (Fourth International). The EIN ceased to function in 2013, but in 2020 a new initiative also involving a number of FI activists was launched by John Molyneux, the Global Ecosocialist Network.

Resolution on climate change (2010): the FI in favour of a reduction of material production

In 2010, the World Congress of the Fourth International (FI) adopted a resolution entitled "Capitalist Climate Change and Our Tasks." For the first time, the movement discussed and adopted a document devoted to a specific aspect of the "ecological crisis." This flowed from a decision taken a year earlier by the FI's international committee. In support of the draft resolution, a wide-ranging document on the science of climate change had been translated into several languages and distributed in the sections. Daniel Tanuro drafted these two texts and introduced the resolution to the congress (Fourth International, 16th World Congress 2010).

The resolution consists of five chapters. The first blames climate change on the capitalist system – though recalling the dismal record of the USSR, its allies, and China as guilty of replicating productivism. The second chapter explains why capitalism is unable to stop the catastrophe. Firstly, because the necessary reduction in greenhouse gas emissions is only achievable by substantially reducing energy consumption and, therefore, also material production. Secondly, because North-South justice demands that the lion's share of this reduction take place in the imperialist countries in order to give others some room to manoeuvre for the satisfaction of vital human needs. The third chapter warns against barbarian Malthusian policies as a bourgeois response to the combination of the economic, climatic, and food crises. The fourth underlines the twofold urgency of a world socialist alternative and of a radical break of the socialist vision from productivism. "The emancipation of the workers is no longer conceivable without simultaneously taking into account the principal natural constraints" (Fourth International 2010, 5). The FI Congress ratified the idea that the objectively new situation justifies an embrace of ecosocialism. The fifth chapter focuses on tasks.

The first Marxist organization to embrace ecosocialism, the Fourth International was also the first to adopt the idea that "the reduction in material production and consumption is immediately necessary . . . because capitalism has led humanity too far into a dead end" (Fourth International 2010, 5). The text clarifies that this decrease "does not in any way prejudge the future possibilities of development, once the climate system has been stabilised" (Fourth International 2010, 5). It also notes that it "constitutes only one quantitative criterion of the necessary transition towards an economy without fossil carbon" and adds:

> [T]his quantitative criterion must be combined with qualitative criteria: in particular, redistribution of wealth, reduction of working time without loss of wages, development of the public sector. If these criteria are satisfied, and on condition that it targets useless or harmful production, the reduction in material production will actually be synonymous with an increase in the well-being, wealth and quality of life of the vast majority of humanity, through . . . the re-conquest of the free time necessary for self-activity, self-organisation and democratic self-management on all levels.
>
> *(Fourth International 2010, 5)*

"It is not enough to affirm that socialism must take ecological questions on board," reads the resolution.

> The real challenge consists rather of creating the conditions so that the socialist project is compatible with the global ecology of the terrestrial super-ecosystem. Development cannot only be conceived of with the aim of satisfying real democratically determined human needs, but also according to its sustainability by the environment. . . . The concept of "human control over nature" must be abandoned. The only really possible socialism from now on is one that satisfies real human needs (disentangled from commercial alienation), democratically determined . . . taking care to carefully question ourselves as to the environmental impact of these needs and the way in which they are satisfied.
>
> *(Fourth International 2010, 5–6)*

The resolution expands on the idea that technologies are not neutral, mentioned in earlier documents.

> The capitalist energy system is centralised, anarchic, wasteful, inefficient, dead-labour intensive, based on non-renewable sources and characterised by a tendency towards overproduction of commodities. The socialist transformation of society requires its progressive destruction and its replacement by a decentralised system, planned, economical, and efficient, living-labour intensive, based exclusively on

> renewable sources and directed towards the production of durable practical values, which can be recycled and reused.
>
> *(Fourth International 2010, 6)*

The socialist revolution should be conceived

> not only as the destruction of the power of the bourgeois state . . . but also as the beginning of a process of destruction of the old capitalist productive apparatus and its replacement by an alternative apparatus, utilising different energy sources, different technologies and different structures in the service of democratically decided objectives.
>
> *(Fourth International 2010, 6)*

In its conclusion, the resolution sets out 12 tasks ranging from raising awareness regarding the seriousness of the situation to developing popular disaster relief practices and fighting neo-Malthusianism in defense of the rights of the poor and women; devising a comprehensive anti-capitalist plan for social and ecological reconstruction, concretely linking the fight for the climate and for social rights, in particular the right to employment; supporting indigenous peoples; opposing the headlong rush into dangerous technologies; establishing links with critical scientists; etc.

FI World Congress of 2018: Capitalist destruction and the ecosocialist alternative

The 17th World Congress of the FI in 2018 adopted a resolution entitled "The Capitalist Destruction of the Environment and the Ecosocialist Alternative" (Fourth International 2018). A long preparatory document had been written by Daniel Tanuro. The ecology commission decided to submit one of its chapters to the congress and to publish the full text as a working paper.

The resolution acknowledges the "extremely alarming level" of the crisis. "Thresholds are already exceeded in some areas, particularly the concentration of greenhouse gases in the atmosphere," which could lead to "a qualitative shift," says the text.

> The Earth System would then enter a new dynamic equilibrium regime, characterised by very different geophysical conditions and an even more marked decrease in its biological richness. At the least, in addition to the consequences for other living creatures . . . this new regime would endanger the lives of hundreds of millions of poor people, especially women, children and the elderly. At the most, it cannot be excluded that it contributes to a collapse of our species.
>
> *(Fourth International 2018, 1.1.)*

The document focuses on the "deep gap between the urgency of a radical ecosocialist alternative, on the one hand, and the relationship of forces and the levels of consciousness, on the other" (Fourth International 2018, 2.).

The measures to take should include "the socialization of the energy and credit sectors," "the abolition of private ownership of natural resources and intellectual knowledge," "the suppression of useless or harmful products," and "the common and democratic management of resources at the service of real human needs" (Fourth International 2018, 2.3.)

But these measures are obviously beyond reach in the present context. The resolution says that

> this gap can only be closed by concrete struggles of the exploited and the oppressed. . . . By winning immediate demands, broader layers will radicalise and their struggles will converge. They will formulate . . . demands incompatible with capitalist logic." The text lists demands that fit into this strategic framework. They are applicable separately, but . . . form a coherent whole, incompatible with the normal functioning of the capitalist system. This program is not exhaustive . . . and will continue to be enriched by concrete struggles. In an ecosocialist perspective this enrichment must be guided by . . . environmental and social justice, common but differentiated responsibilities, the fight against inequality, . . . the end to green colonialism and environmental racism, prioritising collective solutions, internationalism and the precautionary principle.
>
> *(Fourth International 2018, 2.4., and 2.5.)*

"Above all, the exploited and the oppressed must develop their empowerment by democracy, decentralization, control and the collective appropriation or re-appropriation of the commons" (Fourth International 2018, 2.7). Renewable-energy technologies can help promote self-management because "the physical nature and the difficulty of storage of electricity make it is easier to manage in a decentralised, combined and complementary system" (Fourth International 2018, 8.2). The resolution says that "along with food sovereignty this field of struggle is particularly important for the South, as part of an alternative development model to the imperialist model" (ibid.).

A major contribution of the resolution is the way it unifies the vision of capitalism:

> Capitalism relies not only on the appropriation of nature and the exploitation of the labor force through wage labor but also on the patriarchal invisibility of the labor of care and reproduction of the labor force. Added to these three pillars of capitalism is a fourth, exploitation based on race. All have a common denominator that is the appropriation of natural resources, in which the human workforce is a part.
>
> *(Fourth International 2018, 4.3)*

The resolution points to the prominent roles of Indigenous peoples, peasants, women, and young people, providing a materialist explanation of their vanguard positions. Obviously, the key question is the role of the working class. Capitalist relations of production "make it both very difficult and of decisive importance to mobilise the labour movement in the ecological struggle" (Fourth International, 3.1). There is a challenge around job loss: Even if "the transition will create a growth of employment in [new] sectors . . . a global reduction of material production is necessary." Hence the decisive importance of "a reduction of working hours without loss of wages" (Fourth International 2018, 3.2). The resolution concludes that "the struggle for an ecosocialist transitional programme is ultimately achievable only through the emergence of political alternatives for a comprehensive plan of structural anti-capitalist reforms that satisfies both social needs and environmental constraints" (Fourth International 2018, 7.5).

Concrete examples of FI sections' ecosocialist activities

At the World Congress in 2018, debates on ecology reflected the heavy involvement of many activists in concrete struggles, especially in the countries of the South, which are the most affected by climate change.

The section of the Fourth International in Mindanao (Philippines) has long been involved in the defense of communities against increasingly violent typhoons. These comrades are engaged in the development of farming methods based on food sovereignty, the exclusion of genetically modified seeds, and the production of organic food for local communities.

In Bangladesh, one of the countries that is most vulnerable to the effects of climate change, the FI organization is deeply involved in major peasant struggles against climate change. With Via Campesina and other organizations, the comrades' campaign for food sovereignty, the rights of small farmers, and the redistribution of land, along the lines of the Brazilian MST. Since 2011, they have been heavily involved in the organization of climate caravans campaigning in Bangladesh, Nepal, and India.

In Pakistan, too, FI comrades have been at the forefront of the fight against climate change. In 2010, devastating floods submerged one-fifth of the country and left millions homeless. Twenty million people were affected, and 2,000 people lost their lives. Five comrades were imprisoned for defending villagers after a landslide that blocked the Hunza River in the Gilgit-Baltistan region, sweeping away houses and killing 19 people. The landslide formed a 23-kilometre-long lake that submerged three villages, leaving 500 people homeless and 25,000 stranded. They were still in prison seven years after their trial, despite campaigns for their release.

In Brazil, the comrades are involved in building the climate justice movement. In 2015, in Fortaleza, they organised the largest climate march in the country's history. They also demonstrated in 2016 as part of the 350's Break Free campaign in front of the largest Brazilian coal-fired power plant and participated in the Water March in 2017. They are involved in the defense of the Amazon and against the disastrous REDD (Reduction of Emissions due to Deforestation and forest Degradation) treaty, alongside Indigenous peoples, local communities, and environmental groups, violently put down by Jair Bolsonaro's fascist government. FI organizations in other parts of Latin America were involved in the mobilizations around the People's Summit convened by Evo Morales in Cochabamba, Bolivia (2010).

In Europe and North America, comrades of the Fourth International are increasingly involved in climate mobilizations – around the COPs in Copenhagen (2009), Paris (2015), and Madrid (2019) and in the huge youth demonstrations that took place in different countries in 2019. They also participate in local struggles against fossil fuel projects (what Naomi Klein calls "Blockadia"), such as fracking in Great Britain, the exploitation of the tar sands in Canada, the building of the Keystone Pipeline in the United States and the Notre-Dame-des-Landes airport in France, and the Ende Gelände campaign against brown coal mining in Germany. In cooperation with other organizations and trade unions, FI members were also part of the initiative to hold European ecosocialist conferences, bringing together several hundreds of people – in Geneva (2014), Madrid (2015), Bilbao (2016), Lisbon (2018), and Basel (June 2020).

References

Commoner, Barry. 1971. *The Closing Circle. Nature, Man, and Technology*. New York: Knopf.

Engels Frederick. 1934. *The Role Played by Labor in the Transition from Ape to Man*. Moscow: Progress Publishers.

Engels Frederick. 1947. *Anti-Dühring. Herr Eugen Dühring's Revolution in Science*. Moscow: Progress Publishers.

Foster, John Bellamy. 2000. *Marx's Ecology. Materialism and Nature*. New York: Monthly Review Press.

Fourth International. 2003. "15th World Congress, Ecology and Socialism." https://internationalviewpoint.org/IMG/pdf/ecology-and-socialism_a178.pdf

Fourth International. 2010. "16th World Congress, "Capitalist Climate Change and Our Tasks." https://internationalviewpoint.org/IMG/pdf/capitalist-climate-change-and-our-tasks_a1834.pdf

Fourth International. 2018. "17th World Congress, the Capitalist Destruction of the Environment and the Ecosocialist Alternative." https://internationalviewpoint.org/spip.php?article5452

Hennicke, Peter, and Michael Müller. 1989. *Die Klimakatastrophe*. Bonn: Dietz.

Mandel, Ernest. 1972. "La Dialectique de la Croissance. A Propos du Rapport Mansholt." *Mai* 7–14, November.

Mandel, Ernest. 1973. *Spätkapitalismus*. Frankfurt-am-Main: Suhrkamp Verlag.

Mandel, Ernest. 1975. *Late Capitalism*. London: New Left Books.

Mandel, Ernest. 1990. "L'Avenir du Communisme." *Inprecor* 305: 25–30.

Mandel, Ernest. 1992. *Power and Money: A Marxist Theory of Bureaucracy*. New York: Verso.

Marx, Karl, 1981. *Capital*. Vol. 3. New York: Vintage.

Rothmann, Harry. 1972. *Murderous Providence*. New York: Harper Collins.

Trotsky, Leon. 1940. "Dialectical Materialism and Science." *New International* 6 (1): 24–31.

Trotsky, Leon. 1941. "Socialism Will Bring Giant Advances for Mankind." *The Militant* 5 (34): 5.

Trostky, Leon. 1962. "Culture and Socialism." *Labour Review* 7 (3): 101–113.

INDEX

For Product Safety Concerns and Information please contact our EU representative GPSR@taylorandfrancis.com
Taylor & Francis Verlag GmbH, Kaufingerstraße 24, 80331 München, Germany

www.ingramcontent.com/pod-product-compliance
Lightning Source LLC
LaVergne TN
LVHW080805011026
849160LV00015B/1541

* 9 7 8 1 0 3 2 1 3 4 9 4 9 *